Donald C. Leete
Sept. 14, 1982
went to look at Pawling
Center.

The Power Guide

Compiled by Peter Fraenkel

A CATALOGUE OF SMALL SCALE POWER EQUIPMENT

Charles Scribner's Sons New York

First U.S. edition published by Charles Scribner's Sons 1980

Library of Congress Catalog Card Number 80-52102
ISBN 0-684-16739-5

1 3 5 7 9 11 13 15 17 19 V/P 20 18 16 14 12 10 8 6 4 2

Contents

Acknowledgements

This publication would not have been possible without the help of Christine Marsden, Max Ewens and Roger Doyle. My thanks also to my other colleagues in the Intermediate Technology Development Group who have been involved in this work.

The co-operation of the numerous manufacturers and other bodies who supplied information for this guide is much appreciated, and it is hoped this collaboration will continue in the compilation of future editions.

On behalf of the Group, I would like to thank the Ministry of Overseas Development for providing the funds for the project and a loan for the printing of this publication. Their assistance is gratefully acknowledged.

Peter Fraenkel
Power Project Officer
Intermediate Technology Development Group

Foreword

If other countries are to attain the standard of living enjoyed by the industrialised nations, then availability of energy is an essential requirement to enable them to do so. If industrialised countries are to enjoy the same per capita growth and the same increase in the standard of living as hitherto achieved the supply of energy must continue to grow.

Now that signs of shortages in the supply of fossil fuels are apparent there is an accelerated and widespread move towards exploring other possible sources of supply, not only of fossil fuels but also of renewable and natural sources. These range from wind and water to solar and biomass, as well as towards recycling the waste from our cities which could be converted to useful energy.

While research investigation will obviously take time, an examination of the state of the art in regard to all the various small scale sources of power available at the present time is a vital first step in the development of alternatives. In this respect this volume could not be more timely, providing as it does an account of both conventional and unconventional small scale energy converters and a catalogue of commercially available equipment in this range.

It is important to appreciate that in the developing countries the problem is not one of substituting one source of energy for another but of increasing the total per capita energy availability. Currently biomass, mainly in the form of wood and dung, constitutes the major source of energy in many developing countries and even in world terms is of great importance, amounting to around 10-15% of the total world energy use. The internal combustion engine, both diesel and petrol, is widely used in the developing world and will continue to be required for some time to come. Solar energy, either in the form of direct radiation or indirectly in the form of wind, water power or biomass, has a number of attractive features. It is continuously renewable, free, universally available, and does not have harmful ecological effects. Although convertors using renewable energy are not cheap, fuel is free, and there are already a number of applications where renewable energy sources are economically competitive with conventional plant. These include windmills for water pumping, small hydro-generators for electricity supply, methane generation using dung or vegetation to provide gas for cooking, and solar crop driers.

This book lists manufacturers of such equipment, provides a brief summary of the characteristics of the equipment and the criteria which should be borne in mind in its selection. There is no doubt that this well-researched and wide-ranging catalogue of small scale power equipment will provide a powerhouse of information for anyone wishing to find a source of energy to suit his purpose. *The Power Guide* is an important and timely contribution to the literature of appropriate technology.

Professor P.D. Dunn
Department of Engineering
University of Reading

Introduction

SCOPE OF THIS GUIDE
ENERGY, POWER PRODUCTION AND DEVELOPMENT
ECONOMICS OF ENERGY CONVERTERS
NOTE ON UNITS AND TABLES OF CONVERSION FACTORS

Scope of Guide

This guide is intended to help those seeking to buy small-scale power equipment, particularly for use in remote and underdeveloped parts of the world, by indicating a selection of appropriate commercially-available power sources. It gives basic information on the pros and cons and the criteria for choosing a variety of different energy conversion systems, together with a selection of internationally available items of equipment, and the names and addresses of their manufacturers and agents. Additional material on related consultancy and information services, plus a listing of organisations concerned with research and development in small-scale power production is also included.

The product information is based entirely on that supplied by manufacturers. The Intermediate Technology Development Group, (ITDG), has no facilities for testing machinery and therefore cannot take responsibility for manufacturers' claims — the object of this Guide is for it to be used as a means to identify sources of supply. *Therefore the inclusion of any particular item of equipment or manufacturer does not signify any specific recommendation by ITDG.* We should of course be interested to have from readers either endorsements or criticisms of companies or equipment in this edition and would value information on useful products that are missing for possible inclusion in future editions.

It should be explained that ITDG has no commercial interest in any of the equipment listed, since we are a non-profit research organisation, registered as a charity in the UK. The preparation of this Guide was financed by a grant from the UK Ministry of Overseas Development. All the entries were provided at no charge to the manufacturers concerned and the choice of the entries was at ITDG's discretion. The Group exists to help people, particularly in underdeveloped parts of the world, to advance and improve their level of self-sufficiency through the use of technically and economically appropriate equipment. We do this by seeking to diffuse information on equipment that has proved useful in the field in many parts of the world, and in some cases where appropriate devices are lacking, we develop, test and encourage others to manufacture new devices. Further details of our activities in the field of energy and power production are given in Section 10).

This book represents an attempt to satisfy the flood of enquiries we (and similar organisations) continuously receive on where one can obtain small-scale power producing equipment. Some of the entries, particularly for the engine manufacturers, refer to world-famous multinational corporations, others are quite small and little-known specialised companies in many parts of the world. In particular we have included a wide selection of manufacturers of renewable energy conversion devices (that use energy from the sun, wind, rivers or bio-mass) since the increasing cost of fossil fuels, particularly in remote underdeveloped areas, has caused a great deal of interest world-wide in such devices. Manufacturers of renewable energy systems are often small, and the merits and weaknesses of their products and often not as well understood as they might be. We have of course included a large section on conventional diesel and petrol (gasoline) engines, as these will remain the most effective power source for many applications for the nearer future, but have limited the descriptive details on these to a basic specification on the assumption that potential users will write for precise and up-to-date information to the manufacturers.

Energy, Power Production and Development

What is energy?

Energy is required, by definition, to do any kind of work; the rate at which it is used is measured as power, (see page 12 for units and their relationships). We can do a certain amount of work slowly, using little power or quickly using more power, involving the use of a similar total amount of energy in either case.

Virtually all human activity requires work to be done; ploughing fields, lifting water, transporting ourselves, mining for minerals, manufacturing, cooking food, and so on. Even lazing around and doing nothing involves "work" in the technical sense, as our bodies consume a minimum amount of energy to keep our heart, lungs and other vital organs functioning at a rate thermally equivalent to burning about a pound (500gm) of coal a day. So there is an obvious minimum energy requirement for human survival, represented by a minimum food need to prevent malnutrition and starvation.

Our own muscles, or the muscles of domesticated animals, were for centuries the primary source of power behind human industry, and this remains true to this day for the majority of the human race (of the order of 2,000,000,000 people). However, man or woman power alone is hopelessly unproductive as a primemover – our brains are far more impressive than our muscles – for example, one man-year of hard labour represents a mere 150kWh, which is the energy in about 15 litres of kerosene (3½ gallons). You can buy 150kWh as electricity, one of the most expensive forms of power, for under £5 ($9.00) in most countries, and who would offer a year of hard labour, even in the poorest of countries, for that? Muscle power alone is therefore unproductive and very expensive compared with even the costlier sources of artificial power.

Energy in under-developed regions

It is quite clear that all human productive activity can benefit from a certain amount of power-assistance, so much so that energy conversion or power production is an almost essential prerequisite for progress. Unfortunately this belief has stretched so far that in certain quarters it is accepted that energy consumption and wealth are tied together by a rigid correlation as binding as a law of nature, a belief that has led certain industrialised countries to extrapolate historical energy "growth" curves and then plan enormous nuclear power programmes, while less developed countries, fearful of being left even further behind in a world of unbalanced wealth and unbalanced energy consumption often seek to do the same thing on a smaller scale. However, energy consumption per capita in the industrialised countries and to some extent in the urban, developed sectors of less developed countries too, has in most cases passed the level at which further additions are important to improve the quality of life – diminishing returns have set in to such an extent that in some places more energy consumption even spells a decline through the effects of added pollution. Today, the United States uses more power just to run the nation's air conditioners for a minority of its population, than is consumed by all 800 million Chinese for all essential and non-essential purposes put together. Air conditioning is of course desirable (although the human race evolved successfully without it until quite recently) but the industrial nations will obviously cease to air-condition rooms which are simultaneously solar heated through enormous picture windows once the price of energy reaches a level where alternative less energy-intensive "solutions" become more economically attractive. Similarly, much of the energy consumption in poor tropical countries is to air-condition government offices – increasing that kind of consumption increases energy per capita as well as GNP per capita, but does nothing for the underdeveloped predominantly rural areas.

It is often thought that rural electrification might be the solution, since the most efficient form of electricity production (and most easily manageable from a centralised administration) is to use large power stations generating hundreds of megawatts. However it is not economical to distribute electricity to small distant consumers. Large thermal stations and the huge dams in Africa and Asia, such as Kariba, Cabora Bassa, Aswan, Volta, Mangla and so on, are only capable of feeding power to a few urban and industrial areas – in most cases the rural areas in the vicinity of such schemes are still without electricity because the potential market is too diffuse to allow economic distribution. Rural electrification has proved expensive and of limited success in most less developed countries. For example, only 2% of the electricity generated in Latin America goes to the rural areas, while India, with one of the most extensive rural electrification programmes, which has been in hand for thirty years, today supplies only 10% of its electricity to the rural areas inhabited by 80% of the population.* In the end, such schemes, useful as they may be to the general economies of their host countries, serve to boost the already-developed sector of the economy rather than the under-developed parts. As a result, centralised electrification schemes almost inevitably lead to a further increase in the differential between the developed and under-developed areas, with further polarisation between rich and poor, more migration from the countryside to the cities and all the usual consequences of continued rural lack of productivity leading to chronic poverty in rural areas and urban wealth and squalor in close juxtaposition.

Most technology has been developed in response to market forces, in other words to be economic and marketable at a price level which suits the already wealthy, (and it is apparent that this observation applies equally in socialist as well as in capitalist economies – since technology has followed similar patterns and trends throughout the world). As a result, power-production machinery, in common with most other types of machinery, is generally not ideal for use in under-developed areas. It is usually too expensive in capital requirements, too expensive in fuel costs (fuel is nearly always priced higher in under-developed areas due to transport costs) and above all, often requires skills that have to be imported at a higher than normal cost in order to provide maintenance and continuity of operation. Also, due to the small initial market for power in such regions, it must be generated on a small scale, generally from systems of under 100kW, which as a result often involves the use of less efficient or less long-lasting equipment. Therefore it would be wrong to pretend that decentralised, small-scale power production can be easily and economically arranged – but without it any significant improvement in the productivity and standard of life in the least developed parts of the world will be no more than a vain hope.

This Guide gives an indication of the variety of energy converting machines that can be considered for small-scale decentralised applications. At the present time, half

**Energy Policy for the Rural Third World* by Arjun Makhijani, International Institute for Environment and Development, London, 1976.

a century of declining petroleum costs followed by less than half a decade of reversal has placed the internal combustion engine in such a dominant position for most applications that it is likely to remain the most commonly used small-scale mechanical prime-mover for the immediately forseeable future. The i.c. engine is unlikely to improve significantly in future as such tremendous efforts have already been devoted to it in connection with its development for motor vehicles. As a result, many diesel and petrol engines for use as stationary power plants are thinly disguised adaptations of car or lorry engines. In some respects such engines are not ideal (a motor vehicle spends much less of its life with its engine running than the average power plant) but on the other hand, the mechanics of such engines are well-understood everywhere that motor vehicles are to be found and the infrastructure to provide spares and maintenance tends to accompany that for motor vehicles. First costs can also be quite low due to high volumes of production. Other power plants have been specially developed to run for the much larger number of hours normally expected of a stationary plant — such machines are invariably heavier, slower and more expensive in first cost (as is discussed more fully in the introduction to the section on i.c. engines).

Increasing petroleum costs (which have hit hardest at many of the least developed countries lacking sufficient hard currency to pay the higher prices), have caused a revival of interest in machines that can utilise non-petroleum-based energy sources; steam and hot air engines that can burn wood or coal and, of course, devices such as windmills, small hydro-turbines and solar heaters that use freely available natural energy. Since these energy resources are more diffuse than oil or petrol, and in some cases are only intermittently available, the conversion devices tend to be larger and more expensive in terms of first cost per kilowatt. They have also had more limited development (especially in recent years), their volume of production is smaller and they are generally not so well understood by prospective purchasers and users. However, even with all these handicaps, many are economically competitive with internal combustion engines at current prices and increased usage and development combined with further escalations in the price of oil will move the economic factor further away from the i.c. engine towards the windmill and the solar heater or the steam engine and the photocell. It is hoped that this Guide will indicate something of the variety of equipment that is available on the world market and that the introductory sections will provide some guidance on their pros and cons and how the different devices can most appropriately be applied.

In the end the ultimate criterion for any energy converting machine is not its power output, its life or even its efficiency — but its economics. How much does it cost to guarantee sufficient power for the required end-purpose. Whether it is intended to make decentralised rural energy production a "paying proposition" (an aim which historical trends even in developed industrial countries, let alone in the less developed ones, indicated as being unlikely to be successful) or whether rural energy is subsidised or even given away for nothing in the interests of rural development or aid, it is obviously desirable to minimise energy costs and to maximise its effectiveness.

To maximise the effectiveness with which energy is applied, it is important to assess energy end-uses carefully. It is likely that power used for pumping irrigation water or driving workshop machines is going to be more useful for the local economy than power used for lighting or for television sets, (although the latter two applications are not unimportant in terms of redressing the balance of attractions between the rural and urban areas).

Variation of demand and variability of load are other important considerations; for example, a large generator is inherently more efficient at full load than a small one, but a number of small generators may be cheaper to run than one big one that is on part-load for most of the time. Also, a number of small generators would be unlikely to fail simultaneously whereas a single large one may need a standby unit. Small generators are also likely to need a smaller and simpler power distribution system and they are also often more readily repaired and maintained by local people. But smaller generators are, of course, more expensive per unit of output at a given load factor.

Sometimes power must be available at any time or all the time, for example to drive a lathe or milling machine or to run a refrigeration plant, but sometimes the need is not very time-dependent — for example to pump water into a reservoir or storage tank or for heating water in an insulated tank. The latter uses can easily make use of intermittent and less predictable energy sources such as the wind or the sun, but the former will generally need a more controllable prime-mover.

Another important consideration is energy *grade.* Just like currencies, energy can be exchanged or converted and in this case natural constraints, defined under the laws of thermodynamics, dictate that every transaction causes a percentage of the energy converted to be lost as waste heat. Some transactions are inherently more efficient than others — it is more efficient to convert from a higher grade of energy to a lower, while considerable losses are involved in going from a low grade to a high. As a result, electricity or mechanical shaft power are said to be of high grade, because they can be efficiently converted to lower grade heat, but heat is a variable but lower grade of energy — the lower the temperature of the end-use the lower the grade. Therefore, raising steam from a boiler demands a higher grade of heat than is needed for domestic water heating. As a result, low grade solar heat can be used quite efficiently and cheaply for domestic water heating but it is more difficult and less efficiently utilised to run a steam engine. The diagram below shows typical energy exchange rates and it can be seen how high-grade electricity converts. For example 1050W of electricity yields 1000W of heat, but in reverse it takes 20,000W of heat to produce 1000W of electricity directly, or if heat is converted first to mechanical and then to electrical power — 3,600 watts of heat to produce 1000 watts of electricity. This shows why motor-generator sets are more popular than thermo-electric converters, despite their greater complexity and maintenance requirements.

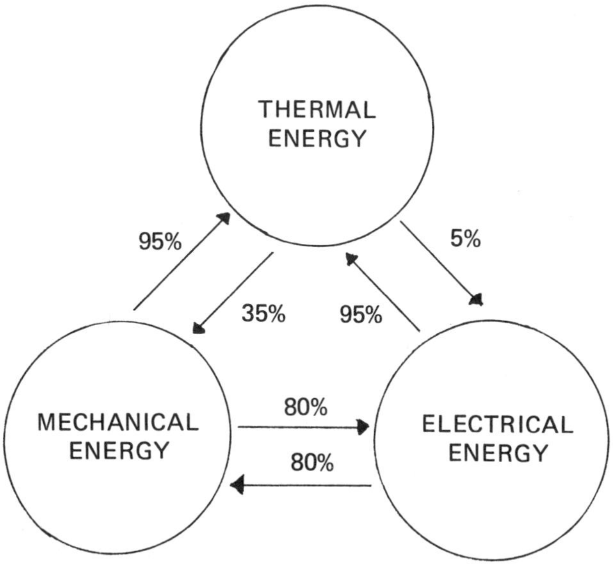

Figure 1. Some typical energy exchange rates

It follows from consideration of energy exchange rates that it would be wasteful to use, say, a diesel generator to produce electricity for heating water, and quite ridiculous to use a thermo-electric convertor for such a purpose! Although convenient, electricity is hard-won energy and should not normally be used for low-grade applications (except perhaps in conjunction with hydro-electric plants when it can sometimes be very cheaply generated). Indeed, since a lot of normally wasted heat is associated with any type of heat engine, including diesel generators (in the cooling water or cooling air) it may be worth considering ways that the heat could be used as a by-product (particularly in the case of larger fixed installations). This is very rarely done, although waste heat could readily be used for crop or hay drying, water heating, desalination or space heating. It may become possible to utilise waste heat for cooling purposes in hot countries once suitable refrigeration and air-conditioning equipment is developed to make use of low grade heat inputs (it is currently technically feasible to do this but suitable equipment is not readily available).

Economics of Energy Converters

Before considering the factors that affect the cost of energy conversion using different types of machine, or what might be called the micro-economics of energy conversion, it is worth keeping in mind the macro-economic effects of any larger-scale rural energy policy. An individual device has an operating cost determined by the *de facto* first cost, fuel price and other variables. However, in situations where the choice is not for one or a few but for a larger number of energy converters, it would be a mistake to ignore how existing costs might be deliberately changed sufficiently to favour a different energy conversion process. For example, local manufacture might become feasible and greatly reduce both first costs and foreign currency requirements – so suitability for local manufacture may be a valid criterion when considering widespread usage. (And of course, small scale local manufacture offers other benefits to an under-developed area). Just as with centralised power production, centralised machinery production gives low unit costs (at the power station bus-bar or the factory gate respectively), but distribution costs can lead to diseconomies compared with small-scale local manufacture just as much as with decentralised power production. Increasing energy prices will lead to increased distribution costs both for mass-produced engines and for centralised, grid-distributed electricity, so this effect should become increasingly important in future.

However, most readers of this Guide will not be in a position to influence the criteria that effect the economics of energy conversion. They will have to weigh up the relative merits of whatever happens to be available to deliver the required power to their envisaged end-use, and hopefully, future editions of this Guide will help to keep them aware of any newly available devices that may be worthy of consideration.

Assessing the annual costs of different devices

The main elements in costing any energy converter are:-

1. Cost of purchasing and replacement finance
2. Cost of fuel
3. Cost of maintenance and repairs

In order to evaluate these costs, it will be necessary to obtain quite a number of figures from manufacturers' literature or from local manufacturers' agents, as follows:

1. Cost of purchasing and replacement finance

A number of formulae are conventionally used to assess the annual charges relating to a fixed capital investment, most of which require knowledge of the capital sum or first-purchase cost, C; the expected life of the machine (amortisation period) in years, n; and at least a notional interest rate based on the value of the interest to be gained if the same sum were invested or the actual interest to be paid if it is borrowed, r. If r is expressed as a decimal percentage (i.e. 0.10 represents ten per cent, 0.08 is eight per cent, etc.), then:-

$$\text{Annual charge } R = C\frac{r(1+r)^n}{(1+r)^n - 1}$$

2. Fuel costs

These depend on:

i. the type of machine (obviously fuel costs will be zero with a windmill and will vary depending on fuel type with steam, petrol or diesel engines);

ii. the rated power output, P – ideally chosen so as to operate at as high a power rating as possible without unduly overloading it and shortening its life – (a heavily loaded machine runs efficiently but wears out fast and vice-versa);

iii. efficiency of conversion which governs the fuel consumption per unit of power output (specific fuel consumption), c; here it is important to assume a fuel consumption in line with typical load factors and not the manufacturer's proudly quoted optimum fuel consumption. A typical figure for a small diesel engine will be around 0.3 litres/hp.h or .07 gall/hp.h (which is 250 gm/hp.h or 0.6 lb/hp.h) under fairly well controlled conditions;

iv. the number of hours of operation per year, h;

v. and finally, the unit price (per litre or gallon or kg) of fuel including delivery and storage, f; (here the trade-off is between size of storage and frequency of delivery).

Therefore P.c.h litres or gallons (depending on units) are needed per year so, fuel costs F = (P.c.h) f.

3. Maintenance and repairs

This depends very much on the type of machine. In general, the frequency of maintenance and repairs tends to increase in relation to the operational speed of the machine; small fast machines need more maintenance than big slow ones; while non-moving devices like thermo-electric generators and photo-voltaic cells need practically no maintenance and rarely go wrong. However, while small, high-speed engines need rather a lot of maintenance, the skills required are often locally available and spares are mass-produced and relatively inexpensive, while a larger, more specialised engine, a gas turbine or any unconventional machine or device may need less attention, but the maintenance and repair skills may be rare and spares may also be harder to obtain and more expensive.

A rough rule of thumb with small engines is that the lubricant consumption rate is 2% of the fuel consumption rate and that servicing and routine replacements will cost an additional 2% or more of the capital cost of the machine per annum.

Typical worked examples

a. *5 hp diesel engine and pump set* (for irrigation). Let the first cost be £700, the annual utilisation 1000 hours, and on the basis of an assumed 10,000 hour useful life, the amortisation period will be a maximum of 10 years. Lastly let us assume a notional interest rate of 10%, which gives:-

$$\text{Annual charge } R = £700 \frac{0.1\,(1+0.1)^{10}}{(1+0.1)^{10}-1} = £114 \text{ p.a.}$$

Let us assume that the engine is derated by 60% and delivers an average of 3 bhp (i.e. 3000hp-h/y). The specific fuel consumption is likely to be around 0.3 litre/hp-h, or 900 l/y. Suppose the delivered fuel cost is £0.21/l (£0.90/Imp. gallon) — (i.e. around double a typical refined product production cost), then:-

Fuel costs F = 3000 x 0.3 x £0.20 = £180
in the first year

If we assume lubricant consumption at 2% of fuel, or 18 l/y and lubricant cost (typically) to be around three times fuel cost (£0.60/l) and maintenance and repairs cost an additional 2% of the cost of the engine (£14), we must add:-

Maintenance and repairs M = (18 x £0.60) + £14 = £24.80 in the first year

However, it would generally be unreasonable to suppose that fuel and maintenance costs will remain constant throughout the life of the engine. If for example inflation is taken as being 10% per annum, then these costs will increase over the ten year life of the unit to such an extent that the average annual cost for the period will be the first year's cost multiplied by a factor of 1.59, (see Table 1 for multiplying factors for other inflation rates and periods).

As a result, combined fuel and maintenance costs of £205 in the first year would average £326 over ten years at 10% inflation. (At 5% inflation the average annual fuel and maintenance cost would be £258, while at 15% inflation the annual average would be £416.) Obviously inflation is neither constant nor predictable, and due to inflation the inflated values are not worth the same by the time they occur as present day currency — however it is worth noting the effect as it is a factor when comparing with devices that do not require fuel and which in real terms become cheaper after inflation.

Therefore, for the diesel engine, the following costs apply:-

	1st year	*average*	*10th year*	
annual running costs:	£320	£440	£560	(10% inflation)
average cost per bhp-h:	10.7p	14.7p	18.7p	

b. *16ft diameter (5m) water pumping windmill* (for irrigation). Let the first cost be £3000 including tower, storage tank and all accessories. Assuming a life of 30 years, then the annual charge for the investment will be:-

$$R = £3000 \frac{0.1\,(1+0.1)^{30}}{(1+0.1)^{30}-1} = £318 \text{ per annum}$$

Fuel costs are zero.

Maintenance involves one oil change per annum, say £10.

Therefore total average annual cost = £328.

Average cost per hp-h depends on the wind regime of the location. Calculations based on an average wind speed of 8 mph (12 kmph) suggest that such a windmill would pump 500 million ft. gallons per annum, which is roughly equivalent to 2500 hp-h.

Hence, average cost per hp-h is approximately 13p in the first year.

On the same basis as before, at a 10% inflation rate for maintenance.

	1st year	*average*	*10th year*
annual running costs:	£328	£334	£341
average cost per hp-h:	13.1p	13.4p	13.7p

On the face of it, there is little to choose between these two power sources, at present day prices with the chosen wind regime. In a windier location the windmill would be cheaper and vice-versa.

What is illustrated is that certain energy converters are predominently influenced by day-to-day running costs, which are subject to inflation, while others involve one large capital investment, so that the annual charges are substantially fixed.

In practice, high capital-cost devices with low or zero running costs tend to be less economical where high interest rates apply, while low capital cost devices (such as engines) with relatively high running costs become progressively less competitive in times of high fuel and maintenance cost inflation.

Table 1
Variation of annual average fuel and maintenance costs over various periods at various inflation rates

Multipliers to apply to current annual average price to obtain mean annual average over the periods shown

inflation @	*5 years*	*10 years*	*20 years*	*30 years*
1%	1.02	1.05	1.11	1.17
5%	1.10	1.26	1.66	2.22
10%	1.22	1.59	2.86	5.47
15%	1.35	2.03	5.12	14.49
20%	1.49	2.60	9.33	

e.g. average annual cost over a ten year period with 10% inflation will be 1.59 times present annual cost.

formula:

average annual cost F_{av}
present annual cost F_1
interest rate i%
period n years

$$F_{av} = \frac{100.F_1}{n.i.} (1 + i/100) - 1)$$

Notes on Units and Tables of Conversion Factors

Unfortunately a wide variety of different measuring systems are in current usage internationally (and the UK is almost like the world in microcosm in this respect, since we are in the midst of a slow conversion process from our own traditional system which is still generally used in modified form in the USA, to the SI (Système Internationale) metric system). As a result the information supplied by manufacturers can be in all kinds of units.

We have generally sought to show SI units (or sometimes other non-SI metric units), in conjunction with British units where possible. However it proves very difficult to abandon some non-SI units such as the ubiquitous horsepower, which usually means more to people in terms of mechanical shaft power than kW. Although SI prefers either mm or m, we have given engine capacities in cc, which should of course be written as cc^3! Again, we would plead that this is what people are generally familiar with and it is certainly the most common engine capacity unit used by manufacturers.

The following tables of units and conversion factors may help in sorting out any confusion which may arise from the selection of different units quoted in this Guide.

A. Methods used for indicating multiples of ten and their standard prefixes

name	*numeric*	*exponent*	*prefix*	*standard symbol*
millionth	$\frac{1}{1,000,000}$	10^{-6}	micro-	μ
thousandth	$\frac{1}{1,000}$	10^{-3}	milli-	m
hundredth	$\frac{1}{100}$	10^{-2}	centi-	c
times thousand	x 1,000	10^{3}	kilo-	k
times million	x 1,000,000	10^{6}	mega-	M
times billion	x 1,000,000,000	10^{9}	giga-	G

B. Principal units in common usage; names and symbols (alternatives in brackets)

	Metric		*British/US*	
length	*millimetre	mm	inch	in (")
	centimetre	cm	foot	ft (')
	*metre	m	yard	yd
	kilometre	km	mile	m.
area	square metres	m^2	sq.ft.	ft^2
	hectare	ha	acre	acre
volume	cubic cm	cm^3 (cc)	cubic inch	in^3 (cu in)
	litre	l	gallon	gal
	cubic metre	m^3	cubic feet	ft^3 (cu ft)
mass	gramme	g	ounce	oz
	*kilogramme	kg	pound	lb
	tonne	t	ton	ton
velocity	*metres/second	ms^{-1} (m/s)	feet/second	ft/s
	kilometres/hour	kmh^{-1} (km/h)	miles/hour	mph (m/h)
			knots	kt
rotation/frequency	*herz	Hz	cycles/second	c/s
	revolution/min.	rev/min (rpm)	revolution/min.	rev/min (rpm)
	radoam/sec.	Ω	radian/sec	Ω
flow rate	litre/minute	l/min	gallon/minute	gal/min (gpm)
	cu.metre/second	m^3s^{-1} (m^3/s)	cu.ft/second	ft^3/s (cusec)
			cu.ft/minute	ft^3/min (cfm)
force	*newton	N	pound force	lbf
	*kilonewton	kN	ton	ton
	kilogrammeforce	kgf		
	tonne	t		
torque	newton-metre	Nm	pound-feet	lbf ft
	kilonewtonmetre	kNm		

	Metric			British/US	
work/heat/energy	calorie	cal		British Thermal Unit	BTU (B.Th.U.)
	kilocalorie	kcal		Therm	therm
	*joule	J		footpoundforce	ft lbf
	*megajoule	MJ		horsepower-hour	hp h
	gigajoule	GJ			
	watthour	Wh			
	*kilowatthour	kWh			

power	*watt	W		foot-pound/second	ft lbf/s
	*kilowatt	kW		horsepower	hp
	*megawatt	MW		brakehorsepower	bhp
	metric horsepower	CV (PS)			

electrical	*amps	A	(current)	volt-amps	VA
	milliampère	mA		kilovolt-amps	kVA
	ampères/hour	Ah			
	*volts	V	(potential difference)		
	ohms	Ω	(resistance)		

pressure	*pascal	Pa		pounds/square inch	
	megapascal	MPa		gauge	psig
	*bar	bar		pounds/square inch	
	kilogram per	$kgcm^2$		atmospheric	psia
	square centimetre	(kg/cm^2)		atmosphere	atm
				foot water	ftH_2O

*Preferred units are indicated by an asterisk.

C. Conversion factor tables

length	*mm*	*m*	*km*	*in*	*ft*	*mile*
	1	.001	10^{-6}	.0394	.0033	$5.4x10^{-7}$
	1000	1	.001	39.4	3.28	$5.4x10^{-4}$
	10^6	1000	1	39360	3280	.539
	25.4	.025	$2.5x10^{-5}$	1	.083	$1.4x10^{-5}$
	305	.305	$3.0x10^{-4}$	12	1	$1.9x10^{-4}$
	$1.6x10^6$	1609	1.609	63360	5280	1

area	m^2	*ha*	km^2	ft^2	*acre*	*sq. mile*
	1	10^{-4}	10^{-6}	10.76	$2.5x10^{-4}$	$3.9x10^{-7}$
	10000	1	.01	$1.1x10^5$	2.471	$3.9x10^{-3}$
	10^6	100	1	$1.1x10^7$	247.1	0.386
	.0929	$9.3x10^{-6}$	$9.3x10^{-8}$	1	$2.3x10^{-5}$	$3.6x10^{-8}$
	4047	.4047	$4x10^{-3}$	43560	1	$1.6x10^{-3}$
	$2.6x10^6$	259	2.590	$2.8x10^7$	640	1

volume	*l*	m^3	in^3	*gal (US)*	*gal (Imp)*	ft^3
	1	10^{-3}	61.02	.264	.220	.0353
	1000	1	6102	264	220	35.31
	.0164	$1.6x10^{-5}$	1	$4.3x10^{-3}$	$3.6x10^{-3}$	$5.8x10^{-4}$
	3.785	$3.8x10^{-3}$	231.1	1	.833	.134
	4.546	$4.5x10^{-3}$	277.4	1.201	1	.160
	28.32	.0283	1728	7.47	6.23	1

mass	*g*	*kg*	*t*	*lb*	*ton*
	1	.001	10^{-6}	$2.2x10^{-3}$	$9.8x10^{-7}$
	1000	1	.001	2.205	$9.8x10^{-4}$
	10^6	1000	1	2205	.984
	453.6	.4536	$4.5x10^{-4}$	1	$4.5x10^{-4}$
	10^6	1016	1.016	2240	1

velocity	*m/s*	*km/h*	*ft/s*	*mph*	*kt*
	1	3.60	3.28	2.237	1.768
	.278	1	.912	.621	.539
	.305	1.097	1	.682	.592
	.447	1.609	1.467	1	.868
	.566	1.853	1.689	1.152	1

rotation	Hz (or c/s)	rpm	rad/s
	1	60	6.283
	.0167	1	.1047
	.159	9.549	1

flow rate	l/min	m^3/s	(Imp) gal/min	ft^3/s
	1	1.7×10^{-5}	.220	5.9×10^{-4}
	60000	1	13206	35.315
	4.546	7.6×10^{-5}	1	2.7×10^{-3}
	1699	.0283	373.7	1

force	N	kN	kgf	t	lbf	ton
	1	.001	.102	1×10^{-4}	.225	1×10^{-4}
	1000	1	102	.102	225	.100
	9.807	.010	1	.001	2.205	9.8×10^{-4}
	9807	9.807	1000	1	2205	.984
	4.448	.004	.5436	4.5×10^{-4}	1	4.5×10^{-4}
	9964	9.964	1016	1.016	2240	1

torque	Nm	kNm	lbf.ft
	1	.001	.738
	1000	1	738
	1.356	1.4×10^{-3}	1

work/heat/energy	cal	J	Wh	BTU	ft.lbf	hp.h
(smaller units)	1	4.184	1.2×10^{-3}	3.9×10^{-3}	3.088	1.6×10^{-6}
	.239	1	2.8×10^{-4}	9.4×10^{-4}	.7376	3.7×10^{-7}
	860.4	3600	1	3.414	2655	1.3×10^{-3}
	252	1055	.293	1	778	3.9×10^{-4}
	.324	1.356	3.8×10^{-4}	1.3×10^{-3}	1	5.0×10^{-7}
	6.4×10^5	2.6×10^6	745.7	2546	2.0×10^6	1

(larger units)	kcal	MJ	kWh	BTU	hp.h
	1	4.2×10^{-3}	1.2×10^{-3}	3.968	1.6×10^{-3}
	239	1	.2778	947.8	.3725
	860.4	3.600	1	3414	1.341
	.252	1.1×10^{-3}	2.9×10^{-4}	1	3.9×10^{-4}
	641.6	2.685	.7457	2546	1

power	W (or J/s)	kW	CV	ft.lbf/s	hp	BTU/min
	1	.001	1.4×10^{-3}	.7376	1.3×10^{-3}	.0569
	1000	1	1.360	737.6	1.341	56.9
	735	.735	1	558	1.014	41.8
	1.356	1.4×10^{-3}	1.8×10^{-3}	1	1.8×10^{-3}	.077
	746	.746	.9860	550	1	42.44
	17.57	.0176	.0239	12.96	.0236	1

power flux	W/m^2	kW/m^2	hp/ft^2
	1	.001	1.2×10^{-4}
	1000	1	.1246
	8023	8.023	1

calorific values	cal/gm	MJ/kg	BTU/lb
	1	4.2×10^{-3}	1.80
	239	1	430
	.556	2.3×10^{-3}	1

1

Solar Energy

SOLAR ELECTRICITY
SOLAR ENGINES
SOLAR HEATING

The sun is indirectly responsible for most of our existing energy production, millions of years of solar radiation having produced the fossil fuels which we are so rapidly consuming. It is also the heat of the sun which evaporates enough sea water to produce the streams and rivers from which we tap hydro-electric power and of course it is the radiation from the sun that is essential to the photosynthesis process which is the basis for plant growth on which all life on earth ultimately depends.

Despite the obvious importance of the sun, recognised by mankind from earliest times, the direct use of solar energy (other than for naturally occurring processes such as crop-growing), has proved difficult. This is partly because of the intermittent nature of sunlight and also partly due to its diffuseness as an energy source compared with the concentrated energy in fossil fuels.

The sun emits about 3.8×10^{23} kW and is expected to continue doing this for 4 billion years or so to come (according to current scientific thinking). A minute proportion of this strikes the Earth, averaging around 1.7×10^{14}kW, of which nearly a third gets reflected back into space, giving an average daily energy supply from the sun of about 2.9×10^{15}kWh reaching the earth's surface. Total world daily energy consumption during 1974 averaged around 1.8×10^{11}kWh – so our current energy needs are only around one sixteenthousandth (1/16000) of what the earth receives every day; put differently – the earth receives every six seconds more energy from the sun than the entire human race consumes in 24 hours. Clearly there is no shortage of solar energy if we can find effective ways of harnessing it.

However the problems of using low-grade heat efficiently and of storing energy during the night or cloudy periods, economically, have made it difficult to tap much of this enormous energy supply. Hopefully, the combination of increasing technical skill and the fact that fossil fuels are being depleted will lead to a great increase in the use of this benign and pollution-free energy source. Certainly, if the equipment listed in the following section provides any indication, we are at the beginning of a mushrooming international solar energy industry, as very many of the devices listed (and their manufacturers) have only appeared within the last few years. Even the oldest established manufacturers of solar water heaters have only been in business for 25 years or so, (solar water heaters were in fact used in California and Florida during the early part of this century but these "prehistoric" designs disappeared when cheap natural gas killed the industry off until the recent revival of the 1970s).

This is not the place to go into any great detail on the technicalities of using solar energy, but a number of basic principles apply and are worth bearing in mind when reviewing the equipment listed.

Some characteristics of solar energy

Obviously the economics of any solar energy utilisation system will depend on the available solar energy at the proposed location. Just under 1kW/m^2 is the maximum solar energy received at ground level on a surface perpendicular to the sun's rays. If the atmosphere is clear, the humidity low, the energy flow per unit area of a surface perpendicular to the sun's rays received near the North Pole can be similar to those near the equator. The main reason why the tropics are so hot is that the ground is more nearly perpendicular to the sun's rays, so the average energy intensity as it hits the ground is considerably greater. Also, of course, the number of hours of unobstructed sunlight available at any location will greatly influence the viability of any solar-powered device. As with wind-power, the availability may vary considerably from day to day and with the seasons, but the total annual energy received at any particular place will not vary greatly from year to year. Therefore the outputs from solar-powered devices over a period of time such as a season are reasonably predictable.

For example, the UK, not one of the sunniest parts of the world, receives a maximum of about 900 W/m^2 on a horizontal surface at midday in mid-summer in clear weather, while at midday in mid-winter in clear sunny weather this figure falls to 200 W/m^2. Hence the best that can be expected from a wholly clear summer day in the UK is around 8.5kWh, while the worst possible result on an overcast, short winter day might be around 0.8kWh. Britain in fact receives a total of around 900kWh/m^2 per year, which averages a power supply of 100W per square metre if calculated over every hour of an 8760 hour year. By contrast, one of the more favourable areas for collecting solar energy, such as the desert regions straddling the Tropics of Cancer and Capricorn, receive some 2000kWh per year, giving an all-the-year-round hourly average of 230 W/m^2. These figures of course apply only to a horizontal surface – if a surface is inclined to track the sun and remain perpendicular to its rays, greatly increased amounts of energy may be received. Since tracking the sun continuously involves complicated and expensive mechanisation, a good compromise is to site solar collector surfaces at an angle close to that of the latitude of the location, facing south in the northern hemisphere and vice-versa in the southern. This gives the optimal direction from which to receive the sun's rays as near to perpendicular as possible for the maximum possible time. In practice it is usual to set fixed collectors at an angle more like the latitude plus 10 or more degrees to make the optimisation relatively better in winter when greater efficiency of collection is more important. In some cases it is worth adjusting the angle seasonally in order to gain a slight increase in received energy per unit area of collector.

Utilisation of solar energy.

The main problems of using the sun's energy are its rather low concentration and intermittent nature. We have processes to convert solar radiation directly into heat or into electricity, or we can use the heat to power a heat engine to produce mechanical power. This section therefore considers the available solar energy conversion devices in three subdivisions – solar electricity, solar engines and solar heaters.

It is possible to overcome the problem of the diffuse nature of sunlight by concentrating it with systems of lenses or, more often, with mirrors, which are cheaper for larger applications. Just as with the well-known example of a magnifying glass, this creates a more intense energy flux which can achieve the temperatures commonly reached with fossil fuels. However, there are major disadvantages inherent in solar concentrators – firstly they can only focus direct sunlight and they therefore cease to function effectively under cloudy conditions and secondly, they need to track the sun in order to be kept in the correct alignment with the rays of sunlight, which inevitably introduces mechanical problems that can greatly add to the cost of such a system compared with a passive non-concentrating solar collector. As a result, most solar energy conversion systems are set up to use unconcentrated solar energy at the optimum angle for the location. Many of these can benefit from diffuse sunlight received through clouds. For example, if a clear sky yields 800 W/m^2, slight haze can reduce this to 600 W/m^2, light overcast conditions to 300 W/m^2 and heavy overcast would typically be 150 W/m^2.

The intermittent nature of sunlight raises different problems. Obviously it can readily be applied for end-uses which are not very time-dependent – for example solar distillation – but a solar cooker may be inconvenient if its owner wants a hot lunch on a cloudy day and it is useless for cooking supper. It is technically feasible to provide energy storage to permit energy derived from sunlight to be available on cloudy days or at night, but in many cases it is difficult to do this at an economic price compared with the energy stored in fossil fuels. This is why solar cookers have been generally unsuccessful, even in hot sunny climates, but simple solar distillation systems where the product can be cheaply stored, have been economically applied for many years in a number of countries. Similarly, solar water heating is readily economic even in quite cold countries, as hot water can be stored at low cost in an insulated tank, but solar space heating in a cold country is more difficult (as is solar air conditioning in hot countries) as much larger quantities of energy require storage to cover nights and extended cloudy periods. One of the most common applications for solar-mechanical outputs is for pumping water for irrigation, since the need is greatest under sunny conditions, and water can be stored at low cost. Solar generated electricity, on the other hand, can only conveniently be stored in batteries, which is only economic for low output applications where limited quantities of stored electrical energy are sufficient. These problems will be discussed in more detail in the introductory passage to each of the categories of solar energy in which this section is sub-divided.

Solar energy economics

Obviously, the energy that can be converted from solar radiation at a particular location by a particular device will be proportional to the area of collector surface (or of concentrator surface) (assuming the collector is at the optimum alignment to maximise the received solar radiation). The efficiency of conversion and of storage will then govern what is usefully available from the system. The collector tends to be the most critical part of a system in terms of achieving reasonable efficiency and it is also generally the largest and most expensive component. Almost all solar energy conversion devices involve compromises aimed at minimising the cost per square metre of collector area while maintaining acceptable efficiencies. The most efficient energy converters are generally less cost-effective in most applications due to a disproportionately high cost per unit area, while very inefficient devices are also not cost-effective as they require excessive areas of collector to produce a given output. Some of the most economic solar energy systems are ones where the collector serves a dual purpose – for example where it forms the actual roofing for a building (rather than being placed on top of the roof to convert an existing building).

Another important aspect of any solar-powered device is its likely life expectancy. Since the running costs are very low, consisting of nothing more than whatever minor maintenance might be needed, (the sun being free), the main cost elements are whatever notional interest charge is payable on the capital investment involved in the installation and of course a depreciation allowance to permit eventual replacement. Hence the life-expectancy, capital cost and the consequent interest charges are the primary elements of the cost of converting solar energy. Therefore a low cost system may not be economic if it has too short a life, while on the other hand a system with an almost indefinite life might be feasible, but it may not be economical as its high cost would never be compensated by its value in the distant future (and it may outlive its usefulness and be overtaken by improved lower-cost technology within its long lifetime, raising the question of whether to replace prematurely). Therefore, much of the design decision-making in solar energy technology consists of finding appropriate compromises between cost and efficiency and cost and life expectancy to suit the technical requirements as well as the economics of the end use.

These elements will be briefly described in conjunction with some of the technicalities of the different solar energy conversion processes in the three subsections dealing with solar-electric, solar-mechanical and solar-thermal conversion.

Solar Electricity (photo-voltaics)

A physical phenomenon known as the "photo-voltaic effect" allows light energy to interact with electrons present in suitably arranged semi-conductor combinations, and produce a voltage across two terminals. If an electrical load is connected across the terminals, a d.c. electricity flows and light energy is converted into electrical energy. Devices that produce this effect are called photocells, and a variety of different materials have been found to be active in this respect.

The most commonly produced type of photocell is based on silicon (one of the most abundant elements, being a primary constituent of sand – but used in an extremely refined form for photocells). Others commonly available are based on cadmium sulphide, gallium arsenide, and a number of other materials. Probably one of the best known applications of photocells is in the exposure meters of photographic cameras – since most photocells generate a voltage proportional to the intensity of the light falling on them, they are very suitable for measuring light intensity in conjunction with a voltmeter.

Unlike most other electrical generators, solar cells are normally not damaged by being short-circuited or open-circuited – they simply develop a voltage to suit the load and the availability of electrons motivated by the amount of incident light energy. Their output is roughly proportional to the received light energy intensity and the active area of the cell. Typical conversion efficiencies are of the order of 5 to 10% of received solar energy, depending on cell type and construction and assuming that the load and photocell array are suitably matched to permit output at the optimum voltage to maximise the electrical power.

As there are no moving parts and it is possible to encapsulate the delicate cell surfaces and contacts in a robust synthetic resin, very good reliability coupled with a long active life, of the order of 20 years or so, is normal. Also, if one solar cell fails it does not normally affect the functioning of its neighbours, and the output can be easily adjusted by adding or removing cell modules.

The main problem with photocells is their high price. This results largely from the high cost of developing and manufacturing them rather than from the intrinsic value of their materials (particularly in the case of silicon solar cells). Originally the only "power" applications were for milli-watt sized loads such as the voltmeter part of a light-meter – but the development of space satellites gave a great boost to solar cell development as they were applied to quite large power applications to run electronic systems in space. Some of the large space satellite installations involved 100 watt plus outputs, and of course the high cost of the solar arrays in relation to their output was unimportant in that application as there were no "conventional" alternatives that could be applied. In the early 1960s, photocells cost hundreds of pounds per peak watt. Increased production and the improvement of manufacturing techniques had brought the price to large buyers down to around £15/peak watt in 1977 (i.e. peak watt means the output in bright sunlight with a solar radiation of around 1kW/m^2) and £10/peak watt by 1978.

Just as the price of electronic pocket calculators and transistor radios dropped enormously once large-scale production began, so it is anticipated that photo-voltaics will come down to around £1/peak watt during the early 1980s and it is thought they may be down to prices of the order of £0.25/peak watt within 20 years. As their price falls, the range of economic applications increases, which allows a higher production rate with accompanying further economies, and so on.

A few years ago photocells were only viable for ultra-low power applications (such as cameras) or for "expense no object" applications like space satellites. Today they are available in modules offering outputs in the 1 to 100 watt range, where their reliability, long-life and freedom from any maintenance requirements make them economic for a variety of specialised remote applications such as keeping batteries charged in inaccessible places, (e.g. for microwave repeaters, navigational aids for ships and aircraft, remote emergency equipment, telephones, railway signals, etc.). A typical module rated at 10 watts peak (in full sunlight), will on average produce 270 watt hours per week under a Mediterranean type of climate – this would be down to an average of only 160 watt hours per week in the UK, but might be as high as 400 watt hours per week in a favourable Middle East location. Its 1979 price would be under £100, and it could be expected to last up to 20 years and involve virtually no running costs or maintenance.

Once the price falls by another factor of ten, as is possible during the next decade, it will become economic to use much larger arrays of solar cells to produce power outputs of several kilowatts peak in order to pump water, power remote dwellings, hospitals or schools and to generally replace small petrol and diesel engines in remote areas.

A price drop of another magnitude to a probable minimum of the order of £0.20 (1977 values), possibly by the late 1980s or 1990s may allow larger scale electricity generation from solar energy – perhaps using decentralised local electrical systems in preference to centralised systems.

Energy storage with solar cells would normally be through the use of batteries; (this subject is discussed in more detail under the section dealing with electrical equipment). Sealed lead-acid batteries or nickel-cadmium cells tend to be better suited for most solar-electric applications, as their low (or zero) maintenance requirements match the similar characteristics of photo-voltaic cells, and so long as solar cells are so expensive and they are only used for rather low powered applications, the battery cost element is quite small compared with the photocell costs.

Optimisation of solar-cell and battery combinations in order to ensure sufficient storage capacity combined with sufficient charging capability is a specialised skill for which most of the manufacturers have developed computer programmes. Accurate records of sunshine hours and intensity are an important input requirement if system costs are to be minimised. Several manufacturers market monitoring devices, often based on a single solar cell, which can be set up in order to measure the radiation received over predetermined periods of say one month, which aid the correct specification of photo-voltaic power systems.

Solar engines

In the nineteenth century some of the first attempts to utilise solar energy actively involved the use of small steam engines, with boilers heated by solar concentrators. One of these successfully ran a printing press at the great Paris exhibition of 1878. A solar steam engine producing 50hp (37kW) ran successfully in Egypt as long ago as 1913. Yet, ironically, today little is commercially available (as can be seen from the following section) although a number of interesting devices are under development (one or two of which are indicated in section 10 on R&D).

The main problem when using solar energy to drive a heat engine is to achieve optimum economics for the system. Heat engines work most efficiently from a high temperature source, while solar collectors are most efficient supplying heat at low temperatures. Early solar engines tended to run at high temperatures so as to use reasonably conventional engines and consequently needed large areas of solar concentrator which had in turn to be constantly oriented to track the sun. Modern developments, such as that by Sofretes, involve the use of special low temperature engines running on heavy organic vapours rather than on steam which can operate at lower temperatures that permit the use of non-tracking, more efficient, flat-plate collectors.

One of the primary applications for solar-powered engines is for pumping irrigation water, where the needs are closely matched to the availability of solar energy. Other applications might be to drive solar cooling (air conditioning or refrigeration) systems, where again the needs are closely matched to the energy availability. It seems likely that there will be considerable development in this field during the coming years, although in the long run it may be difficult for solar heat engines to compete with photo-voltaics driving electric motors once the price of solar cells comes down for small-scale applications. Various large-scale solar power stations are being investigated, including one based on a space satellite in stationary orbit which could beam its power to earth as microwaves and which would operate 24 hours/day. Other systems using indirect solar heat, such as is present in ocean thermal gradients (the difference in temperature between the surface and the depths of the oceans) are under evaluation — obviously such developments are well outside the scope of this guide, but they do indicate how important solar energy is likely to be as a primary source of energy in the future.

Solar Heating

It can be seen from the large number of solar heating designs in the following section that this is one of the most common applications for solar energy.

Solar water heating is quite widely in use in some of the sunnier parts of industrial countries, such as the southern USA, northern Australia and Japan and it is becoming a growth industry even in less sunny regions such as northern Europe.

Solar space heating is more difficult to arrange as the needs tend not to coincide with the times when solar energy is most available, hence heat storage becomes a major element of solar space heating systems and these are best designed into the structure of a building rather than added to a conventional building.

As a result, it is solar water heating that offers the best prospects for being economic. In many cases solar water heaters may be used in conjunction with thermostatically controlled electric water heating in order to save electricity and reduce the running costs. In any case it is normal to evaluate solar water heating economics in terms of what it would cost to heat the same amounts of water using fossil fuels or electricity or whatever is being substituted for by using solar energy.

Solar water heaters are passive devices; the only mechanical component needed, and then only in some cases, is a small electric pump needing very little power. As this is not an essential, solar water heating can be used in locations away from an electricity supply. Unlike photocells or solar engines, solar water heaters can readily be manufactured on a small-scale with low cost conventional workshop equipment, and as a result, many small manufacturers all over the world are now making them. There are probably many more manufacturers, particularly in the USA, than have been included in this edition. It is also encouraging that a number of engineering companies in Third World countries are also beginning to manufacture solar water heaters. It is hoped particularly to include other Third World manufacturers in future editions and any who are not included are invited to get in touch with us. It is well worth seeking a nearby manufacturer, because solar water heaters are relatively bulky and it is not very economic to import them in quantities from distant industrial countries due to high transport costs. Hence it is likely that a local manufacturer, even producing quite a small output, can compete with an overseas supplier quite successfully. In view of this a number of established manufacturers in industrial countries expressed interest in negotiating licenses to permit overseas manufacture of their products. Also a number of small engineering companies in non-industrial tropical countries have found it worthwhile to carry out some research and development to develop their own designs. Both these trends can be expected to continue.

Technical aspects of solar water heaters

As with most solar energy systems, the solar collector is generally the largest and most expensive element of a water heating system. In general it consists of a blackened absorber plate which is heated by the sun's rays. The absorber carries water pipes or channels which are either integral with it or attached in such a manner as to obtain good thermal contact to allow heat to flow from the absorber into the water.

A blackened absorber with water flowing through it, or containing water is the simplest form of solar water heater. If it is intended to heat the water to temperatures significantly above ambient, and to keep it warm once it has been heated, even after dark, then various steps must be taken to prevent heat loss. This is because the temperature attained by the absorber will cease to increase once the rate of heat loss equals the rate of heat input. Obviously, the more the heat losses can be cut down the hotter and more efficient the absorber will get. Therefore it is normal to insulate the back and sides of the absorber to minimise heat losses, and to have a glass or plastic transparent cover over the side facing the sun. The latter serves two purposes — it traps solar energy by the so-called "green-house effect" under which glass readily passes short visible-light wave lengths but is opaque to longer heat wave lengths. As a result, a glass cover is like a one-way valve — it lets in light-energy easily, but obstructs the departure of heat-energy. The other function of the glass is to trap a layer of air near the absorber to prevent the cooling effects that would be present if wind or convection could cool the surface. Multiple layers of glass allow higher temperatures to be attained. So-called "selective surfaces" consisting of various metallic oxides can be applied to metallic absorber surfaces which have a similar function to a glass cover in that they absorb light energy readily but do not readily allow heat to radiate — certain solar absorbers have this feature which improves the overall efficiency of the system.

The solar collector is normally connected by two pipes to an insulated storage tank — one pipe leads hot water to the top of the tank while the other, running from the bottom of the tank, lets cooled water return for reheating in the absorber. Sometimes the hot water from the absorber does not mix with the water in the tank, but is circulated through a copper (or other metal) heat exchanger in the tank. Then the hot water to be drawn from the tank does not actually pass through the solar heater but is indirectly heated, and the liquid in the solar heater remains unchanged — which is an advantage as it allows anti-freeze and anti-corrosive additives to be added to the water in the solar heater. In some cases liquids other than water may be used in certain types of solar water heater offering indirect heating of the water in the storage tank.

If the storage tank can be sited at a point higher than the top of the solar water heater, it is possible to take advantage of the fact that hot water is less dense than cold water to allow the hot water to "thermo-syphon" up to the tank and displace colder water down to the solar water heater via the second return pipe. Systems such as this do not require any pumps or mechanical moving parts, as when there is no sun, thermosyphoning stops and the hot water remains at the highest part of the system inside the insulated tank. However, in some cases it is not easy to arrange for the tank to be above the solar water heater, in which case small electric circulating pumps of the kind used for central heating systems are generally used — these have to be controlled by an electronic temperature sensing system which switches the pump on automatically whenever the temperature at the absorber exceeds the temperature at the top of the storage tank, and vice versa. Most of the smaller low cost systems try and use the thermosyphon effect to keep the costs down.

A few types of solar water heater consist of a complete combined solar collector and tank. The capacities of such systems are usually quite small, although they can of course by inter-connected in many cases to increase the availability of hot water.

A few systems heat air rather than water at the absorber. Systems using air do not suffer from the same corrosion and freezing problems that can affect water-filled absorbers, but for water heating an air to water heat exchanger is needed and fans are also usually needed to drive the air.

Many of the solar water heaters marketed in industrialised countries are intended for heating swimming pools, as the fuel bills for heating even quite a small swimming pool with electricity or oil are very high due to the large volumes of water involved. Swimming pool heaters usually need to heat a lot of water with only a small temperature rise, so they are designed to be of low cost per square metre and less efficient than the heaters designed for domestic hot water supplies. Some of these might be useful for domestic water heating in hot countries where the temperature increase required is not as great as in the colder parts of the world and a lower efficiency would therefore be acceptable. They would also be applicable to preheat water for other systems.

Many different types of absorber are offered by different manufacturers, some being aluminium, some copper, some steel and some plastic. Technically copper has a lot of advantages, being both a very good conductor of heat and also resistant to corrosion; unfortunately it is by far the most expensive absorber material generally used and may in some cases be "too good" to be economic. Aluminium "roll-bonded" absorber panels have been marketed by a number of firms, but they cannot operate with ordinary water inside as they corrode too quickly — instead inhibitors have to be used in the water or else special oils are used as heat carrying fluids instead of water. Steel does not conduct heat as well as copper or aluminium, but is much cheaper than either. It is less corrosion prone than aluminium, particularly if it is used in an indirect system with a corrosion inhibitor. Alternatively thick steel pipes or stainless steel panels may be used. Composite systems in which copper pipes are mounted into an aluminium absorber are also sometimes used. Plastic absorbers have been quite widely introduced, particularly for lower temperature applications — care must be taken with most of these that they cannot accidentally be drained of water while exposed to sunny conditions as this can lead to overheating which may distort the plastic. Obviously plastic is completely corrosion resistant, but it may eventually degrade under strong sunlight.

Other uses for solar heaters are solar distillation, crop-drying, timber seasoning, etc. As most of these applications require installations that are constructed on site, complete systems are not normally available ready-built, but advice or plans for their construction may possibly be obtained from some of the organisations who offer consultancy services (listed in section 10).

Solar heater economics

In common with most devices using renewable energy resources and consequently not having any significant "running costs", the economics of solar water heaters depend on achieving the optimum compromise between the conflicting requirements of achieving good efficiency, low cost per unit area of energy absorbing surface and an adequate life expectancy for the system. The other important element in the economic calculations is the value of the energy produced, which generally depends on the cost of the energy being substituted for, such as fuel oil or electricity.

The following table indicates approximate payback periods when solar heating is valued against different fuel costs. The calculations assume unchanging fuel costs during that period (which is most unlikely) so they are rather pessimistic. Typical UK solar energy figures amounting to 900kWh/m^2 per year are compared with a hotter country receiving 1400kWh/m^2 per year. The payback periods are referred to a 4m^2 collector having three possible initial costs for the system, namely £200, £300 and £400; the lower value would relate to a thermosyphoning simple system installed on a do-it-yourself basis, while the higher level relates to an electrically pumped system, with electronic temperature sensing, that is professionally installed. For convenience, £5 per annum is deducted in all cases to cover pumping costs, whether or not a pump is used, making the figures for a thermosyphon system even more pessimistic.

This table shows how the unit cost of the heating source being replaced greatly affects the pay-back period. Obviously if inflation of fuel costs is taken into account, shorter pay-back times would be achieved.

Although solar water heaters can pay for themselves in a reasonable period with electricity prices at the levels that they are in many countries, and they can also save valuable fossil fuels in many countries where mains electricity is produced by diesel or thermal power stations, it has often needed political action to help establish them in the few countries where they are widely used. This is because the capital cost of an electric immersion heater is much lower than that of a solar heating system and the decision as to which system to install is usually in the hands of the building contractors, rather than of the future residents who would need to pay for the electricity that will be used. A solar heated house tends to be slightly more expensive to build but is a lot cheaper to run, both for the residents and for the economy of the area in which it is located. Solar heating can also offer significant local job creating possibilities since the equipment lends itself to small scale manufacture.

Payback periods in years for systems with different first costs

Fuel costs p/kWh	*a) Received energy 900kWh/m²/yr*			*b) Received energy 1400kWh/m²/yr*		
1p = £0.01	*£200*	*£300*	*£400*	*£200*	*£300*	*£400*
0.5	65	97	129	26	39	53
1.0	18	27	36	10	15	20
1.5	10	16	21	6	9	12
2.0	7	11	15	4	7	9
2.5	6	8	11	3	5	7
3.0	5	7	9	3	4	6
3.5	4	6	8	2	4	5

Solar Electricity

BRIAU S.A.
B.P. 43
37009 Tours Cedex
France

Telex: 750729 F
Phone: (47) 61 38 17

Suppliers of solar pumping installations driven by photo-voltaic modular array. The number of solar collector modules, inclined to the optimum position for the latitude where the unit is installed, can be varied to suit the required power output. Units can raise water from 50m or more and can discharge to any required height. Automatic start/stop units are incorporated to guard against no load pumping and water loss through over filled tanks. The inclusion of a battery bank allows continuous running even during periods of zero insolation.

SOLAPAK PRODUCTS
14 Brooksfield Rd
London W4 IDQ
U.K.

Telex: c/o 53374 Victor Wallsend
Phone: 01-994 8173

School House
Covent Usworth
Washington
Tyne & Wear
NE37 1NU
U.K.

Telex: 537681 ALWRLD
Phone: (0632) 464646

Large selection of modular arrays (Solarex USP) producing up to 300 watts/module. The constituent Solapak panels come in a variety of sizes from 1 to 9 watts of 1.5 to 14V.

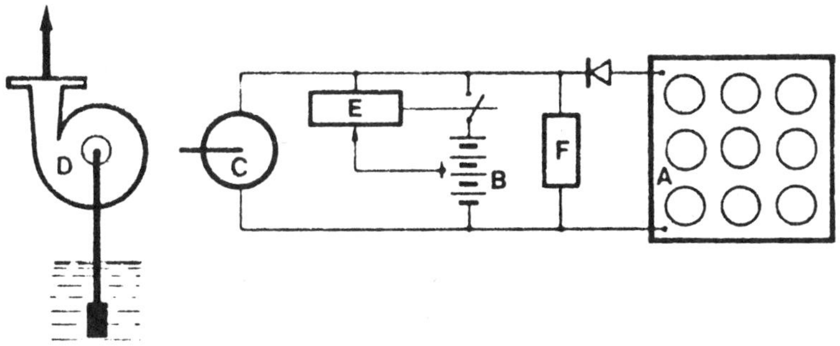

A: Solar Panels
B: Accumulator battery
C: Motor
D: Pump adapted to needs of installation
E: Electric control cabinet
F: Limit switch for charge

Briau Photopile Generator for Pumping

LUCAS SERVICE OVERSEAS LTD
Thame House
Windmill Road
Haddenham
Aylesbury
Bucks HP17 8JB
U.K.

Telex: 83615
Phone: Haddenham 291681

Supply silicon solar cells encapsulated in silicone rubber. Various modular arrays are available to suit different demands from 1 to 30W and 2 to 12V. The basic modules available are:

Model	*Charging point data*			*Peak output data*			*No. of cells*
D0206	–	–	–	0.6A	2.2V	1.3W	5 of 55mmØ
E1215	1.53A	13.5V	20.7W	1.45A	17.5V	25.4W	36 of 90mmØ
E1206	0.61A	13.5V	8.2W	0.56A	17.0V	9.5W	36 of 55mmØ

Series "E" Solar Array

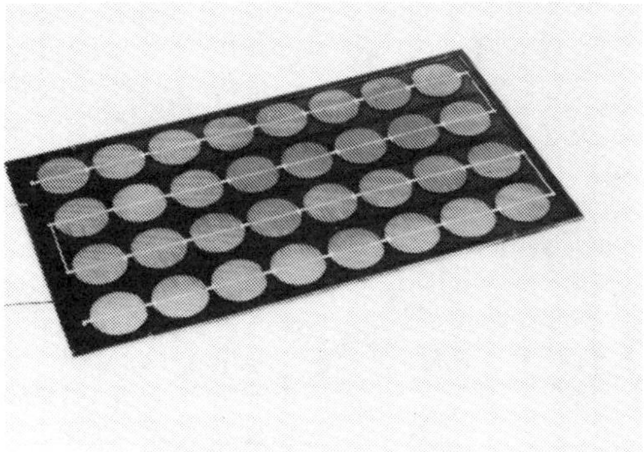

Solar Power Module MST 102

FERRANTI LTD
Gem Mill
Fields New Road
Chadderton
Oldham OL9 8NP
Lancs
U.K.

Telex: 668038 Ferem G
Phone: (061 624) 0515

Manufacturers of silicon solar cell modules suitable for charging 12V batteries. The cells are encapsulated in silicon resin with an optional polycarbonate cover for harsh environments. Two models are available:

Model No.	*Maximum power point*			*No. of cells*	*Shipping weight*
	Volts	*Amps*	*Watts*		
MST 102	13.0	0.54	7.00	32	1kg
MST 103	–	–	–	36	1.1kg

A new range based on larger cells is under development.

SOLAR ENERGY COMPANY
World Wide Energy Systems
P.O. Box 649
Gloucester Point
Virginia 23062
U.S.A.

Cables: Solenco

This company offers a range of individual silicon solar cells ranging from 2mA at 0.45V up to 2000mA at 0.45V (optimum) for the customer to assemble and in addition there is a wide range of modular cell assemblies as follows:

	Suncharger	236	1.5V	250mA	0.375W
	"	296	14V	650mA	9.1W
	"	346	14V	250mA	3.5W
	"	348	12V	150mA	1.8W
	"	372	12V	250mA	3.0W
	"	421	12V	580mA	6.96W
	"	704	14V	250mA	3.5W
	"	1626	2.5V	250mA	0.625W
	"	1270	6.5V	1.2A	7.8W
	"	2938	14V	650mA	9.1W
	"	2700	28V	750mA	21W
Super	"	3348	12V*	1.8A	21.6W
"	"	6348	12V*	3.5A	42W
"	"	8348	12V*	4.7A	56.4W
Solar power pack		S6348PP		(200Wh/day)*	

*These are available with optional voltage outputs of 12, 24 or 36V to order.

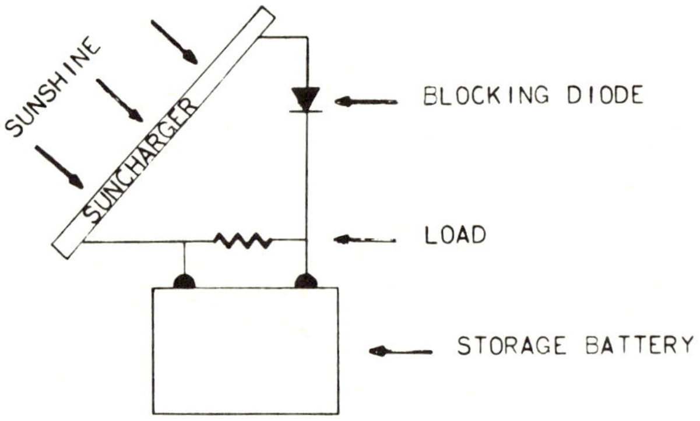

Solar Energy Co. Array

Solarex Unipanels

SOLAREX CORPORATION
1335 Piccard Drive
Rockville
Maryland 20850
U.S.A.

Telex: 710 828 9709
Cables: Solarex
Phone: 301 948 0202

U.K. Distributor: SOLAPAK PRODUCTS, 14 Brooksfield Road, London W4 1DQ (Phone: 01-994 8173).

This company manufactures a wide selection of individual silicon solar cells as round, rectangular and quadrant shapes for customer assembly, plus a range of modular silicon cell panels with the cells encapsulated in silicon rubber, the trade name for the assemblies being "Solarex Unipanels".

Model No.	*Nominal Voltage*	*Output at 100mW/cm²* *Amps*	*Watts*	*Watt hr/week (US average)*	*watt/ft² (min)*	*weight lb*	*kg*
215	12	0.125	1.5	45	4.0	0.7	0.3
230	12	0.25	3.0	100	4.0	1.3	0.59
260	12	0.5	6.0	200	4.0	2.3	1.06
1260	12	0.5	6.0	200	5.0	2.0	0.905
1270	12	0.58	7.0	220	5.8	2.0	0.905
280	12	0.62	8.0	240	5.2	2.3	1.06
615	6	0.25	1.5	45	3.7	0.7	0.3
950	9	0.5	5.0	160	5.3	1.8	0.8
435	14	0.25	3.5	105	4.2	1.4	0.64
470	14	0.5	7.0	220	4.4	2.5	1.15
480	14	0.62	8.0	270	5.0	2.5	1.15
1470	14	0.5	7.0	220	5.0	2.1	0.97
1480	14	0.58	8.0	260	5.8	2.1	0.97
670	7	1.0	7.0	220	4.4	2.5	1.15
770	7	1.0	7.0	220	4.9	2.3	1.06
785	7	1.2	8.5	280	6.0	2.3	1.06

Also a "Solar Electric Monitor" type 203.

Solar Power Corp. Array

SOLAR POWER CORPORATION
5 Executive Park Drive
North Billerica
Massachusetts 01862
U.S.A.

Telex: 710 347 6792
Phone: 617 667 8376

110-111 Strand
London WC2E 0DA
U.K.

Phone: 01-386 8918

This company is a subsidiary of the giant Exxon Oil Group and produces a series of silicon solar cell modules encapsulated in UV stabilised silicon rubber and available with various types of carrying frame or legs incorporating different combinations of the basic modules to give various output voltages. Some examples of basic modules available are:

Model No.	*Charging point*		*Peak power point*		
	Volts	*Amps*	*Volts*	*Amps*	*Watts*
1002	–	–	2.3	0.6	1.4
E4-125-0.6	4.5	0.64	4.9	0.6	2.9
E6-185-0.6	6.8	0.64	7.7	0.6	4.6
E6-365-1.2	6.8	1.29	7.7	1.2	9.2
E6-369-3.0	6.8	3.22	7.9	3.04	24.0
E12-365-0.6	13.6	0.65	16.0	0.6	9.6
E12-369-1.5	13.6	1.62	16.5	1.52	25.0
G12-361	13.6	2.00	16.5	1.90	31.3

The last model, prefixed G, has a tempered glass cover which is claimed to be of lower cost, good weathering resistance and durability and to keep the cells cooler.

This company offers computerised system design to match cells and batteries optimally and a single cell solar insolation monitoring unit is also available.

SOLAR SYSTEMS INC.
Division of Silicon Sensors, Inc.
Highway 18 East
Dodgeville
Wisconsin 53533
U.S.A.

Phone: (608) 935 2707

Solar Systems Array

This company produces a wide selection of individual silicon, cadmium sulphide and selenium photo-voltaic cells suitable for the customer to assemble, and in addition offers the following purpose-built modular arrays of silicon photo cells for power production: (data applies with 100mW/cm^2 and 25°C standard).

Model designation *	*Nominal output data*†			*Alternative voltage at same power*
	Volts	*Amps*	*Watts*	
SPM-75-2	2	.32	.64	2, 4, 8 and 16
SPM-100-2	2	.48	.96	2, 4, 8 and 24
SPM-150-2	2	.72	1.44	2, 6, 12, 18 and 36 (& 3.6 economy)
SPM-200-6	6	.18	1.08	6 and 18 (also derated at 2V)
PCM-510 Power-Mod	2	.044	.09	
PCM-540 Power-Mod	2	.10	.20	

*the final suffix number gives the voltage; for example the 6V version of SPM-75 is SPM-75-6.
†the output data for the lowest voltage version is given as an example.

SPECTROLAB INC.
(Subsidiary of Hughes Aircraft Corp.)
12500 Gladstone Avenue
Sylmar
California 91342
U.S.A.

Telex: 910 496 1488
Phone: (213) 365 4611

Spectrolab produce the "Solectric" series of modular silicon solar cell arrays. The silicon cells are embedded in clear silicon rubber and covered with a glass faceplate with an extruded aluminium "I" beam backing.

Five basic modules which can readily be linked by bus bars are available. Their nominal outputs with 100mW/cm^2 light intensity at 28°C are as follows:

Nominal voltage	*Minimum test performance*			*Length**		*Weight*	
	current	*voltage*	*power*	*in*	*cm*	*lb*	*kg*
4V	1.0A	5.4W	5.4V	29	73	3.6	1.63
6V	1.0A	8.1V	8.1W	42	107	5.2	2.36
6V	0.5A	8.1V	4.05W	22	57	2.8	1.27
12V	0.5A	16.2V	8.1W	42	107	5.2	2.36
14V	0.5A	18.9V	9.4W	49	124	6.0	2.72

*the lengths are to nearest whole inch or cm.

3kW Solectric Array

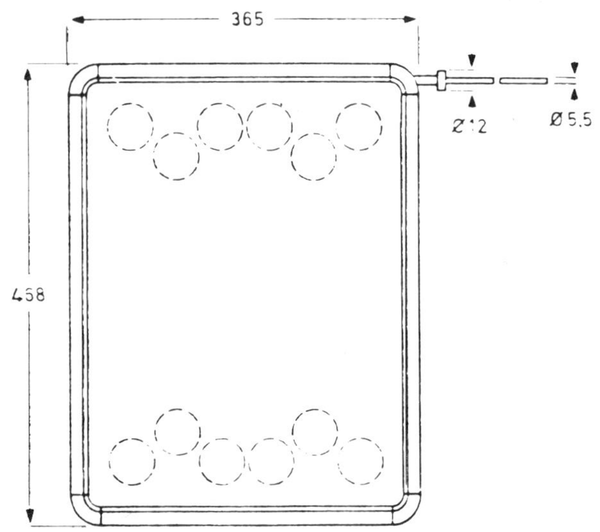

Solar Cell Module BPX47A

Philips Nederland BV
Afd. Elonco
Boschdijk 525,
Eindhoven
Netherlands

Phone: (040) 79 33 33

also:

ELCOMA Philips Electronics Components and Materials, P.O. Box 50, Lane Cove 2066, New South Wales, Australia.

MULLARD LTD., Mullard House, Torrington Place, London WC1E 7HD U.K. Phone (01) 580 6633.

NORTHAMERICAN PHILIPS, 230 Duffy Avenue, Hicksville, NY11802, U.S.A. Phone (516) 931 6200 and numerous other Philips subsidiaries, worldwide.

The Philips solar cell module consists of 34 series connected 57mmØ silicon solar cells, moulded in transparent resin and laminated between two glass sheets.

Performance data at 100mW/cm^2 insolation and 25°C ambient: Optimum power 10.7W with voltage 15.5V and current 0.69A.

Solar Engines

ORMAT TURBINES LTD
Szydlowski Road
Yavne
P.O.B. 68
Israel

Phone: 055 95461
Telex: 31902
Cables: ORMATURB TELAVIV

Manufacturer of a range of solar-powered turbines for a variety of applications including water pumping, electricity generation, etc.

Turbine runs on organic heavy vapour working fluid and has only one major moving part. Over a thousand units have been installed in a variety of continents and these have been in production for some ten years.

Solar powered systems can be supplied with the possibility of an optional fuel burner to guarantee power at night or during cloudy periods.

Ormat Turbines Installation

SOLAR PHYSICS CORPORATION
1350 Hill Street
El Cajon
California 92020
U.S.A.

Phone: (714) 440 1625

Manufacturers of a solar power unit utilising the sun via tracking mirrors to heat a working fluid in a receiving tower. This working fluid is used to power an engine which drives a generator. Electrical output of up to 5kW currently offered, but larger models offering outputs up to 100kW are at the planning stage. This company is well known for its pioneering work in the field of "power towers", which have been postulated as a likely configuration for future large scale power production. Consultancy services in this field are also offered.

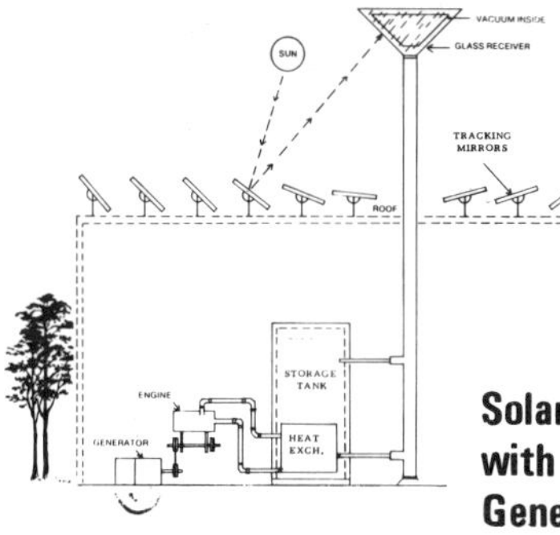

Solar Power-Tower with 5kW Electrical Generating Plant

YAZAKI CORPORATION
International Division
6th Floor Tokyo Ryutsu Center
1-1 6 chome
Heiwa Jima
Ota-Ku
Tokyo
Japan

Telex: J24437, 246 8395
Cables: Yazaki Tokyo
Phone: Tokyo (03) 767 3960

This group manufactures a newly developed solar heated absorption refrigeration unit producing chilled water for building cooling by the absorption cycle.

Yazaki Solar Heated Absorption Refrigeration Unit

RENAULT MOTEURS DEVELOPPEMENT
R.N. 186
La Boursidière
92350 Le Plessis-Robinson
France

Telex: Reno 200288 F
Phone: 630 21 03

Renault Motors are now marketing a novel type of solar heat engine developed by Sofretes in France. Quite a large number of these machines are now in use in countries such as Senegal, Mexico, Upper Volta, Mauritania, Niger, Mali, Cameroon, Chad, Abu Dhabi and Sudan. Since they are relatively expensive machines they depend on intense sunlight for the best return.

Two versions in various sizes are available, these and solar pumping stations designated by the prefix MS and solar power stations prefixed TS, which can also be used as pumping stations.

The following performance data assumes average ambient temperatures over 25°C, average solar radiation of 700W/m^2, pumped water temperature 28°C.

Model No.	*Water output at 25m head*	*Nominal power (pumping)*	*Daily Average running time*	*Nominal power (shaft)*
MS 3	3.5m^3/hr	240W	4-6 hr/day	–
MS 5	5.75m^3/hr	392W	4-6 hr/day	–
MS 7	8.9m^3/hr	545W	4-6 hr/day	–
TS 25	1200m^3/day	17.2kW	5-6 hr/day	25kW
TS 50	2400m^3/day	34.8kW	5-6 hr/day	50kW

Sofretes Panel Collectors for Solar Pump

SPILLING CONSULT AG
Sonnenweg 4
CH-5610 Wohlen
Switzerland

Telex: 57 939
Cables: Spill Ch
Phone: 057 6 73 57

Spilling solar energy plant consisting of a solar collector and a steam motor generator set with outputs from 2kW upwards.

This was all the information available to us at the time of going to press.

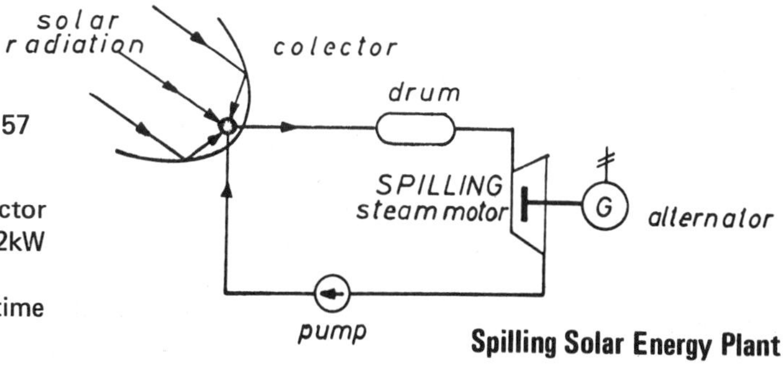

Spilling Solar Energy Plant

Solar Heating

AIR DISTRIBUTION EQUIPMENT (M & W) LTD
64 Whitebarn Road
Llanishen
Cardiff CF4 5HB
U.K.
Phone: 0222 754677

Manufacturers of solar water heating systems for home water heating. Systems incorporate solar panels plus a differential temperature controller to control the flow of water through the panels.

Solar panel module of 2ft x 2ft (0.37m^2).

GRP/aluminium case to glass cover.

Solar Panel

ARC SOLAR CENTRE LTD
176 Ifield Road
London SW10 9AF
U.K.
Phone: 01-370 4804

Manufacturers of a range of solar panels under the tradename "Sunstor":

a. *Sunstor domestic collector*

Size: 1524mm x 610mm x 76mm, weight 28lb (12kg) empty, area of collector 0.92m^2.

Absorber: Copper pipes soldered to copper absorber plate.

Casing: Aluminium sheet (18 swg) with fibreglass wool and polyurethane insulation.

Glazing: 4mm single sheet of glass to be supplied by customer.

Installation: Either thermosyphon for simplicity or electric pumped.

b. *Sunstor industrial collector*

Size: 2000mm x 1000mm x 76mm, absorber area 1.9m^2, weight 17.5kg empty.

Absorber: Copper plate with copper pipes bonded by patented heat fusion process.

Surface coating: 3M Nextel with reduced infra-red emission.

Casing: Aluminium sheet with fibreglass wool and polyurethane foam.

Glazing: 4mm glass.

Installation: Intended for large installation for industrial/commercial premises.

c. *Sunstor "Waterfall 2" low cost, low temperature collector*

Size: 1830mm x 1220mm (2.16m^2).

Construction: All metal unglazed collector with copper waterways clamped to an aluminium absorber plate.

Applications: Preheating cold water for other solar or non-solar water heating systems or for heating swimming pools.

d. *Sunstor miniature demonstration panel*

This is a novel miniature complete thermosyphoning solar collector and storage tank to demonstrate solar water heating or for heating small quantities of water. It is intended mainly for educational purposes or for studies.

Effective absorber area: 0.1m^2 (426mm x 335mm).

Construction: Similar to domestic collector described in paragraph a.

Storage tank capacity approximately 2 litres. It is claimed that it will heat two litres of water to 55°C in strong sunlight quite rapidly.

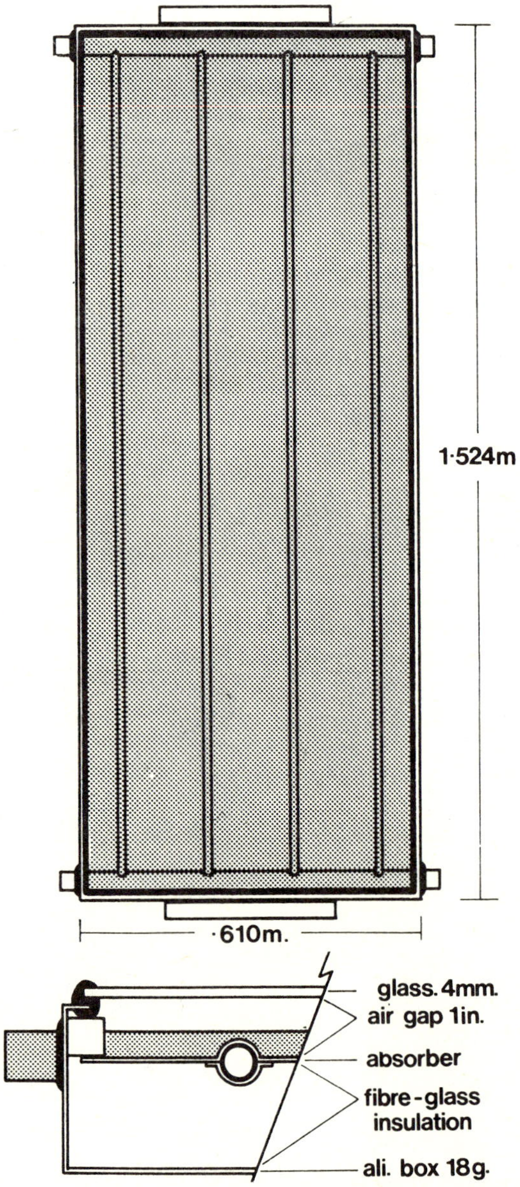

Sunstor Flat Plate Collector

ASAHI TRADING CO LTD
Asahi House
Church Road
Port Erin
Isle of Man
U.K.

Phone: (062 483) 3347/ 3758

Manufacturers of a glazed solar panel using glass tubes as the collector pipes in which the heating water runs.

This is a Japanese system and it is claimed that 2½ million units are in use in Japan.

The unit is quite heavy when filled with water and does not require a separate storage tank as it is for direct rather than indirect systems. Having glass water channels there is no corrosion problem. The channels are claimed to be efficient as collectors, particularly at absorbing solar energy when the sun is not due south or conditions are cloudy.

The case is of steel and the unit is glazed with plastic.

Size: 1460mm x 1670mm x 190mm.
Weight: Empty 75kg; full 255kg.
Capacity: 180 litres.

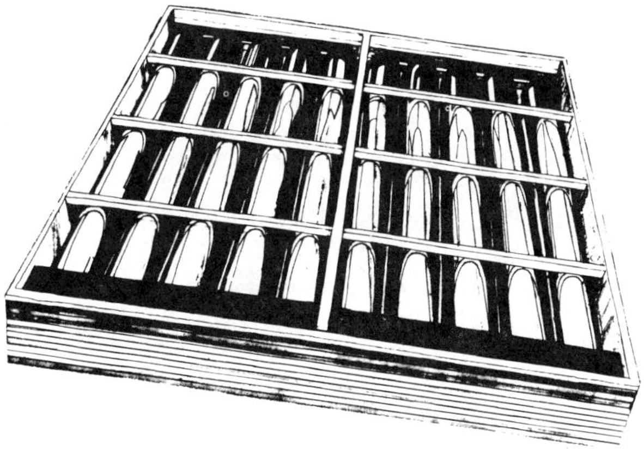

Asahi Solar Panel

BURKE RUBBER COMPANY
2250 South Tenth Street
San José
California 95112
U.S.A.

Phone: (408) 297 3500

A simple unglazed trickle flow solar panel for low temperature gain water heating, (i.e. used in the U.S.A. mainly for swimming pool applications).

Module areas are: 8ft x 8ft, 12ft or 16ft (2.44m x 2.44m, 3.66m or 4.88m).

Construction material: Du Pont "Hypalon" (Chlorosulphonated polyethylene) with a copper sparge pipe.

Surface temperature limits: 40°F to 250°F (–40°C to 121°C).

1. Water Out
2. Installation Grommets
3. Water in at top
4. Copper Distribution Pipe
5. Water Distribution Material Inside Panel
6. Black Hypalon Panel
   8 ft. Wide x 8 ft. Long
   8 ft. Wide x 12 ft. Long
   8 ft. Wide x 16 ft. Long
7. Overflow Relief Valve

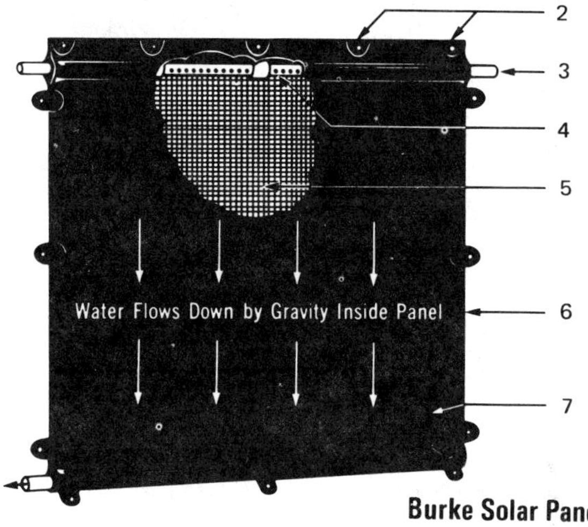

Burke Solar Panel

CALOR SOL LTD
Lancaster Road
Shrewsbury SY1 3NG
U.K.

Phone: (0743) 51578

This company offers a variety of water and air heating solar collectors. The following are all in glass reinforced plastic casing with polyurethane insulation and glass reinforced plastic glazing.

Mark 1: Cascade water heating corrugated metal plate — stove-enamelled matt black, 15mm copper sparge pipe and 25mm UPVC outlet collector. For low temperature applications. 2.92m x 0.97m = 2.83m^2 effective area.

Mark 2: Warm air ventilation corrugated metal plate stove-enamelled matt black, size as above.

Mark 3: For indirect water supply with aluminium roll-bonded absorber, size as mark 1.

Mark 4: Similar size to above units, but with pressed steel radiator absorbers interconnected with copper pipes.

Some further low cost models are also offered, which have 20g galvanised steel cases and thinner GRP glazing and the insulation is expanded polystyrene.

Mark 5: Roll bonded aluminium absorber of 2.05m^2.

Mark 6: Corrugated galvanised cascade collector of 2.3m^2 effective area.

Mark 7: Similar to Mark 6, but designed for air heating rather than water heating and having suitable inlet and outlet ducts.

Calorsol Solar Heat Collector Piate

CSI SOLAR SYSTEMS DIVISION
12400 49th Street
Clearwater
Florida 33520
U.S.A.

Phone: (813) 577 4228/4489

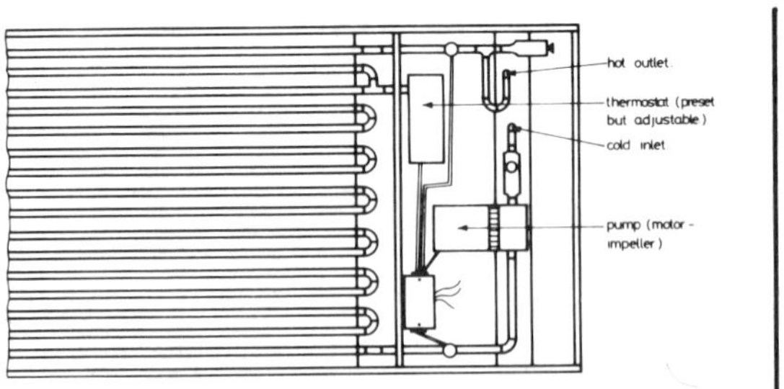

CSI Solar Panel

Manufacturers of a range of solar heater modules with integrated controls and an electric circulating pump.

Model No.	*Size*	*Weight*	*Water Channels*	*Absorber Material*	*Frame*	*Insulation*
250	2ft x 8ft	80lb			Stained	Polyethy-
500	2ft x 11ft	100lb	Copper	Aluminium	red-	lene
1000	4ft x 11ft	220lb	tube		wood	

Champion "Solar Furnace"

CHAMPION HOME BUILDERS CO
5573 North Street
Dryden
Michigan 48428
U.S.A.

Phone: 313 796 2211

Manufacturers of a solar heater for hot air domestic heating. The unit is a free standing unit using a rock fill heat store. The system is easily installed in conjunction with any conventional warm air heating system.

Sizes: Three models of 108, 144 and 180ft^2 collector area (10.1, 13.4, 16.7m).

Absorber: Aluminium with double glass cover and urethane insulation.

Heat store: 9, 12 or 15 cu.yds of gravel.

Fitted with circulating fans and electronic controller.

DIY-SOL INC
P.O. Box 614
Marlboro
Massachusetts 01752
U.S.A.

Phone: 617 877 6018

This company supplies a kit and instruction book to allow a sophisticated and effective hot air solar heating system to be built. The kit includes only the special equipment necessary and ordinary construction materials are supplied from local sources by the customer. This is claimed to be a low cost method of installing solar heating.

Hot water can be provided through the use of a heat exchanger between the solar heated air and the water supply.

The collector is glazed with Dupont's "Tedlar" and incorporates a special type of efficient absorber called "Xsoletherm" – both are included in the kits.

This system might be useful for crop drying or timber dring processes, etc.

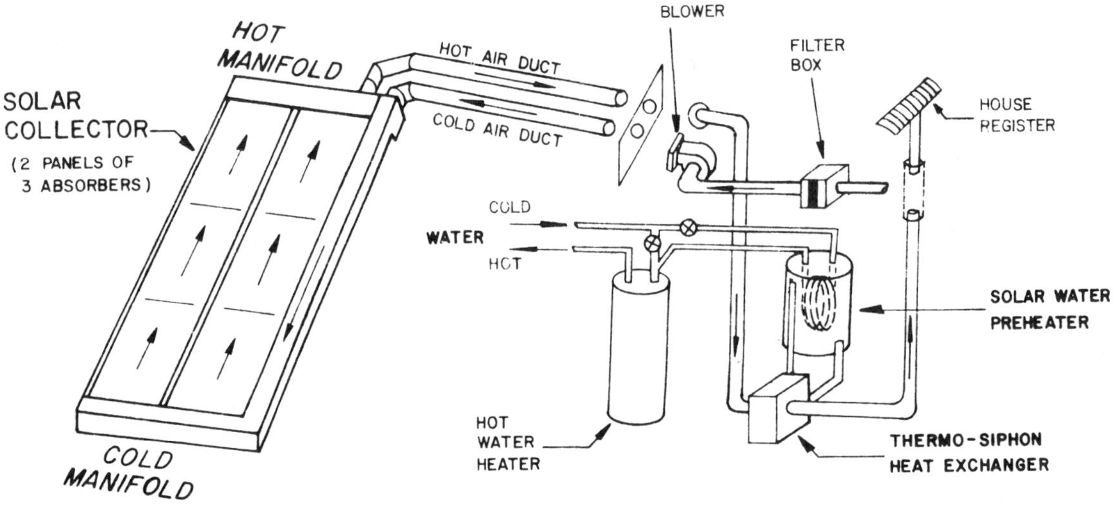

DIY-SOL "Hot Water Plus" System

E.I. DU PONT DE NEMOURS
& CO INC
Film Department
Speciality Markets Division
Wilmington
Delaware 19898
U.S.A.

Manufacturer of a wide range of plastic materials including "Tedlar" PVF film which is particularly suited for the glazing of solar flat plate collectors. "Tedlar" is available in various grades; 100BG30TR passes UV wavelengths and some infra red. "Tedlar" is partially selective in that it blocks infra red in the 7 to 12 micron wavelength range partially.

E.A. HYDRAULICS & Phone: Nairobi
METAL INDUSTRIES LTD 559273
P.O. Box 30730 28918
Nairobi 28907
Kenya

This is a solar water heating panel made in Kenya with a glazed cover and combined galvanised tank of 120 litres capacity. It is claimed to heat the contents of the tank to 70°C after two hours of Kenyan sun.

Overall dimensions of the installation are 5ft x 4ft x 3ft and net weight (empty of water) is 240lb. A 1.5kW electrical immersion heater can be included as an optional extra to guarantee hot water during periods of limited sunshine.

Solar Panel

EDWARDS ENGINEERING Telex: 130131
CORP. Cables: Edengco
101 Alexander Avenue Phone: (201) 835
Pompton Plains 2808
New Jersey 07444
U.S.A.

The Edwards Completely Packaged Solar System provides a complete assembly of related heat transfer units to accomplish the heating and cooling of residential/commercial spaces.

Solar Collectors: The solar collector panels are 2ft or 3ft in width and of any specified length to fit the building design. The panels are so designed that they will take the place of the normal roofing material such as the shingles. The panels consist of 1 inch of aluminium covered fibreglass insulation, heat absorbing aluminium plates, copper or aluminium tubes, one or two layers of transparent plastic sheeting, and an aluminium frame for fastening and holding the parts to the roofdeck.

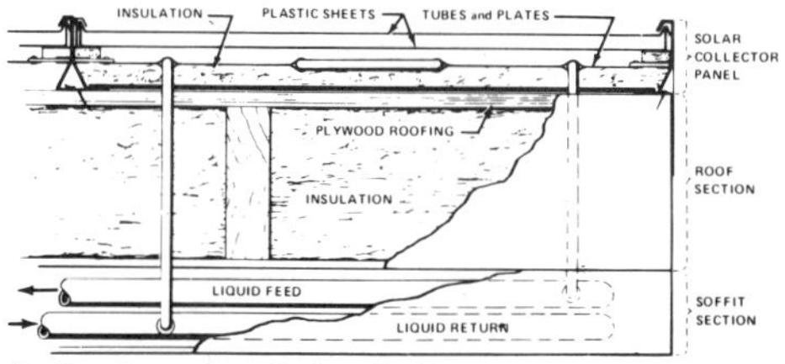

Edwards Engineering Solar Collector

E & K SERVICE COMPANY Phone: 206 486
16824 – 74th Avenue N.E. 6660
Bothell
Washington 98011
U.S.A.

Manufacturers and licencees of an unusual wooden framed solar collector claimed to minimise heat loss by conduction which can be a problem with some metal-cased collectors.

Panel Sizes: 2ft x 4ft, 4ft x 4ft, 4ft x 8ft, 4ft x 12ft.

Absorber: Aluminium sheet with copper tubes.

Glazing: GRP plastic or $^3/_{16}$" glass or can be supplied unglazed.

Weight: Approx 6.5lb/ft^2 (glazed) or 4.25lb/ft^2 (unglazed).

A miniature version of the above panels, size 1ft x 2ft is available for educational purposes or as a sample.

E & K Service Company
SOL-R panel

GRUMMAN CORPORATION Telex: (516) 575
Energy Systems Division 6555
4175 Veterans Memorial Highway Phone: (516) 575
Ronkonkoma 6205
New York 11779
U.S.A.

Manufacturers of direct and indirect solar heating systems which include solar panels and all ancillary equipment for domestic water heating, for use in conjunction with other heating systems or on their own.

Typical unit size: 24.58ft^2 (2.28m^2) of absorber area.

Weight: 87 or 122lbs depending on absorber.

Absorbers: Aluminium Rollbond for closed circuit indirect systems. Aluminium Finplank with copper water passages for direct or indirect systems.

Glazing: Arched acrylic covers or single/double tempered glass.

Insulation: Foil enclosed fibreglass.

Casings: Aluminium alloy with baked enamel finish.

N.B. Finplank absorbers are available separately for building into "home made" cases. Weight 56.5lb for 24.95ft^2 absorber.

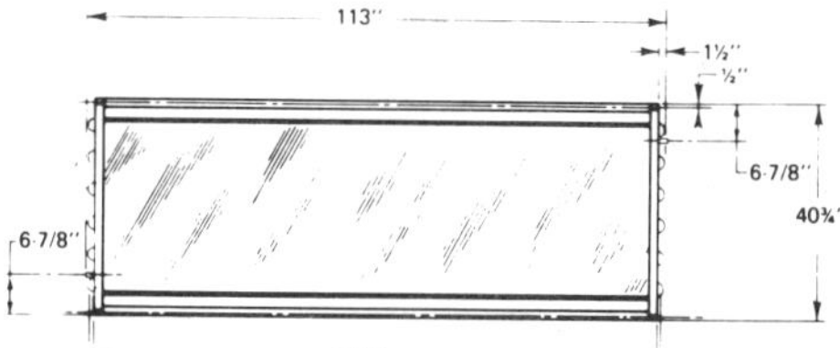

Grumman Solar Panel

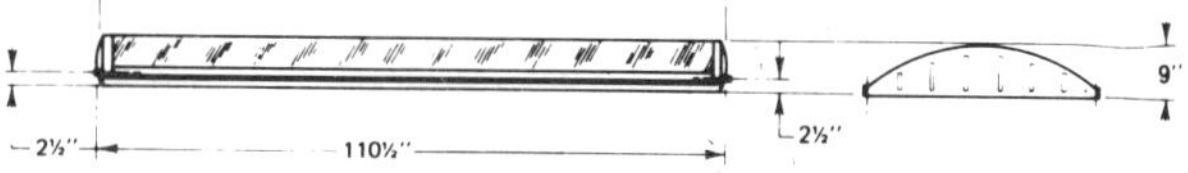

ENERGY DYNAMICS CORPORATION
6062 East 49th Avenue
Commerce City
Colorado 80022
U.S.A.

Phone: 303 321 3314

"Solaris" Collector

Designers and suppliers of solar systems using the "Solaris" solar heating module. Systems include designs for domestic water heating and space heating as well as the incorporation of a heat pump connected system.

"Solaris" collectors are to the following specifications:

	Model	*Width*	*Nominal Height*	*Depth*
Dimensions:	308	4′ – 0	8′	4″
	312	4′ – 0	12′	4″
	316	4′ – 0	16′	4″

Additional space is required at top and bottom for plumbing connections.

Weight: 2.5lb/sq.ft dry: 3.25lb/sq.ft wet.
Absorber: Aluminium.
Absorptive coating: Flat black epoxy coating.
Cover: 1/8″ tempered glass — single layer — transmittance — 88%.
Transfer fluid: Water.
Plumbing connections: Top: 3/8″ OD copper tubing. Bottom: 1″ CPVC pipe.

FAFCO SOLAR HEAT EXCHANGERS
138 Jefferson Drive
Menlo Park
California 94025
U.S.A.

Phone: 415 321 6311

Manufacturers of an unglazed, low temperature solar water heating system using polyolefin (plastic) tubular extrusion absorbers between header pipes.

Standard panel sizes are 4′ x 10′ and 4′ x 8′. Electronic controllers and associated equipment are available.

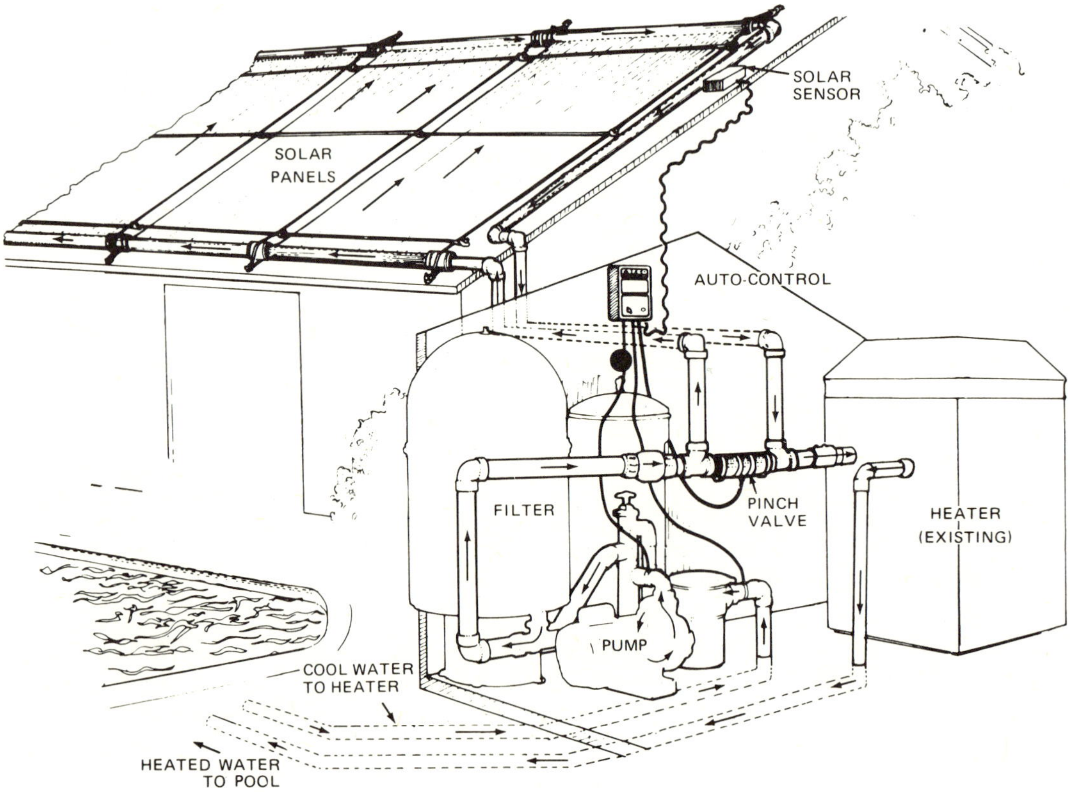

Fafco Solar Heating System

Hadley Home Built Solar Panel

HADLEY SOLAR ENERGY Phone: 215 444 3618
Box 1456
Wilmington
Delaware 19899
U.S.A.

Publishers of a booklet and plans for construction of do-it-yourself solar heating panels for both air heating and water heating.

Specifications:	*Water heater*	*Air heater*
Size	4 x 8ft = 32ft^2 (2.97m^2)	4 x 8ft = 32ft^2 (2.97m^2)
Absorber	galvanised steel	galvanised steel
Water channels	copper pipe	—
Glazing	acrylic or glass	acrylic or glass
Insulation	fibreglass	fibreglass

HELION Phone: 213 367 8291
P.O. Box 4301
Sylmar
California 91342
U.S.A.

Helion solar panels are offered as kits and are claimed as an easily fitted low-cost system.

Overall dimensions: 45 x 96in = 30ft^2 (2.79m^2).
Weight: 55lb.
Absorber: Aluminium sheet with copper tube water channels.
Glazing: Tedlar coated fibreglass.
Insulation: Polyurethane foam/fibreglass.
Casing: Aluminium.

FILON DIVISION, VISTRON CORPORATION Phone: (213) 757 5141
12333 Van Ness Avenue
Hawthorne
California 90250
U.S.A.

Manufacturers of fibreglass reinforced (GRP) panels for use in solar collectors and passive glazing installation. Units are guaranteed to retain 90% light transmission for 15 years. Licenses for manufacture apparently granted in Africa, Asia and South America.

Helion Solar Panels

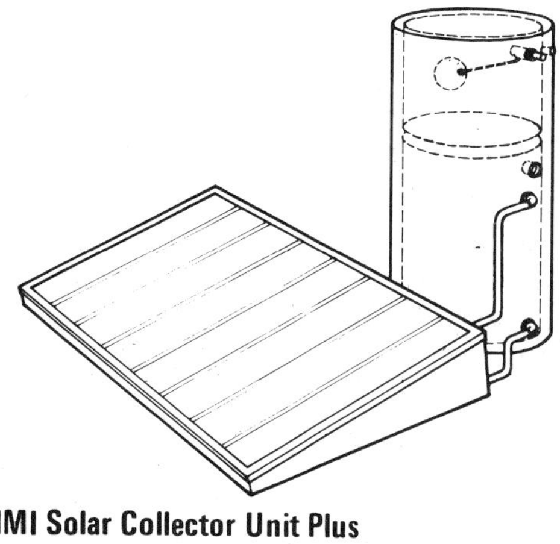

IMI Solar Collector Unit Plus Hot Water Cylinder

IMI RANGE LTD Phone: 061 338 3353
P.O. Box 1
Bridge Street
Stalybridge
Cheshire SK15 1PQ
U.K.

A large company which offers mainly components for trade use or export in quantity. Manufacturers of "Solaric" absorber panels.

Sizes: 3 widths and 3 lengths available giving 9 areas ranging from 0.21 to 2.82m^2. Favoured size is 1.75 x 0.86m = 1.6m^2.
Absórber: All copper consisting of copper water tubes fitted with copper absorbing fins treated with black selective coating.
Casing: The absorbers are recommended to be built into the roof structure of a building and glazed over, but plastic cases for the 1.6m^2 size are available if independent "detachable" units are needed.
Other items: Copper heat exchanger tanks, with optional sprayed on insulated surface are also available.

McKEE SOLARONICS LTD
12 Queenborough Road
Southminster
Essex CM0 7AB
U.K.

Phone: 0621 772477

Manufacturers of a range of plastic solar collectors for water heating:

Industrial/Domestic heat exchanger: collector 2050mm x 1230mm ($2.4m^2$). The absorber is a polypropylene extrusion in translucent plastic with a metalised reflecting sheet behind it and this system is novel in that it uses a dark fluid to transfer heat from the collector which is claimed to give good efficiency. The unit is single glazed with a tough polycarbonate plastic window. It is claimed that these units will withstand temperatures in the range from −40°C up to +120°C without damage (and can tolerate the effect of water freezing within the channels). Maximum water pressure is 0.6MPa (90 psi). Insulation is by transparent cellular plastic either side of the absorber panel. In situations where direct solar heating of water is required, the panel can be obtained with the internal polycarbonate absorber painted matt black to avoid the need for using a dark liquid.

Swimming pool heat exchanger: collector 1886mm x 1240mm ($2.4m^2$). This is to a similar specification to the above system, but is slimmer as it does not include the insulating voids filled with cellular plastic since it is not intended to achieve such a high temperature increase above ambient. Applications such as pre-heating or low temperature increase processes such as crop or wood drying may be appropriate in addition to the intended application of swimming pool heating.

McKee Solaronics Industrial/ Domestic Solar Heater

METALCO (HEATERS) LTD
P.O. Box 1307
Nicosia
Cyprus

Telex: 2776 Metalc-Cy
Cables: Metalco-Nicosia
Phone: 32101/2

Manufacturers of a complete water filled solar domestic hot water system. System is free standing and relies on the thermosyphon principal for its operation. An electric heating element may be incorporated to provide hot water during sunless periods.

Absorber area: approximately 131 x 92cm (2x) = $2.4m^2$.
Construction material: not specified.

Metalco Solar Hot Water System

DISTRIMEX LIMITED
88 The Avenue
London NW6 7NN
U.K.

Cables: Imexhouse London NW6
Phone: 01 459 1391

Distrimex produces the Miromit solar energy collector, which was developed in Israel over 20 years ago and incorporates a unique selective surfacing known as "Tabor Selective Black" after its inventor, the well known solar energy researcher, Dr H. Tabor.

Collector module surface area: $1.428m^2$.
Collector absorber material: Mild steel electroplated with "Tabor Selective Black".
Insulation: Rockwool 40mm thick.
Outer casing: 0.6mm galvanised steel 1850mm x 960mm x 95mm.
Glazing (not supplied): Recommend 3mm or 4mm glass.
Weight: 60kg.
Claimed life expectancy: "Well in excess of 15 years".

Other products from Distrimex include the "Sunsource Solar Controller" which operates at 240V (or 110V by request) to switch on a pump for a pumped solar heating system.

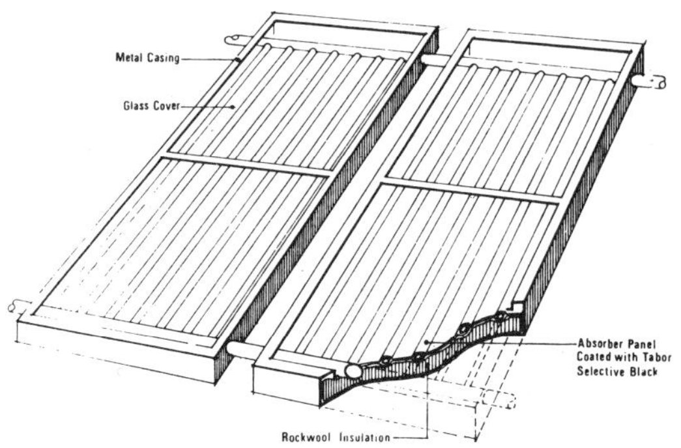

Miromit Solar Energy Collector

NATURAL ENERGY SYSTEMS INC
1654 Pioneer Way
El Cajon
California 92020
U.S.A.

Phone: 714 440 6411

Manufacturers of solar panel heating systems for domestic space heating or water heating installations. The company also produce a solar distillation system (illustrated) for purifying water.

Precise technical specifications were not known at the time of going to press.

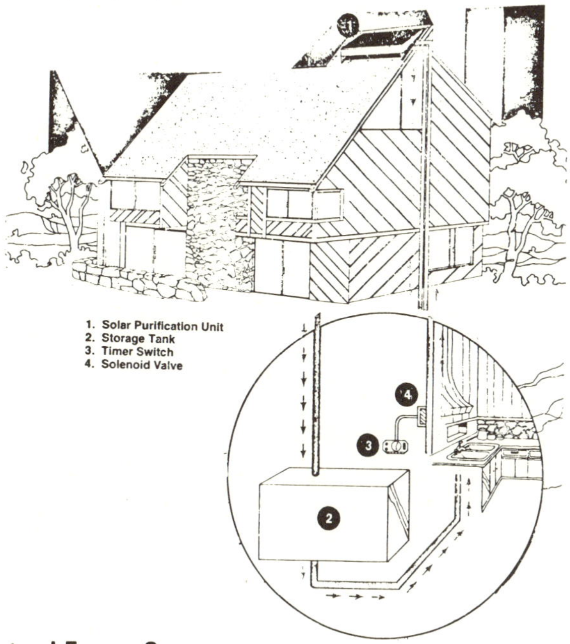

Natural Energy Systems Solar Water Purification System

THE PEOPLE/SPACE COMPANY
Division PAC
49 Garden Street
Boston
Massachusetts 02114
U.S.A.

Phone: 617 261 2064

Manufacturers of an open flow collector in which the water flows over the fibreglass collector surface beneath a translucent fibreglass cover. Due to its simplicity, this system is claimed to be of low cost and the construction materials are certainly long-lasting and durable.

One module size is available 4ft x 8ft x 4ft ($2.97m^2$) which can be obtained either double glazed or single glazed.

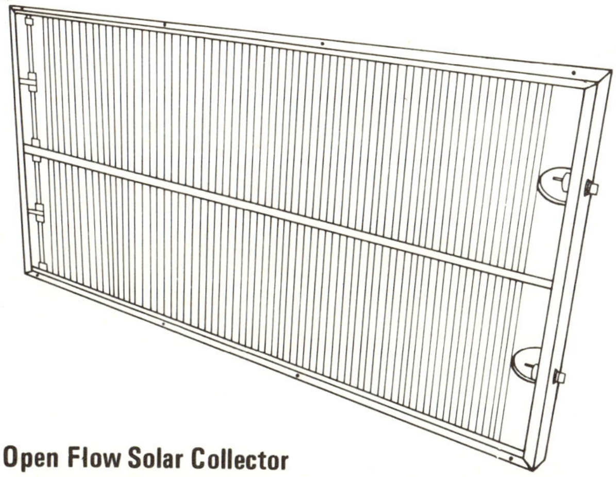

Open Flow Solar Collector

NORTHRUP INC.
302 Nichols Drive
Hutchins
Texas 75141
U.S.A.

Phone: (214) 225 4291

This company is a major solar energy developer and claims the world's largest solar heating/air conditioning installation of $16,000ft^2$ ($1487m^2$) at Trinity University in Texas.

Main product is a concentrating solar collector which tracks the sun using electric drive motors and concentrates the sunlight through an acrylic linear fresnel collector lens onto a copper selectively coated absorption tube; this system is claimed to be of high efficiency in terms of energy collection per unit area compared with flat plate collectors. It is claimed that it can supply water at 200°F (93°C) under 90°F ambient conditions with an efficiency of 65%. Effective collecting area per module is 9.7ft ($0.901m^2$).

Other systems available include a low cost (claimed) integral thermosyphon water heater and storage (which has no moving parts at all). This has a $21.78ft^2$ absorber of extruded aluminium with copper water tubes coated with 3M Nextel selective surface and with a Tedlar plastic cover. It has a 40 gallon (150 litre) storage tank with anodic corrosion protection and insulation is a fibreglass blanket.

This collector is also available without the storage tank as a normal flat-plate collector module and glass can be used (customer supplied) for glazing as an option. Storage tanks, pumps and other components are also available.

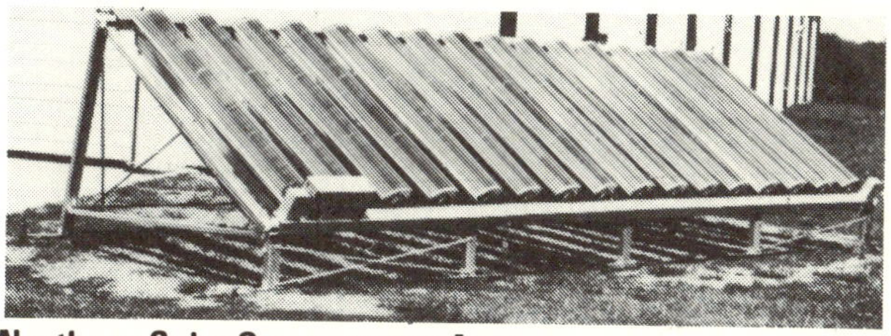

Northrup Solar Concentrator Array

RAYOSOL
Apartado 21
Torremolinos
Malaga
Spain

Phone: 320432

Manufacturers of a glazed solar water heater using a copper absorber with copper water tubes. The case is of fibreglass/polyester with unspecified insulation material behind the absorber. Standard absorber area is approximately $1.5m^2$.

Rayosol Solar Panel

PIPER HYDRO INC
2895 East La Palma
Anaheim
California 92806
U.S.A.

Phone: (714) 630 4040

Designers and manufacturers of complete solar water heating systems for individual dwellings or industrial estates. Units are solar/gas heated for constant hot water and space heating.

This company is one of 36 U.S. manufacturers whose products were approved by ERDA in 1976 for use in the Federal Solar Heating and Cooling Demonstration Programme.

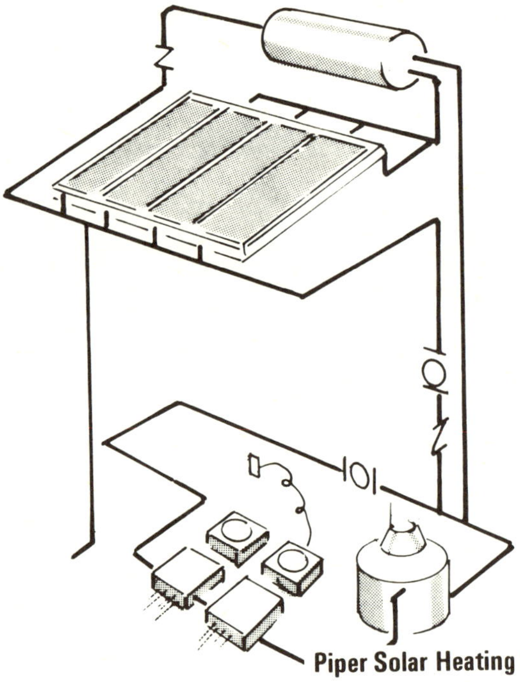

Piper Solar Heating System

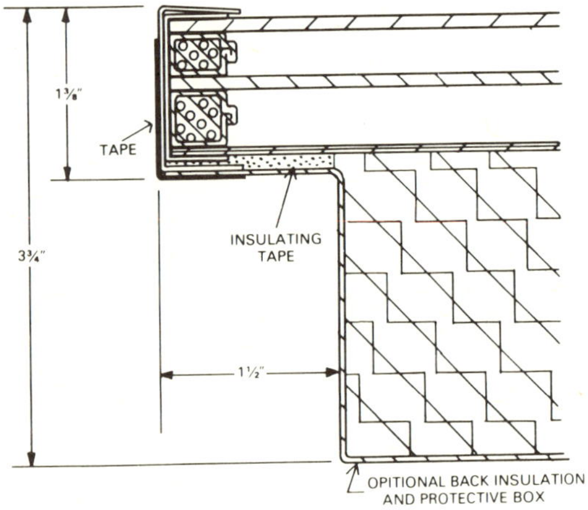

PPG Solar Collector

PPG INDUSTRIES INC
One Gateway Center
Pittsburgh
Pennsylvania 15222
U.S.A.

and:

BOMERT, TEVES & BLANKLEY LTD
(Agents)
Pembroke House
44 Wellesley Road
Croydon
Surrey CR9 9PD
U.K.

Telex: 866570
Cables: Glaspit
Phone: 412 434 2045

Manufacturers of solar collectors for water and space heating. These systems can include heat storage units for periods of little sunshine. Standard collectors are as follows:

Size	*Absorber*	*Casing*	*Glazing*
$34\frac{3}{16}'' \times 76\frac{3}{16}'' \times 1\frac{5}{16}''$	Copper alloy bonded to copper water tubes with semi-selective PPG Duracron black coating.	Painted galvanised steel with fibre-glass insulation.	Single or double pieces of $\frac{1}{8}''$ tempered glass

---

REYNOLDS ALUMINUM EXPORT CORPN.
6603 West Broad Street
Richmond
Virginia 23261
U.S.A.

Telex: 827-358
Cables: Reyfor Rch
Phone: 281 2000

Manufacturers of an all aluminium water heating solar panel using one piece extruded water tubes and fins enclosed in an aluminium frame. A double layer of Tedlar glazing is shrunk onto the glazing frame over the unit. Standard panel is 4ft x 8ft = $32ft^2 = 2.97m^2$.

CROSS SECTION THROUGH FRAME

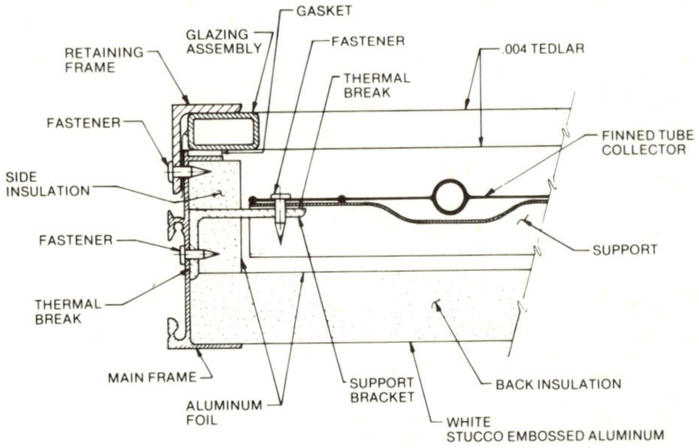

Reynolds Solar Collector

RAYPAK INC
31111 Agoura Road
Westlake Village
California 91359
U.S.A.

Telex: 67 3308
Phone: 213 889 1500

Manufacturers of a range of solar panels for high, medium and low temperature increase applications as follows:

Model	*Size*	*Area*	*Absorber*	*Glazing*	*Case*
SG18P	950 x 2095mm	1.99m^2	Copper tubes in aluminium	Single glass	Enamelled steel
DG18P	950 x 2095mm	1.99m^2	" " "	double glass	" "
SK800	1090 x 2440mm	2.66m^2	" " "	unglazed	no case
SK1000	1090 x 3050mm	3.33m^2	" " "	"	" "

Also supply controllers, pumps and associated equipment.

Raypak Solar Panel

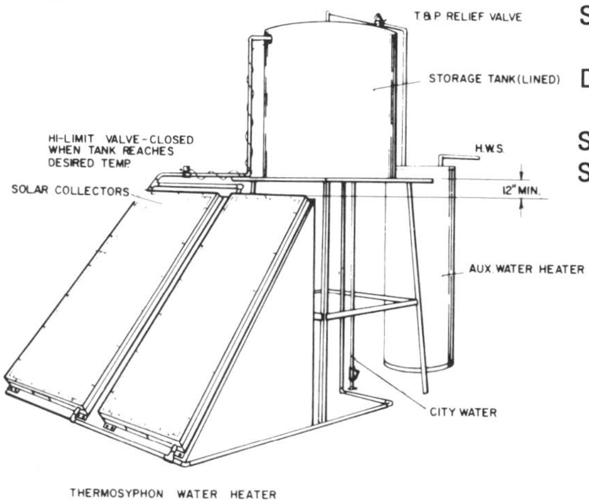

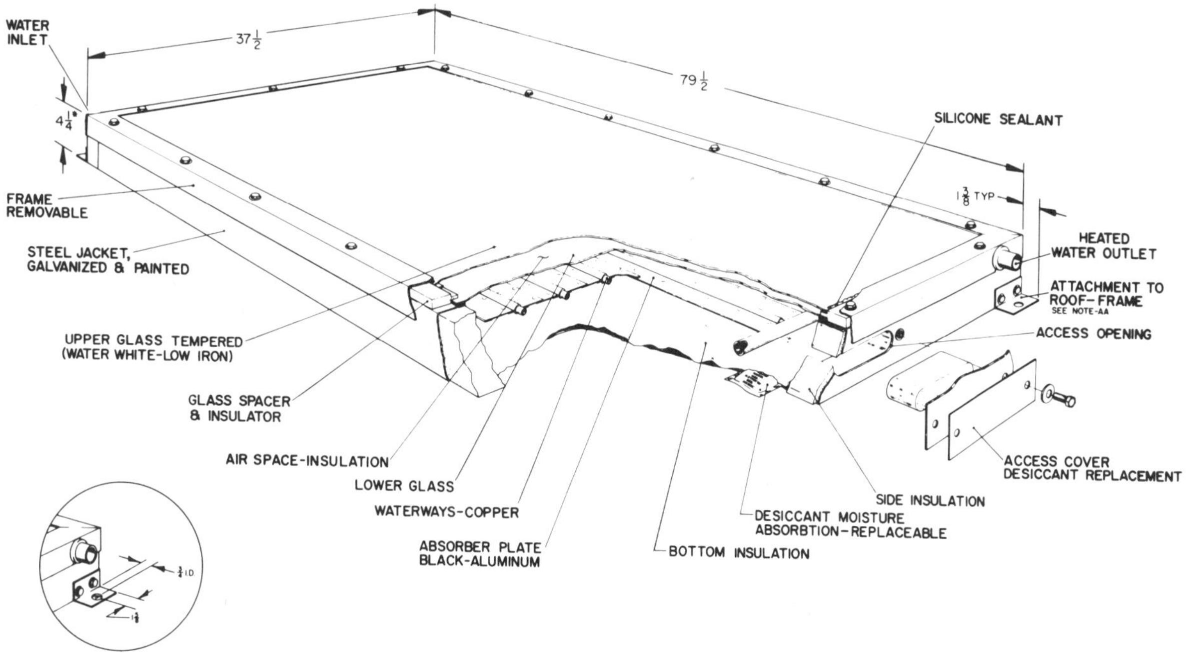

SATA Solar Collector and Tank

SATA/BALAJU YANTRA SHALA (P) LTD
(Swiss Association for Technical Assistance)
P.O. Box 113
Kathmandu
Nepal

Cables: SATA Kathmandu
Phone: 21205

This group produces solar water heating systems and provides detailed plans for home construction of units. Designed for conditions existing in Nepal using low cost materials; similar conditions probably prevail in a number of other developing countries where solar heaters might be useful.

Size of typical module: 180 x 89cm (1.6m^2).

Absorber: Aluminium sheet or galvanised mild steel sheet with steel water pipes.

Cover: Glass.

Insulation: Fibreglass.

Case: Mild steel sheet 1.2mm thick.

REDPOINT ASSOCIATES LTD
Lynton Road
Swindon SN2 2QN
Wiltshire
U.K.

Telex: 449543
Phone: 0793 28440

Manufacturers of a 1.75m^2 solar collector module. The unit is designed for fast response to variations in insolation.

Absorber: Selectively surfaced with refrigerent-filled heat pipes to copper heat exchanger to water.
Frame: Light alloy.
Glazing: 1 glass sheet.
Insulation: Fibreglass back.

This is an unconventional system in that the absorber uses heat pipes to transfer heat to the water. It is available as an absorber and heat exchanger alone for overseas manufacturers who would like to fabricate the case and glazing locally.

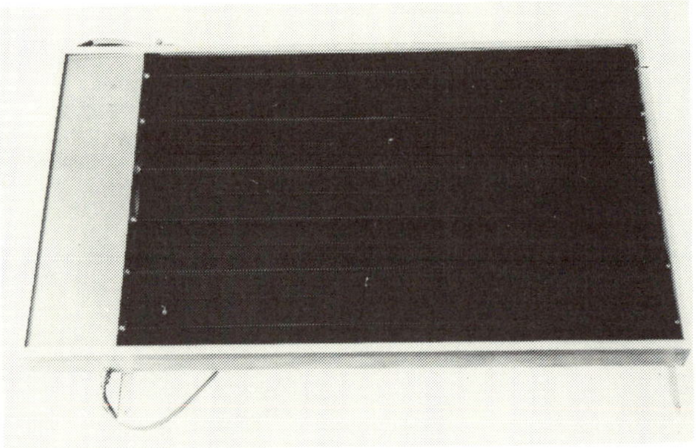

S175 Solar Panel

FRED RICE PRODUCTIONS INC.
P.O. Box 643
48-780 Eisenhower Drive
La Quinta
California 92253
U.S.A.

Phone: 714 564 4823

Publishes quarterly *Sunergy* newsletter. Offers consultancy services on a variety of solar energy applications and markets Sav Cylindrical Solar Hot Water System.

Sav Cylindrical Solar Hot Water System

ROBINSON OF WINCHESTER LTD
Robinson House
Winnall Industrial Estate
Winchester SO23 8LH
Hants.
U.K.

Telex: 477149
Cables: Robinsons Winchester
Phone: (0962) 61777

Manufacturers of polypropylene plastic solar panels under the trade name "Suncell". These units are unglazed and work at temperatures up to 70°C. Other versions with single glazing are also available.

Two sizes of absorber are available; 1.2 x 2m; 1.2m x 3m.

These units are for low temperature-rise applications and are low cost and of light weight.

Suncell Swimming Pool Solar Heater

SENIOR PLATECOIL LTD
Otterspool Way
Watford
Herts WD2 8HX
U.K.

Telex: 935574 Setl
Cables: Sentherm Watford
Phone: Watford 26091 35571

Manufacturers of a wide range of plate type heat exchangers for process industries and including "Solarmiser" solar collector absorber plates available in carbon steel and in stainless steel from 20swg thickness upwards in various dimensions and arrangements.

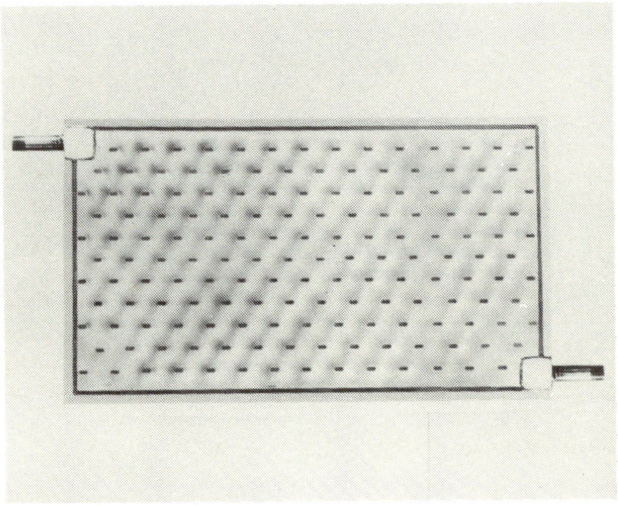

Solarmiser

SOLAHART PTY LTD
112 Pilbara Street
Welshpool
Western Australia 6106

Cables: Hartmet
Phone: 686211

Some 10,000 of these systems are claimed to be in use in Australia. The main system offered is the "Solahart" model 240L (illustrated). This consists of a complete solar hot water system of two collectors and an insulated storage tank. The collectors have copper water tubes in an aluminium absorber panel. The effective area is 1.9m^2 each (3.8m^2 for the complete system). The cover is aluminium with single glass glazing and a combination of polystyrene foam and glassfibre insulation. Tank capacity is 240 litres (54 gallons) with 75mm of polyurethane insulation. An 1800 watt electric booster element can be fitted to guarantee hot water even under adverse solar conditions. Total weight of this system is approximately 160kg. A booster system with only eight gallons of storage (36 litres) and a single 1.74m^2 absorber is also available, construction being similar to that described above.

Solahart 240L Unit

SOLAR EQUIPMENT DIVISION
Monorail Chairlift Ltd
P.O. Box 10368
Wellington
New Zealand

Telex: NZ 31368
Phone: "Opinion"
720 306

Manufacturers of an unusual cylindrical solar water heater with internal water storage. Unit consists of a double clear plastic outer skin with the tubular inner water heater using a double pass system. Units can be connected in series or parallel. It is claimed that the cylindrical collector offers advantages over flat plate collectors by being able to accept solar energy from all directions equally well, including heat reflected from the roof and that less light is reflected back by the glazing. As a result it is claimed to be particularly efficient.

Agents in the U.S.A. for this system are:

Fred Rice Productions Inc., P.O. Box 643, La Quinta, California 92253, U.S.A. (see previous page).

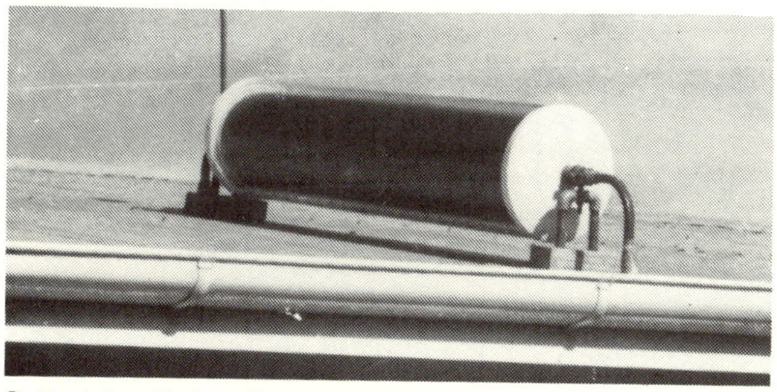

Cylindrical Solar Water Heater

KALWALL CORPORATION
Solar Components Division
Kalwall Corporation
88 Pine Street
P.O. Box 237
Manchester
New Hampshire 03105
U.S.A.

Phone: 603 668
8186

A variety of solar collectors is offered:

"Fes Delta Focussing Collector" – Model 54A. 54ft^2 (5m^2) collector area with two dimensional focussing reflecting surface and aluminium "Roll-Bond" absorber.

"Aquarius I" all copper absorber with copper oxide selective coating and double glazed with "Sunlite" GRP. System will raise 42 gallons of water (160 litres) approx. 60-70°F per day above ambient.

"Solar-Kal Airheater" is a very simple but effective flat plate air heater with aluminium (flat or 'V') absorber panel and "Sun-lite" glazing in GRP.

Sizes of standard module:
47⅜" x 95½" (1.2m x 2.42m).
59⅜" x 107½" (1.5m x 2.73m).

This company manufactures a selection of materials that can be bought in bulk for constructing solar panels as integral parts of buildings or in other purpose-built applications. Products include:

"Sun-lite" collector glazing material. This is thin translucent fibreglass plastic sheet, which is strong, light in weight, easily cut and of low cost and transmits 85-90% of incident solar energy. Available in rolls 4ft x 10ft, 20 or 50ft and 5ft x 50ft. Thicknesses .025" and .040" (0.635mm and 1.02mm).

Various absorber panels – e.g. black coated aluminium sheet, flat and 'V' corrugated. Copper "Roll-band" panels for solar water heaters and aluminium "Roll-bond" preferably used with non-corrosive liquids as heat exchange media.

Insulation materials, sealants, control equipment, absorber paints, air blowers, water pumps, insulated storage tanks, etc. also available.

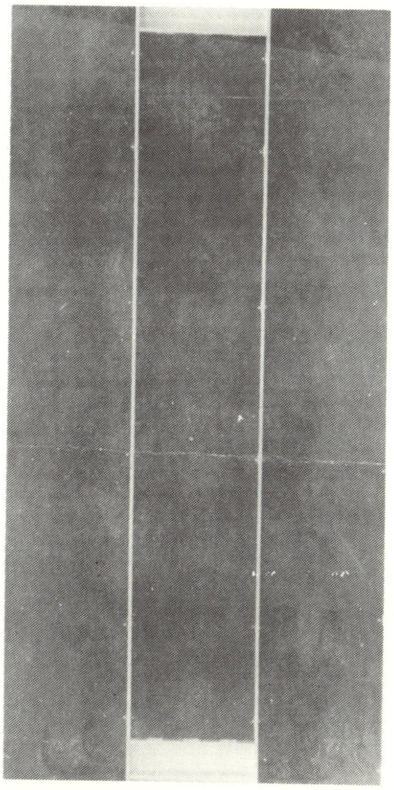

Solar-Kal Air Heater

SOLAR CORPORATION OF AMERICA
P.O. Box 399
Warrenton
Virginia 22186
U.S.A.

Phone: (703) 347 0550

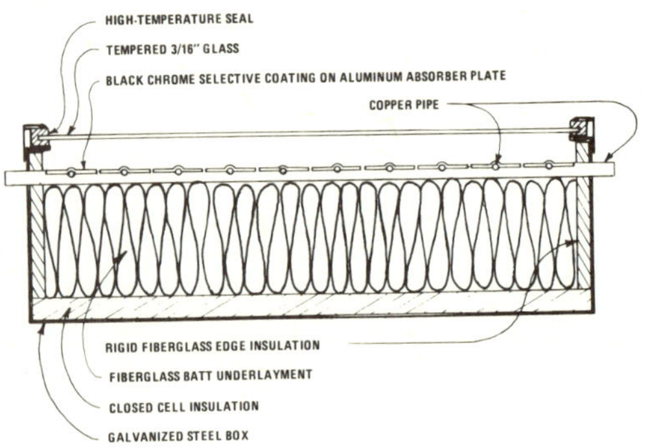

Solar Collector

Main product is a high efficiency solar collector module, as follows:

Absorber area: 32.16ft^2 (2.99m^2).

Absorber Spec: Eleven aluminium extrusions enclosing copper water tubes and coated with black chromium selective surface.

Glazing: Single 3/16" tempered glass.

Insulation: Fibreglass blanket plus unspecified plastic foam.

Outer casing: Galvanised steel plate.

Another model to a similar specification has optional double glazing and is about half the area of the above unit in size.

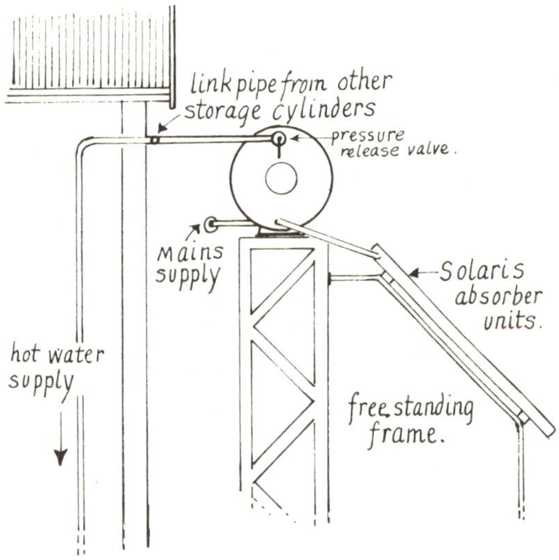

SOLARIS INDUSTRIES LTD
Bruce McLaren Road
Henderson
Auckland
New Zealand

Cables: Solaris
Phone: HSN 68 408
HSN 68 470

Manufacturers of two sizes of solar panel of 40ft^2 and 21ft^2 (3.72m^2 and 1.95m^2). The absorber is of copper sheet coated with "thermal black" and has copper water tube arranged in a serpentine embossed into it. The outer case is of fibreglass with fibreglass wool insulation and the unit is glazed with a single pane of 5.5mm float glass (supplied by customer). The system follows the Australasian popular and effective pattern of using a horizontal water cylinder which collects the hot water by thermosyphon effect. Cylinders insulated with urethane and including facility for electric standby heating are available in a choice of sizes from 136 to 546 litres. A single panel with a 60 gallon (270 litre) tank is claimed to yield an average of 8.69kWh per day in Sidney, Australia (highest monthly average per day was 12.43kWh and lowest was 4.32kWh).

Solaris Panel

SOLAR WATER HEATERS LTD
153 Sunbridge Road
Bradford
Yorkshire BD1 2PA
U.K.

Telex: 51170
Phone: (0274) 24664

This system is based on the Solarex polypropylene plastic absorber panel consisting of plastic moulded sandwich structure containing a host of integral water channels. Each absorber unit is 10ft x 4ft (3.05m x 1.2mm) giving a total of 40ft^2 (3.72m^2). These have been widely used in the U.S.A. mainly as swimming pool heaters.

The absorbers are available unglazed for low temperature applications or pre-heating water for other systems. They are also available in an aluminium frame with polyurethane insulation and double glazing. Pumps, supplementary tanks, automatic control units, etc., are also available.

Solar Panels

SUNDU COMPANY
3319 Keys Lane
Anaheim
California
U.S.A.

Phone: 714 828 2873

Manufacturers of low temperature unglazed plastic solar panel. The basic modules are assembled from 6in (150mm) wide plastic rectangular sections containing integral water tubes and made from ABS material, similar to that used for many types of plastic water pipe. Two sizes are available of 4ft x 10ft and 4ft x 8ft (3.72m^2 and 2.97m^2).

This system is for low temperature applications such as preheating for other systems of swimming pools.

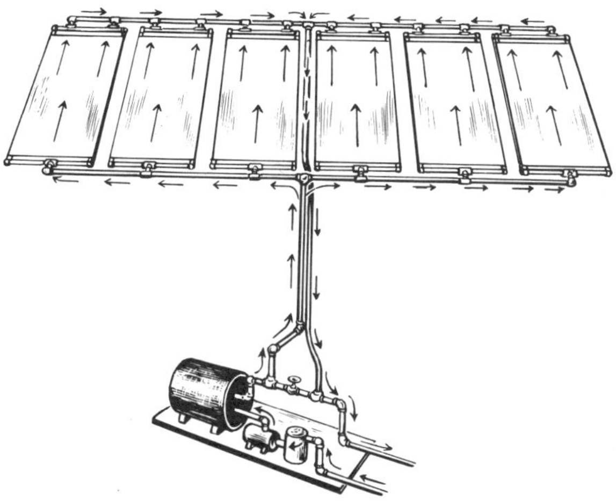

Sundu Swimming Pool Panel Assembly

SONNENENERGIE-HEIZUNGS GENERATOREN PRODUKTIONS UND VERTRIEBS GMBH
7919 Buck
German Federal Republic

Phone: (07343) 493

also:

Intermarketing Paris
25bis Avenue Brezin
92380 Garches
France

Phone: 970 52 95

Designers and manufacturers of a complete solar water heating system for domestic and swimming pool applications. It is a fairly sophisticated modular system involving heat exchangers and temperature-sensing switches, and is normally installed by heating engineers. The manufacturers' information is available in German or French, but not English.

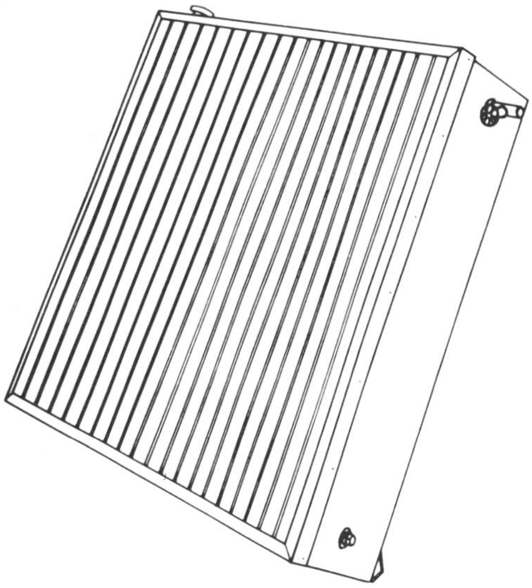

SUNEARTH INC.
Progress Drive
Box 515
Montgomeryville
Pennsylvania 18936
U.S.A.

Phone: 215 699 7892

This company offers two alternative collector modules for water heating. Both have absorber areas of approximately 19sq.ft. (1.77m^2).

Absorber for both in aluminium with copper water passages and selective black surface.

One is single glazed with tempered glass, the other is double glazed with an acrylic outer cover and Teflon (PTFE) inner.

Aluminium frame for both models with fibreglass insulation.

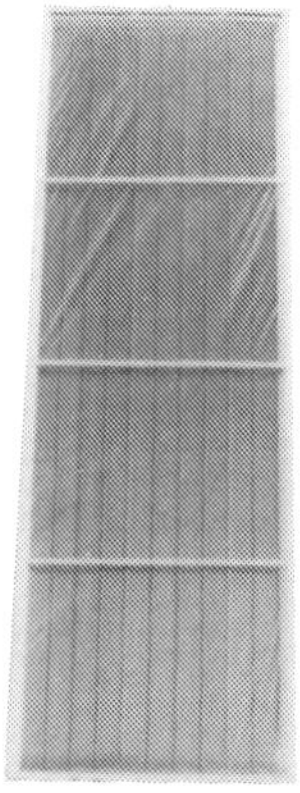

Sunearth Solar Panel

SOLARON CORPORATION Phone: 303 289 5971
4850 Olive Street
Commerce City
Colorado 80022
U.S.A.

This company offers solar heaters using air as the circulating medium with the ability to heat water as well as applications involving space heating. This avoids the problems with liquid filled systems of freezing and corrosion.

Typical size per unit: 3 x 6.5ft = 19.5ft^2 (1.81m^2).
Weight: 153lb (69kg).
Absorber: 28swg steel with ceramic enamel surface.
Glazing: Double glazed, hermetically sealed tempered glass.
Casing: 24swg steel.
Insulation: Fibreglass.

Numerous accessories are offered, including a pebble-bed heat store, air to water heat exchangers, etc.

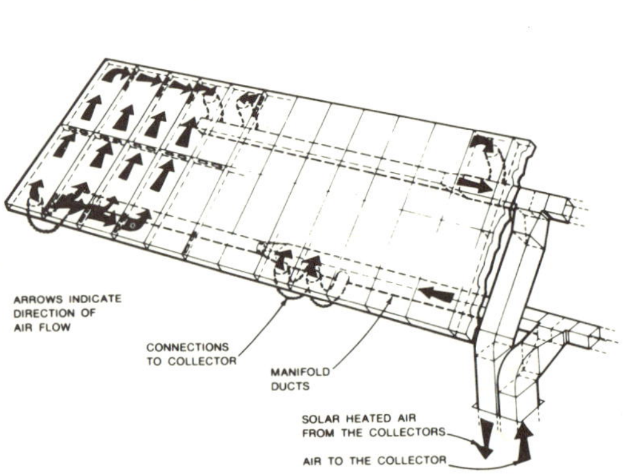

Solaron Air Type Solar Collector Series 2000

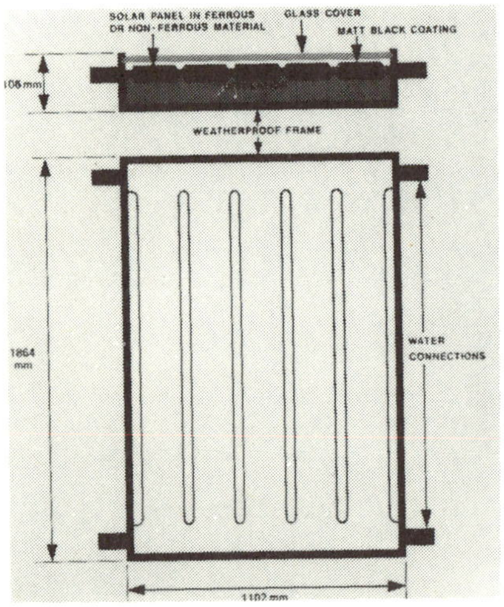

Sundwell Solar Panel

SUNDWELL ENERGY SYSTEMS
Hawker Siddeley —
Washington Engineering Ltd
Solar Heating Division
Industrial Road
Washington
Tyne and Wear NE37 2SB
U.K.

Telex: 53674
Cables: Trafo Washington
Phone: 0632 463001

Collector module surface area: 2m^2 (approx.).
Collector absorber material: Mild or stainless steel.
Back and edge insulation: Polyurethane foam 'Quelflam'.
Glazing: 4mm float glass.
Water content: 8.22 litres.
Weight of panel (drained): 56kg (approx.).
Pressure: Tested to 0.7kg/m^2.
Frame: 0.9mm mild steel, powder coated and stove enamelled.

SUNPOWER DISTRIBUTORS LTD Phone: 446570
P.O. Box 172
Taradale
Napier
New Zealand

Manufacturers of a glazed chlorinated PVC plastic solar collector for domestic water heating applications. Units have a surface area of 1.28m^2 and are encased in galvanised cases. Approximately 1050kWh per year per panel can be collected in New Zealand. Typical installations are indirect by thermosyphon to a special holding tank.

Sunpower CPVC Solar Panel

SUNMASTER Phone: (052 785) 3833 2454
Foster House
Redditch Road
Studley
Warwickshire B80 7AX
U.K.

Manufacturers of flat plate solar collectors that can be delivered to the purchaser in a knocked down form for home assembly. The group also manufactures a controller for electric control of circulation pumps in a solar heating system by sensing the temperature differential between the collector and storage.

Panels are 4ft x 4ft (1.48m^2) or 5ft x 3ft (1.4m^2). The absorber is galvanised steel with a copper water tube serpentine soldered to it.

Sunmaster Solar Collector

SUNRAY SOLAR HEAT INC Phone: (212) 857 0193
202 Classon Avenue 638 6540
Brooklyn
New York 11205
U.S.A.

An 18sq.ft. (1.67m^2) flat plate, single glazed, solar collector.

Absorber: All copper with "Solar Selective Black" selective coating.
Glazing: "Shatter resistant" glass.
Casing: Aluminium.
Insulation: Fibreglass blanket plus polyurethane foam.

Sunray Solar Panel Array

SUNSENSE LTD Phone: 0733
1 Church Street 252672
Northborough
Peterborough
U.K.

A manufacturer with a range of solar products as well as an agency for Elektro wind generators.

Main type of solar panel is supplied unglazed (customer adds own 4mm glass).

Size: 0.675 x 1.46m = 0.98m^2 with a copper absorber, copper water tubes, at 150mm centres, fibreglass wool insulation, aluminium sheet casing, weight empty and unglazed = 12kg per panel.
Manufacturer claims 0.018 to 0.024m^2 of panel is required per litre of hot water storage under British conditions.

Also offer electronic solar panel pump controller.

Sunsense Solar Panel

J.M. SOLAR HEATERS Phone: Gaberone
P.O. Bag 004 4389
Gaberone
Botswana

A range of low cost solar water heaters from this manufacturer has recently been introduced. These generally have galvanised water pipe water tubes attached to a galvanised steel absorber panel. A galvanised steel case has fibreglass insulation and single glass covering. The system can be supplied complete with storage tank, as illustrated and is normally installed to thermosyphon to avoid the need for an electric pump.

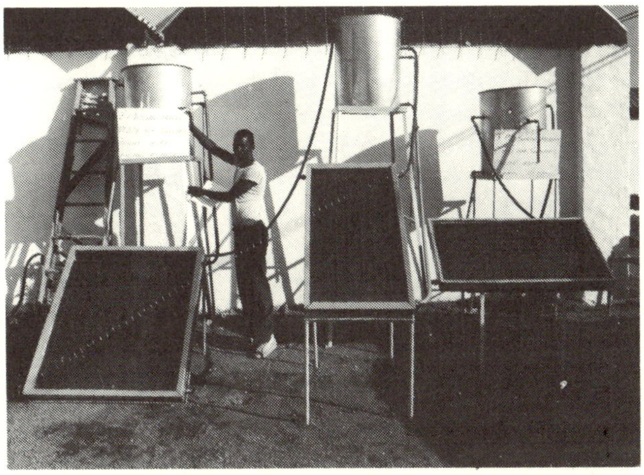

SUNWORKS
Division of Enthone Inc.
P.O. Box 1004
New Haven
Connecticut 06508
U.S.A.

Manufacturers of both liquid and air heating solar panels with mainly copper absorbers. Typical specifications:

Absorber area: 18.68ft^2 (1.74m^2).
Absorber construction: Copper sheet with selective black (Enthone) surface and copper tubes for liquid versions.
Casing: Aluminium with fibreglass blanket insulation.
Glazing: Tempered glass, $^3/_{16}$".

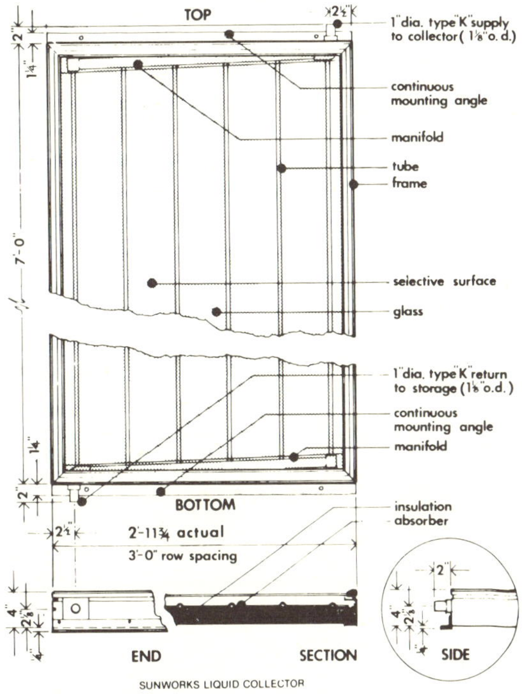

Sunworks Liquid Collector

SUNSHINE MANUFACTURING COMPANY Phone: 503 643 6172
4870 SW Main Highway
Beaverton
Oregon 97005
U.S.A.

Manufacturers of a tracking solar concentrator for water heating, or for producing steam at up to a temperature of 550°F. Tracking is automatic by differential solar heating of gas which works a servo system.

The company also produces a heat store that stores up to 100,000 BTU (29.3kWh) at 400°F (205°C) for use in conjunction with the collector.

Specifications:

Physical dimensions: 48″ x 96″.
Collection area: 25 sq.ft.
Concentration ratio: 6.
Operating temperature: 550°F.
Operating pressure: 25 psig.
*Power output: 6kWh/day.
Shipping weight: 180lbs (82kg).
Features:
Automatic sun tracking: Electronic temperature control: Weatherproof construction: 15 year minimum useful life.

*(thermal).

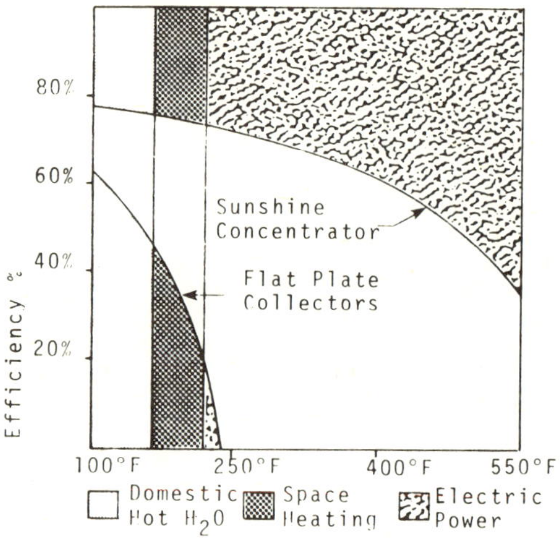

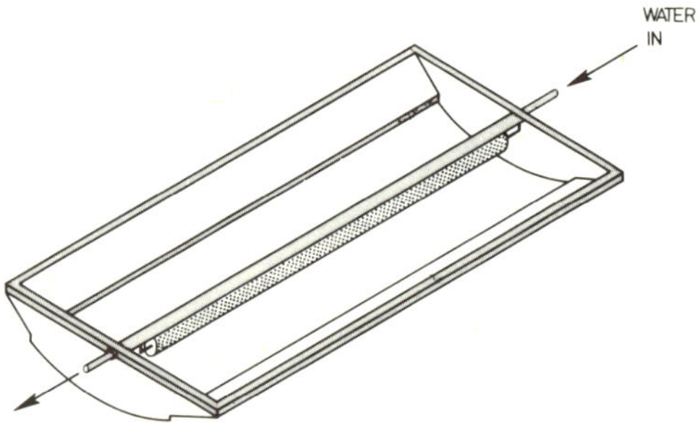

Instantaneous Collector Performance

Also manufacturers of a solar power cell for storage of solar heat for use in periods of low insolation and to smooth out the variations in solar output due to weather conditions.

Physical dimensions: 60″ x 60″ x 60″ (1.5m x 1.5m x 1.5m).
Thermal storage capacity: 550kWh.
Charge rate: 360kWh/day (input at 500°F = 260°C.)
Maximum discharge: 250kWh/hour (output at 400°F = 204°C).
Leakage rate: 25kWh/day.
Shipping weight: 1200lb (544kg).
Minimum life (claimed): 15 years.

TUNNEL COMPANY LTD Phone: Fort Ternan 7Y7
P.O. Fort Ternan
Kenya
East Africa

This company has been manufacturing a low cost type of solar water heater for a number of years, based on a corrugated steel absorber panel treated with suitable internal coatings to prevent corrosion. The glazing consists of a uv-resistant polyester envelope which surrounds both sides of the absorber; this needs replacing at intervals from 1½ to 4 years, but is of relatively low cost. Water temperatures of 65°C are claimed for sunny days in Kenya, while 40°C can be obtained on overcast days. The standard absorber module is 10ft x 2ft 3in. (3.05m x 0.69m) giving an area of 22.5ft^2 (2.09m^2).

Complete systems including hot water storage tanks, with optional electric, cylinder gas or charcoal booster heating capability are available to provide hot water for 5 or 10 showers per day or for 1, 2, 3 or 4 baths per day.

A larger absorber element 20ft x 2ft 3in is also offered for institutional or industrial use.

Hutchinson Solar Shower

2

Wind Power

WIND-ELECTRIC SYSTEMS
WIND-PUMPING SYSTEMS

The energy in the wind has been used throughout man's history in many parts of the world and was developed to such an extent that up to the time of the industrial revolution it was the most powerful and widely used energy source under human control. The fossil-fuel-burning heat engines that came with the industrial revolution led to the decline of windpower due to a better power/weight ratio and more predictable availability of power from first, the steam engine and later, the internal combustion engine.

However, even during the huge growth of fossil fuel based energy consumption in the industrialised areas, wind has continued to be used — primarily for pumping water — in remote regions where fuel supplies or maintenance facilities are difficult to arrange for internal combustion engines. The cattle and sheep raising industries of the arid central USA and even more so in central Australia still rely today on water-pumping windmills. As many as a million of these windmills are still in use in the world, predominantly in Australia and the USA. Indeed, it is true that the windmill, rather than the six-gun, was more important, if less dramatic, in "opening up the Wild West" for ranching. Windmills were also quite widely used in large numbers for generating small outputs of electricity in the 1920s and 1930s in the more remote parts of the industrialised countries, until subsidised rural electrification allowed their replacement by mains electricity.

Today, with greatly increasing oil prices, there is a considerable revival in interest in wind power, both for water pumping and for electricity. Major research programmes are also in hand to develop large windmills with megawatt-sized outputs for feeding electrical grid systems and saving fossil fuels. The few surviving manufacturers of traditional small wind pumps and generators are enjoying a revival of sales and have been joined by a number of new competitors offering in some cases novel designs incorporating recent technical innovations. It can be expected that both the quality and the variety of wind machines appearing on the market will continue to grow during the next few years, as we hope will be indicated by future editions of this publication.

The wind

It is not intended to go into a great deal of detail on the complex subject of wind behaviour and wind energy extraction; however there are a number of important factors which can greatly affect the output of a particular wind machine and which consequently have an important bearing on whether the use of a windmill for a particular application in a chosen location will be economic or not. Just as the economics of running an engine will be closely tied to the cost of fuel, so the economics of windmills depend closely on the availability of wind energy at a particular site.

The energy available in the wind is not at all proportional to its speed — in fact the output of a windmill will vary with the *cube* of the windspeed. In other words, doubling the wind speed has the effect of increasing the available energy by a factor of *eight* ($2^3 = 8$), while a halving of the windspeed reduces the energy to one eighth. So the energy availability is very much more variable than the windspeed. The result of this is that it is generally not practicable to make use of winds with speeds lower than about 5 to 7 mph (8 to 11 km/h) while winds higher than about 30 mph (or 50 km/h) tend to be too powerful to be conveniently used. Therefore, most windmills are designed to make use of winds in the range mentioned; they do not function in lower windspeeds and invariably either furl themselves or deliberately shed a lot of the available power in higher windspeeds with the aid of an automatic governing or furling system to prevent any damage. Hence it is obviously important that there are reasonably frequent winds available at a proposed windmill site in the range preferably above 10 mph or 16 km/h and certainly above 5 mph (8 km/h).

Although the wind is variable in an unpredictable manner from day to day (not even the most well-equipped forecaster could predict the exact windspeed at a certain time in a particular location on a particular day), the total wind energy passing over a particular site remains surprisingly constant if measured over the years. It may vary considerably seasonally in many places, but the total for any particular year generally only varies by a moderate amount. Obviously some sites are very much better than others and of course there are many places where wind speeds are so low on average that it would be quite unrealistic to try and use windmills. On the other hand, some only moderately windy locations offer the prospect of wind energy for certain applications at a lower cost than any other alternative.

Wind behaviour is governed by a combination of global, regional, seasonal and diurnal weather patterns. Globally there are a series of wind belts of varying intensity and prevailing direction, that used to be of great importance in the days of sailing ships, and which are described in detail in most school geography books and world atlases. There are mobile regional effects caused by atmospheric depressions (cyclones and anti-cyclones), and more constant regional effects caused by hills or mountain ranges and the effect of lakes or sea shores. Seasonal changes such as Monsoon or Harmatan winds are caused by the heating and cooling of a large continental land mass in summer and winter. Lastly there is commonly a diurnal wind pattern caused by temperature differences between night and day. All these factors combine to produce varying winds which particularly on land are modified further by surface irregularities and "hot spots" which give the wind its irregular gusty character with which we are all familiar.

In general the latitudes within 10° either side of the equator tend to have lower average wind speeds than most other regions. On the other hand, the coastal regions of hot countries or areas around large lakes in hot sunny regions tend to have marked diurnal winds caused by the temperature differences between the air over the land and that over the water. So, very approximately, coastal regions (by the sea or a large lake) and islands in a marine trade-wind belt often offer the best wind conditions for windmill operation, while inland regions near the equator tend to be poorer.

Choice of site in a given area is also important. Wind speed increases with height above the ground according to the formula $\left(\frac{H_1}{H_2}\right)^{0.17} = \left(\frac{V_1}{V_2}\right)$ where V_1 is the wind speed at height H_1 and V_2 at height H_2. The exponent 0.17 is an approximate figure for "average" ground surfaces – over a smooth surface like water the variation of wind speed with height is less marked and a lower exponent applies, while any obstructed surface covered with trees or buildings creates vertical eddies which have the effect of interfering with the wind motion higher up. Because the power in the wind is related to the cube of the wind speed, putting a windmill on a higher tower can have a more impressive effect on its output than might be expected – for example the wind speed is generally recorded by weather stations at a standard height of 10m above the ground (33 ft) – therefore if a windmill is placed 20m (66 ft) above the ground it will feel a wind speed $(2)^{0.17} = 1.125$ times that at 10m, an increase of 12½%. Bearing in mind that the power available is related to wind-speed cubed, there will be $(1.125)^3 = 1.424$ – in other words 42% more power available than at a wind speed recorded at 10m. Obviously placing the windmill higher still will improve the wind power availability still further, but this has to be paid for in terms of a more expensive tower. In the end a compromise is generally sought in which the minimum tower height which will give the rotor unobstructed access to the wind is used – most manufacturers specify that the lower part of the rotor disc should be at least 6m (20 ft) above the highest level of any obstructions such as trees or buildings within 200m (600 ft) of the windmill, if possible. If there are any obstructions nearer to the windmill than this, it is better to allow an even greater clearance than 6m. In practice, this ideal cannot always be achieved and some obstructions may be tolerated.

Sizing of windmills

There are of course two main applications for windmills; pumping water and generating electricity. This section of the Guide has been divided into two parts dealing with the two main windmill types. A glance at each part will show that, in general, water pumping windmills have multi-bladed rotors, while most electricity generating machines have two or three bladed rotors more like an aircraft propellor in appearance. The reason for this is that a high starting torque (turning force) is needed to get a water pump started, and the provision of many blades eases starting against a heavy load in light winds. Unfortunately a multi-bladed rotor is less efficient at converting wind energy once it gets going, so that many of the multi-bladed water pumpers are actually only about half as powerful in a given wind as an "airscrew" type electricity generating windmill of the same diameter. Also, it is of course generally more expensive to build a multi-bladed rotor due to its greater material content. The other main characteristic of the two main types of rotor is that multi-bladed rotors run slowly in a given wind while propellor-like ones run fast – a good analogy is to think of water pumpers as "low geared" in their interaction with the wind, while electricity generators are "high geared" – this is another reason for the choice of rotor, as reciprocating well pumps need to be driven at quite low speeds (up to around 60 strokes per minute) while electricity generators need to run at 1500 rpm or so, so that the faster the rotor speed the less gearing is required to drive the generator at the optimum speed.

The principles involved in using the wind are generally very similar for both major types of windmill. The output of electricity generators would be proportional to wind speed cubed, but the output of pumping units would correspond to windspeed itself. The relative power of different sized windmills in a given wind is proportional to the area of their rotors, or the *square* of their diameters (since the area of a disc is $\pi d^2/4 = 0.785d^2$ – where d is the diameter). An efficiency factor will also apply which is a measure of the effectiveness of the rotor at using wind power – with most water pumpers the overall efficiency or *power coefficient* (C_p) is around the 0.10 to 0.25 level (10 to 25% of the energy in the wind intercepted by the rotor usefully applied) while many electricity generators can achieve C_p values of between 0.3 and 0.4. A further complication is that the C_p varies considerably with any windmill at different wind speeds, usually being best in low windspeeds where it is most important to use the wind as efficiently as possible, and tailing off to a low level in higher windspeeds.

Manufacturers generally give the outputs of their windmills at various wind speeds in their sales literature. Electricity generators are often quoted as giving a certain rated output at a certain *rated speed* — this is usually quite a high windspeed above which the windmilll is governed to give no increase in output, but below which the output is related to windspeed cubed. Wind-pumps on the other hand are sensitive to the size of pump fitted — fitting a smaller pump will allow the windmill to start in a lower windspeed, but it will produce less water in a high windspeed than with a bigger pump. Good judgement is required to fit a pump which will maximise the output from the windmill in a given wind régime.

Care must be taken when reading manufacturers' brochures not to be fooled by "Rated Power Outputs". If a windmill is rated by the manufacturer for a high wind speed, say 25 mph or 40 km/h, then it may appear to be much more powerful than a rival rated at a more realistic speed of say 15 mph or 24 km/h, because there is $(5/3)^3$ = 4.6 times as much power in the wind at the higher of the two rated speeds. In other words, a wind generator rated at 25 mph, all other things being equal, should be expected to be 4.6 times more powerful *at its rated speed* than one rated at 15 mph. What matters more is the output at average speeds prevailing in the proposed wind generator location and the most important thing is the total amount of energy at all windspeeds that can be converted — i.e. the number of kWh per year that a system can produce. It is to be hoped that manufacturers will stop the misleading practice of publishing "rated output" figures for unrealistically high windspeeds, but will publish actual amounts of energy converted in typical wind régimes — in the meantime the reader is cautioned to note carefully the windspeed at which any quoted rated power output is deemed to apply and to make the necessary allowances.

Generally, to obtain optimum matching of a wind-rotor and generator, the correct rated wind speed shall be not much more than twice the mean windspeed of the chosen location.

To allow the reader to arrive at an approximate estimate of the size of windmill likely to be needed for any application, the following table has been compiled to illustrate the relative outputs from various sizes of electricity generating windmill. The electrical outputs are estimated at a Cp fixed at 0.3 and given purely in kW for different windspeeds. A certain amount of rounding of numbers has been done, for clarity, as these tables are of course only intended as an approximate guide.

An indication of the output of various sizes of wind-pump can be obtained by studying the figures given by manufacturers in several of the entries that follow.

Table 2 shows typical figures for a 16 ft diamater wind-pump. Because Cp declines as wind speed increases, the output only goes up in proportion to V rather than V^3. The wind-pump starts at 5 mph.

Windmill specification

In order to arrive at reasonably accurate predictions of likely windmill performance in a given location, regular wind records are needed, ideally for several years, from as close a meteorological station as possible. There are usually three main levels of wind data recorded by different categories of weather station. The most elaborate (and useful) consist of continuous or hourly records of wind speed; such detailed information as this is only usually recorded at major weather stations such as those attached to international airports. The next level consists of "run of wind" figures (that is the number of miles or kilometres of wind that have gone past a cup anemometer), recorded twice or three times each twenty-four hours; this information allows the average windspeed to be estimated for each day and each night, or in some cases for the morning, afternoon and night. Lastly, the crudest wind data consists of run of wind figures for longer periods, such as daily or monthly; these give no indication of diurnal variations in wind speed, but do give an indication of the relative windiness of a place compared with other places, which can allow at least a vague assessment of the suitability of the wind regime.

If run of wind figures taken regularly, three times per day are available, or, better still, hourly figures, it is possible to produce a velocity/duration chart from which the annual output of a windmill may be predicted. What is required is the frequency with which a particular windspeed occurs. If records are taken only three times a day, say at 0800, 1200 and 1800 — suppose the wind run

The formulae used to calculate Table 1 are:-

$P = 0.00000125 d^2 V^3$ (P in kW, d in ft, V in mph)

$P = 0.00000323 d^2 V^3$ (P in kW, d in m., V in Kmph)

Where: P = Power

d = Diameter

V = Velocity

Table 1
Electricity generating windmills with C_p = 0.3. Output table

rotor diameter		6	8	12	14	16	20	24	30	ft
		1.8	2.4	3.7	4.3	4.9	6.1	7.3	9.2	m
wind	*speed*									
5	8	.005	.010	0.22	.029	.038	.060	0.86	.135	kW
10	16	.043	.077	.173	.235	.307	.480	.691	1.08	kW
15	24	.146	.259	.583	.794	1.04	1.62	2.33	3.64	kW
20	32	.346	.614	1.38	1.88	2.46	3.84	5.53	8.64	kW
25	40	.675	1.20	2.70	3.67	4.80	7.50	10.8	16.9	kW
mph	*kmph*									

Table 2
Typical 16 ft (4.9 m) dia. water pumping windmill

Wind speed		*Output x head*	
5	8	45000	61
10	16	90000	122
15	24	135000	183
20	32	180000	244
25	40	225000	305
mph	*km/h*	*ft.gall/h*	*m.m³/h*

between 0800 and 1200 was 30km, then the average wind speed for those four hours was 30/4 = 7.5 km/h and we would say that a frequency of 7.5 km/h had occurred four times; between 1200 and 1800 suppose the run was 36km, giving an average of 6km/h occurring 6 times, and so on. Then to make use of the windmill data given in Table 2 we would record all wind averages below 5km/h, as being of no use to the windmill, all between 5 and 11 are taken as producing the output corresponding to 8km/h, all between 12 and 20 are taken as producing an output similar to 16km/h and so on. Obviously if hourly figures are available they need not be averaged over a whole morning, afternoon or night, but they would be grouped within the speed bands given for the windmill as just described.

Therefore by knowing from Table 1 or 2 what the windmill will produce in a certain windspeed in 1 hour, and by knowing how many hours in the year that windspeed tends to occur, we can calculate the annual output due to winds of that speed. This can be repeated for each speed range and the results can be added to give the total for the year as indicated in Table 3.

Obviously this calculation can be considerably refined by taking more averages over narrower wind-speed ranges and by averaging figures for several different years (assuming suitable windmill and meteorological data are available). This calculation will tend to under-estimate the windmill output if it is set up under comparable conditions to the anemometer used for recording the wind data, especially if the windmill rotor is considerably higher than the anemometer. The figures used apply to Leicester in the UK, which has quite a low average wind speed (by UK standards) of 6.2 mph (9.9 kmph) but which corresponds quite well to typical annual averages for many tropical locations.

It is interesting that if we take the output per hour corresponding to the annual average wind speed and then multiply it by the number of hours in the year (8760) we obtain a much lower apparent output. This is because a steady low wind (implied by taking an annual average) has much less total power than a fluctuating wind of the same average (the average of the cubes of a series of numbers is greater than the cube of their average).

Obviously the same process can be applied using Table 1 in order to estimate the likely output from an electricity-generating windmill in the same wind regime.

If detailed meteorological data is not available from any weather station that is near enough to the proposed location to be representative of its wind regime, then annual average wind-run figures of the crudest kind can be used as a very rough guide to whether a location could be used for wind powered pumping or generation. In general, places with average annual wind runs less than about 44000 miles or 70000 kilometres, (i.e. average speeds below 5 mph or 8 mph) are unlikely to be very satisfactory, while over 90000 miles or 144000 km suggests a promising location. Anything in between will probably be adequate especially if experience suggests considerable diurnal variability (e.g. sea or lake breezes) and if a good exposed location can be used or the windmill is raised on a high tower.

Where limited meteorological data is available another useful indicator might be the performance of any windmills already in your area, or, if planning to introduce a number of units for the first time, it may be justifiable to set up one small unit initially and use it over a year or so to provide a measure in itself of the possible output. There is nothing better than a windmill for measuring the suitability of the wind regime — although they cost much more than just an anemometer.

In cases were the wind regime was misjudged, providing the error was not too great, it is possible to improve the low wind performance with pumping windmills by reducing the stroke or fitting a smaller diameter pump, and vice-versa.

Table 3
Velocity/duration chart example for a 16 ft diameter water pumping windmill

Wind speed ranges	*No. of times speed within range recorded* (hours p.a.)	*Windmill output at this speed range for head of 100 ft (30m) from Table 2* (Imp.gal/h)	(m^3/h)	*Cumulative total outputs* (Imp.gal)	(m^3)
under 3 mph under 5 km/h	2765	—	—	—	—
3 to 7 mph 5 to 11 km/h	2644	450	2	1189800	5288
8 to 12 mph 12 to 19 km/h	2272	900	4	2044800	9088
13 to 17 mph 20 to 27 km/h	882	1350	6	1190700	5292
18 to 22 mph 28 to 35 km/h	137	1800	8	246600	1096
over 22 mph over 35 km/h	60	2250	10	135000	600
Total for year				4806900	21364

Storage

Due to the variability of the wind, if a supply of electricity or water must be guaranteed, it becomes necessary either to provide a storage capable of covering the longest likely windless periods, or to provide a standby capability using some other energy source such as a small engine. Obviously, the windier the location the smaller the storage needs to be. Water storage is of course considerably cheaper than electrical storage in batteries, and it becomes feasible to provide for windless periods of a week or more in some cases. The capital cost of electricity storage becomes rather high if kilowatt sized loads are to be supplied, and it is often cheaper to use a small standby diesel or petrol engined generator in such situations. The combination of windmill and generator can be considerably more economical in the long term than either system considered alone, since the windmill greatly extends the life of the engine, reduces the costs of maintenance and saves a large proportion of the fuel that would otherwise be needed.

Wind-Electric Systems

AERO-POWER
2398 – 4th Street
Berkeley
California 94710
U.S.A.

Phone: (415) 848 2710

Model "A" wind generator is rated at 1000W and delivers 12V for battery charging. Rated output is reached at 25 m.p.h. (40km/h) and the machine starts at approximately 6 m.p.h. (10km/h).

The rotor is of 8.5ft (2.6m) diameter with three spruce blades with governing by pitch change. A 14.5V 75A maximum alternator is fitted (3-phase rectified to d.c.).

Various towers from 30 to 80ft (9m to 24m) are available and so are voltage regulators, inverters and batteries. This manufacturer also offers a Super-8 film about their wind plant.

Aero-Power Model A Wind Generator

BOSMAN
Waterbeheersing En
Milieuverbetering B.V.
Steegjesdijk 4
Postbus 3518
3364 Piershil (Z.-H.)
Netherlands

Telex: 24475
Phone: (01869) 1316

Manufacturers of a small wind generator set for battery charging operations at 350 to 400W at outputs of 12 or 24V. The twin bladed rotor is made of fibreglass and automatically feathered. Storm protection is effected normally by turning the unit out of the wind. Unit can be supplied on a 7 metre tower.

Bosman Wind Generator Set

ALSTHOM-NEYRPIC-
TECHNIQUES DES FLUIDES
B.P. 75
38041 Grenoble Cedex
France

Telex: 320750
Cables: Neyrpic Grenoble
Phone: (76) 96-48-30

Manufacturers of two electricity generating windmills rated at 15 and 40hp which can be fitted with different rotors to suit different wind régimes:

Wind speed		*15hp Mechanism*		*40hp Mechanism*	
		Rotor of 8m (26ft)	*Rotor of 10m (33ft)*	*Rotor of 13m (43ft)*	*Rotor of (16m (52ft)*
m/s	*m.p.h.*	*hp*	*hp*	*hp*	*hp*
6	13	–	5.8	–	16.3
7	15	5.8	9.3	16.5	25.8
8	18	8.7	14.7	24.5	38.4
9	20	12.5	–	34.9	–
9.5	21	14.7	–	41.1	–

Orientation and speed control are by an auxiliary fan tail rotor and centrifugal pitch change of the rotor blades respectively. Various outputs are possible, typically 110V and this machine is normally supplied with either a 12 or a 15m (40 or 50ft) tower.

Alsthom-Neyrpic Windmill

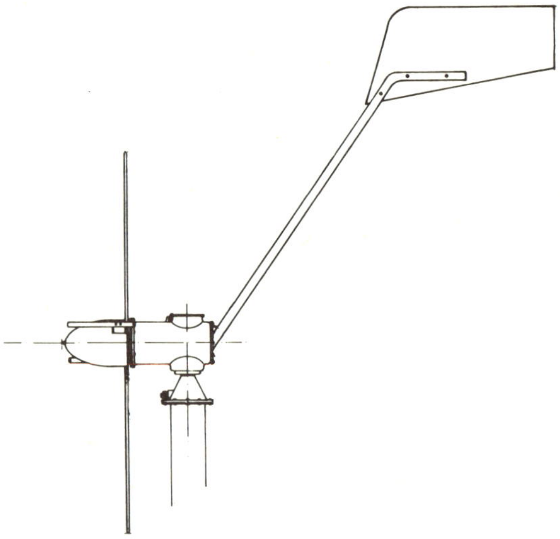

Wind Generator 24 FP7 E

AEROWATT COMPANY
37 rue Chanzy
75011 Paris
France

Telex: Sysna 680160F
Phone: 371 35 78
371 36 21

This is a well established manufacturer in this field with a range of generators offering rated outputs from 24 to 4100W. These machines are conservatively rated and designed to be of high quality and the manufacturer offers special rotors for extreme conditions such as freezing or wind-blown sand.

The principal characteristics of their wind generators are given below.

All these machines run up-stream of the tower with centrifugal governing via blade pitch variation, giving protection automatically in wind speeds up to 125 m.p.h. or 200km/h.

Larger machines are available to special order up to a maximum of 100kW.

*These models can be used to drive a 380V 3-phase, 50Hz submersible pump directly.

Model	*Starting wind speed*		*Rated wind speed*		*Rated power output*	*Rotor dia.*		*Nominal voltage of battery bank*
	(m/s)	*(mph)*	*(m/s)*	*(mph)*	*W*	*(m)*	*(ft)*	*V*
24 FP7	4	9	7	16	24	1.2	3.9	24
100FP5	2	4.5	5	11	100	3.2	10.5	24
150FP7D	3	6.7	7	16	130	2.0	6.5	24
300FP7B	3	6.7	7	16	350	3.2	10.5	24
*1100FP5	2	4.5	5	11	1125	9.2	30	24
*1100FP7B	3	6.7	7	16	1125	5.0	16.5	24
4100FP7	3	6.7	7	16	4100	9.2	30	48

CHLORIDE TRANSIPACK LTD
Stanley Road
Bromley BR2 9JF
Kent
U.K.

Telex: 896071 code Transipack London
Cables: Transipack Bromley
Phone: 01 460 9861

Standard model of wind generator the Transwind 1.1kW machine starts generating at wind speeds of 7.5kt and reaches its rated output at 10kt. Rotor diameter is approximately 2m with variable pitch control to maintain a constant 50Hz 240/110V output.

A similar 3.5m rotor machine rated at 4kW is also available to special order.

See also this company's wide range of static inverter systems.

Transwind 1.1kW Wind Generator

1200 and 2500W Wind Generator Units

COULSON WIND ELECTRIC
RFD 1 Box 225
Polk City
Iowa 50226
U.S.A.

Phone: 515 984 6038

Retailer of reconditioned, second-hand and new wind generators, and ancillary equipment. Offers consultancy service based on 25 years experience particularly with the well known Jacobs generators.

Stocks: reconditioned and second-hand Jacobs Winchargers. New Winco 200W units.

DAVEY DUNLITE
(Division of PYE Industries Sales Pty Ltd)
P.O. Box 120
Oakleigh
Melbourne
Australia 3166

Telex: 32542
Cables: Daylite Melbourne
Phone: 544 6666

The main model is a 2kW rated wind generator (in 25 m.p.h. or 40km/h windspeed, starting speed 10 m.p.h. 16km/h), which has been in production for many years. It has a 13ft (4m) three-bladed rotor with galvanised steel aerofoils. Governing by centrifugally actuated pitch-change and a brake is fitted to stop the rotor manually. A version with a 10ft 6 in (3.2m) rotor is available capable of withstanding 120 m.p.h. winds in very windy locations (some have been used in the Antarctic).

The generator is a brushless machine with an integral gearbox, claimed to be able to run without any maintenance attention for intervals of five years. Outputs of 24V or 110V optional.

A new 5kW model is also available. Specification details were not available at the time of going to press.

U.K. Agents are Pye Telecommunications, Newmarket Road, Cambridge.

U.S. Agents are Independent Power Developers Inc., Box 1467 Noxon, Montana 59853.

2000 Watt Wind Generator

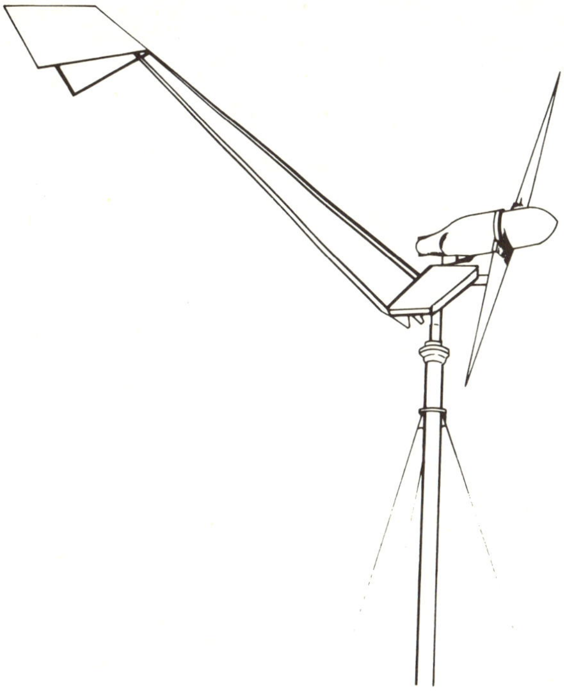

Forces Motrices Neuchateloises Wind Generator

FORCES MOTRICES NEUCHATELOISES S.A.
Rue Pourtales 13
CH-2000
Neuchatel
Switzerland

Telex: Ensa 35 140
Phone: 038 25 77 51

This wind generator is supplied as a complete system consisting of the windmill, tower, servicing console and storage batteries designed to produce an output of 220V, 50Hz, via an inverter.

The wind generator is nominally rated at 5kW and apparently achieves its rated output at 10m/s (22 m.p.h.). It cuts in at 2.5m/s (6 m.p.h.) and governs by turning out of wind above 15m/s (34 m.p.h.). It has a fixed pitch fibreglass two-bladed rotor of 5m diameter and has an overspeed device to prevent speeds in excess of 500 r.p.m.

The alternator is brushless with permanent magnets offering high efficiency through the speed range used, producing 3-phase alternating current at 380V. Its output is rectified to 60V d.c. and a static inverter then produces 220V +5% or −10% at 50Hz ±5% sinusoidal. The normal installation comes with a battery storage of 160Ah consisting of 5 batteries in series.

DOMINION ALUMINUM FABRICATING LTD.
3570 Hawkestone Road
Mississauga
Ontario
Canada

Telex: 06-961482
Cables: Dafcol
Phone: 416 275 5300

This is an unusual turbine, being of vertical-axis configuration known as a "Darrieus" rotor. Two models are available with a choice of voltage for the smaller one:

Rotor dia. (ft)	*Output voltage*	*Rated output Power*	*Rated output windspeed*	*Average monthly output 9*	*11*	*13*	*15 m.p.h. (av. wind)*
15	24V	4kW	23 m.p.h.	110	190	290	420kWh/month
15	110V	4kW	23 m.p.h.	110	210	360	560kWh/month
20	110V	6kW	23 m.p.h.	210	400	680	1070kWh/month

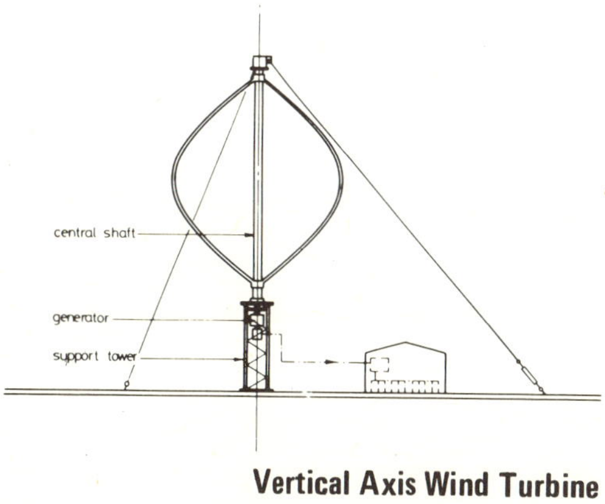

Vertical Axis Wind Turbine

EDMUND SCIENTIFIC COMPANY
101 East Gloucester Pike
Barrington
New Jersey 08007
U.S.A.

Telex: 831-564
Phone: 609 547 3488

Manufacturers of a 12V d.c. portable wind generator producing 196W at 14 m.p.h. and 600W at 25 m.p.h. 3-blade, 9ft diameter rotor on tubular pipe tower driving 12V generator via timing belt. Wind load governing with rotor pivoting against spring-loaded tail. Shipping weight 51 lb (23kg).

Also can supply "Wincharger" generators (9V), (200W @ 12V d.c.).

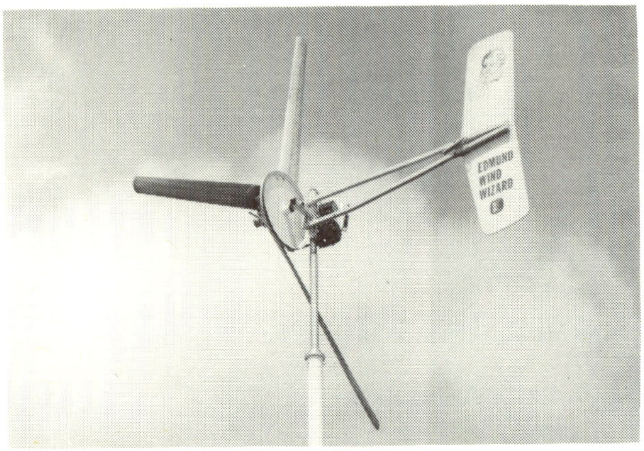

Edmund Portable Wind Generator

GRUMMAN CORPORATION
Energy Systems Division
4175 Veterans Memorial Highway
Ronkonkoma
New York 11779
U.S.A.

Telex: (516) 575-6555
Phone: 516 575 6205

The Windstream 25 Unit is rated at 15kW in winds of 26 m.p.h. Voltage outputs available at 110V, 220V, and 440V a.c. and at 110 or 220 V d.c. Cut-in speed is 8 m.p.h.

Annual energy output from this system estimated by manufacturers as:

Average wind speed for site:	10	12.5	15 m.p.h.
Yearly energy production:	14,000	22,000	37,500kWh

Rotor diameter 25ft (7.6m) with centrifugal overspeed brakes and automatically controlled pitch angle. Rotor weight 340kg nacelle weight 575kg and centrifugally cast concrete tower weighing 4080kg.

Windstream 25 Unit

ENAG SA
Route de Pont-l'Abbé
F 29000 Quimper
Finistère
France

Phone: 95 44 25

Model 1 "Super-Enag": two-bladed governed rotor generating up to 250 watt 12V or 24V output.

Model 2 "Super-Enag": three-bladed governed rotor generating rated output up to 1000 watt at 24V or to 1200 watt at 30V.

Model 3 "Super-Enag" special: similar to Model 1 in output, but fully automatic for remote operation without any kind of supervision.

Model 4 "Super-Enag": three bladed rotor with rated output of 2000 watt at 110V.

Batteries and control equipment can also be supplied, plus towers or stub-towers to suit the various models of 1.1m, 2.2m, and 1.5m.

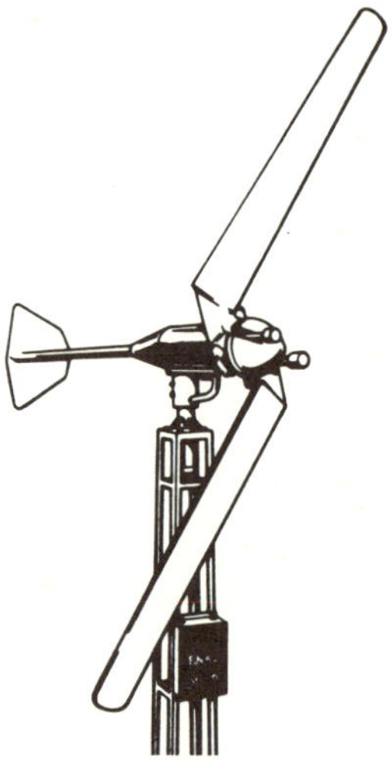

Enag Wind Generator

ÉOLIENNES HUMBLOT
8 Rue d'Alger
Coussey
88300 Neufchateau
France

Phone: (29) 94 09 09

This firm mainly manufactures windpumps but produces one wind generator called Aerogenerateur Ideolec, to the following specifications.

Rotor diameter: 4.20m (13.8ft), two bladed in polyester.

Starting speed: 3m/s (7 m.p.h.) giving 60W output.

Middle range: 6m/s (14 m.p.h.) giving 800W output.

Rated speed: 10m/s (22 m.p.h.) giving rated output of 1100W.

Tower: 9, 12 or 15m (29, 40 or 50ft) angle steel frame.

Generator: 24V, 1100W maximum.

Extras: Regulator, inverter, battery bank, etc.

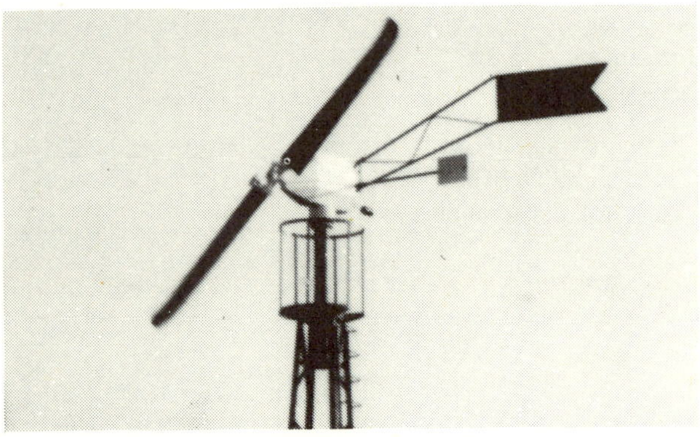

"Ideolec" Wind Generator

Elektro Wind Generator Installation

ELEKTRO GMBH
St. Gallerstrasse 27
8400 Winterthur
Switzerland

Telex: 76-299
Elektro
Phone: 052 22 34 34

Manufacturers of a range of windmill generators (for well over 30 years) with outputs from 50 watts to 10kW, using aerofoil section blades on lattice frame towers.

Model	*Rotor diameter*	*blades*	*Rated output*	*Rated windspeed*		*Available voltages*				*Net weight*
	(m)	*(no.)*		*(m.p.h.)*	*(km/h)*					*(kg)*
W50	vertical axis .45m		50	40	60	6	12	24	–	35
W250	vertical axis .66m		50	20	30	12	24	36	–	70
WV05	2.5	2	750	20	30	12	24	36	48	65
*WV15G	3	2	1200	23	37	12	24	36	48	135
*WV25G	3.6	2	2200	22	36	24	36	48	110	180
*WV35G	4.4	3	4000	23	37	48	60	110	–	235
*WVG50G	5	3	5500	26	42	60	110	–	–	265
*WVG120G	6	3	9000	–	–	110	–	–	–	315

*These models also available with 3-phase a.c. generators for heating applications.

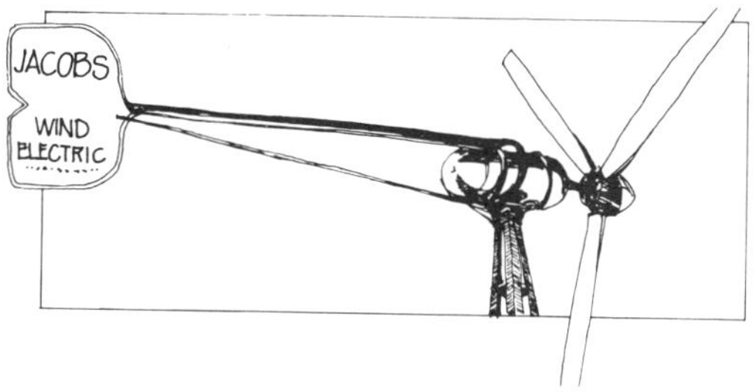

Jacobs/North Wind Generator

NORTH WIND POWER
COMPANY INC
P.O. Box 315
Warren
Vermont 05674
U.S.A.

Phone: 802 496 2955

This company reconditions and rebuilds Jacobs wind generators, a very successful and innovative design in its time (production started in 1931) and still known as one of the most reliable systems available. North Wind have introduced a number of improvements, such as electronic control. All the Jacobs/North Wind generators have three-bladed wooden rotors with aerofoil blades and centrifugal pitch change. Special direct drive generators have been developed which are claimed to match the rotor characteristic particularly well.

The following *Jacobs* machines are offered:

apc J47 3000W @ 120V	apc J46 2800W @ 32V
apc J45 2500W @ 120V	apc J49 2500W @ 32V
apc J50 2000W @ 32V	apc J51 1800W @ 32V

Full rated power is obtained at the relatively low wind speed of 20 m.p.h. (9m/s).

North Wind also offer the following *Aeropower* generator:

Rotor: 6ft diameter spruce, twin aerofoil with pitch change feathering.

Generator: 1000W @ 14V maximum direct drive (max. output in 32 m.p.h. 14.3m/s wind).

The following complete systems are offered:

Jacobs 1.8-30kW 32/120V d.c. or 120V a.c.

Average wind speed for site:	8	10	16	m.p.h.*
Monthly output in kWh	150-250	200-450	500-850	

Aeropower 1kW 12V d.c. or 120V a.c.

Average wind speed for site:	10	12.5	15	m.p.h.*
Monthly output in kWh	33	70	110	

*Note: 1 m.p.h. = 1.6km/h

Also available is a wide selection of different towers in aluminium or galvanised steel, both free-standing and guyed with heights from 20ft (6m) to 100ft (30m). Batteries, inverters, wind measuring equipment, woodstoves and windpumps are also available.

RALPH HOWE MARKETING LTD
New Orchard and High Street
Poole
Dorset
U.K.

Telex: 41495
Cables: Ralphowe Poole
Phone: Poole (02013) 77377/8

Small units for trickle charging electric storage batteries. Intended mainly for marine applications on sailing ships or for remote buoys and harbour beacons, but ideal for any similar application needing small power levels delivered with minimum supervision and maximum reliability.

Model	*Rotor diameter*	*Rated output*	*m.p.h.*	*Cut-in speed m.p.h.*	*Net weight kg*
Selectromarine					
5W12	432mm (17")	5W @ 12V	25	10	10
5W24	432mm (17")	5W @ 24V	25	10	10
10W12	432mm (17") (2 rotors)	10W @ 12V	25	10	15
10W24	432mm (17") (2 rotors)	10W @ 24V	25	10	15
Ampair 50MK3	670mm (26.4")	50W @ 12V	35	12	8.2
Aerocharge 3	432mm (17")	5W @ 12V or 24V	25	10	4

Complete wind-powered marine buoys and navigation aids intended for reliable unattended operation also available.

SENCEBAUGH WIND ELECTRIC
P.O. Box 11174
Palo Alto
California 94306
U.S.A.

Phone: 415 964 1593

This company offers two designs by the well known American wind energy expert Jim Sencenbaugh, as follows:

Model	*Rotor diameter*	*Rated output*	*Maximum output*	*Cut-in speed*	*Weight*
1000-14	12ft (3.7m)	100W at 23 m.p.h.	1200W at 30 m.p.h.	7-8 m.p.h.	300lb shipping
24-14	20in (51cm)	24W at 21 m.p.h.	30W at 30 m.p.h.	8-9 m.p.h.	18lb net.

Also available are Dunlite windmills, towers, storage batteries, wind recording equipment, inverters, etc.

Catalogue $5 – including postage to overseas enquirers.

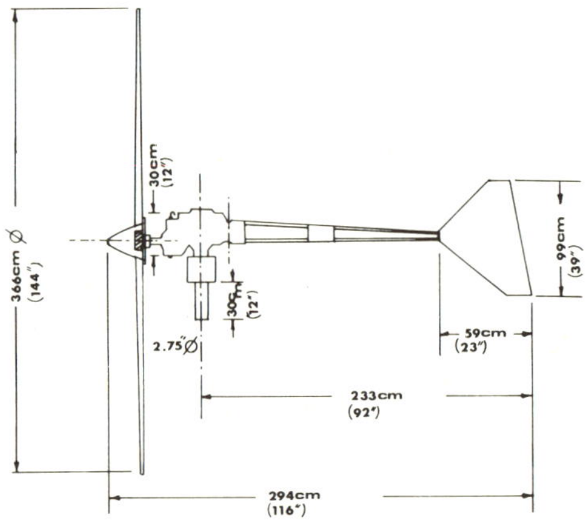

Sencenbaugh Model 1000-14

Kedco 1200 Unit

KEDCO INC
9016 Aviation Boulevard
Inglewood
California 90301
U.S.A.

Phone: 213 776 6636

Kedco offer a range of four aerogenerators based on the Jack Parks design offered by Helion as drawings or by Topanga Power as a kit.

These machines have three-bladed light alloy rotors with variable pitch centrifugally activated governing. The models 1200 and 1600 have 14V (28V optional) d.c. outputs for battery charging, while the 1210 and 1610 have a variable (180V max) voltage permanent magnet alternator rated at up to 2000W. Further details are summarised as follows.

Model	*Rotor dia.*	*Rated power*	*Rated speed*	*Cut-in speed*
1200	12ft 3.7m	1200W	22 m.p.h.	7 m.p.h.
1600	16ft 4.9m	1200W	17 m.p.h.	8 m.p.h.
1210	12ft 3.7m	2000W	26 m.p.h.	11 m.p.h.
1610	16ft 4.9m	2000W	22 m.p.h.	10 m.p.h.

Note that the model 1600 in particular is very conservatively rated compared with most machines on the market. "Tropicalising" is offered as an optional extra and towers and a suitable synchronous inverter are available.

LUBING MASCHINENFABRIK Phone: 05442 625/7
D-2847 Barnstorf
Postfach 110
German Federal Republic

The single Lubing 400W wind generator design results from 25 years experience and is an expensive machine but said to be of good quality. It has six blades made from fibreglass, three smaller ones of fixed pitch to aid starting and three longer variable pitch, centrifugally governed ones which produce most of the power.

The generator is brushless with an output of 24V a.c. which is rectified to 24V (nominal) d.c. via silicon diodes and an electronic regulator. Drive is via an integral gearbox and the windmill comes on a hinged aluminium pole tower which can be winched down horizontal to facilitate servicing.

Performance claimed is as follows:

Wind velocity	*m/s*	4	5	6	7	8	9	10	11	12
	m.p.h.	9	11	13	16	18	20	22	25	27
Output	*Watts*	24	72	136	220	325	375	400	400	400

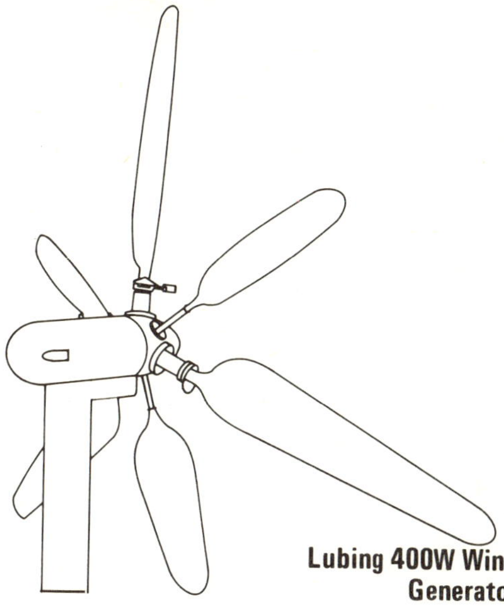

Lubing 400W Wind Generator

PINSON ENERGY CORPORATION
P.O. Box 7
Marstons Mills
Massachusetts 02648
U.S.A.

Cycloturbine Model C2E is supplied to the following specifications:

Rotor diameter: 12ft (3.6m).
Blade length: 8ft (2.5m).
Swept area of rotor: 96ft^2 (8.9m^2).
Overspeed control: Centrifugal.
Starting windspeed: 5 m.p.h. (13.5m/s).
Governing windspeed: 30 m.p.h. (13.5m/s).
Govering r.p.m.: 200
Electrical output: 2kW @ 24 m.p.h. (11m/s). 4kW @ 30 m.p.h. (13.5m/s).

This machine is a very recent development having a vertical-axis configuration (Darrieus) with cyclic pitch variation to achieve self-starting and to limit the speed.

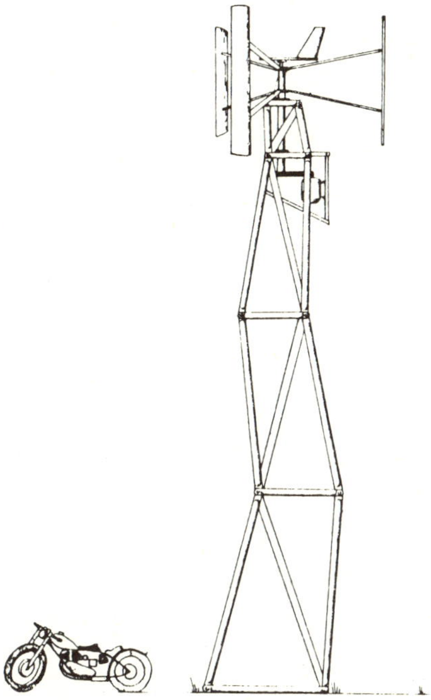

Cycloturbine C2C
Atop 33ft Octahedron Tower

NOAH ENERGIE SYSTEME Gmbh Telex: 889445 d Glahn
Mühlenstr 11 Phone: 02242 2543
D-53 Bonn
W. Germany

&

Wippenhohnerstrasse 31
D5202 Hennef 1
W. Germany

Single and double rotor units with power outputs from 15kW to 130kW depending on size and wind speed, as follows:

Type 30/90kW: 12m diameter, 6 bladed rotor, rated output 30kW and maximum output 90kW.

Type 15/45kW: 12m diameter, 3 bladed rotor with rated and maximum outputs of 15 and 45kW respectively.

Type 45/130kW: 16m diameter, 3 bladed rotor of 45 and 130kW rated and maximum outputs.

Larger double rotor machines with outputs from 200kW up to one with a maximum output of 1800kW are offered as special projects on request.

Noah 30/90kW Plant

P.I. SPECIALIST ENGINEERS LTD
The Dean
Alresford
Hants
U.K.

Telex: 447152 PICOR G
Phone: 096 273 3242
Telegrams: PICOR

This windmill embodies another version of the vertical axis principal often attributed to Darrieus, but has a novel feature of "variable geometry" developed by Dr Peter Musgrove of Reading University. This allows the blades to fold outwards if the turbine overspeeds and thereby effectively governs it.

The manufacturer has just begun production, with a primary application intended as cathodic protection of pipelines and other steel structures. However, the system could equally be employed for any low voltage d.c. application such as battery charging.

In common with other Darrieus-type windmills, it is not self-starting. This machine incorporates an electric starter which is activated by an electronic wind-sensing system whenever adequate wind conditions for operation are felt.

Rotor diameter: 9.5m (14ft 9in).
Blade length: 3m (20ft)
Swept area of rotor: 13.5m^2 (148ft^2)
Overspeed control: Centrifugal.
Starting windspeed: 4m/s (9 m.p.h.) (adjustable).
Governing windspeed: 10m/s (22 m.p.h.) (adjustable).
Operational speed range: 80 to 160 r.p.m.
Electrical output: 50-100W @ 5m/s (11 m.p.h.). 500W @ 7m/s (16 m.p.h.). 100W @ 10m/s (22 m.p.h.).

Electrical output via CAV alternator and diode rectifier to 12, 24 or other voltages. Performance figures given depend on choice of output and alternator.

Work on a 6m (19ft 6ins) diameter self-starting version of the windmill is now well advanced. It is hoped it will be available shortly after publication of this Guide.

4.5m Variable Geometry Wind Generator

QUIRK'S VICTORY LIGHT CO. PTY LTD
33 Fairweather Street
Bellevue Hill
New South Wales 2023
Australia

Cables: Quirklite Sydney
Phone: 36 6630

A 12V or 110V unit supplying 3kW in a 25 m.p.h. wind. Generation starts in a 6-8 m.p.h. wind and plants are designed to withstand wind speeds of 140 m.p.h. Brushless generator with integral gearbox. Centrifugal governor with variable pitch galvanised steel rotor blades. Designed for 5 year operation without maintenance.

Manufacturer also offers batteries, inverters, various towers, etc.

Quick's Windlite Generator

ZEPHYR WIND DYNAMO COMPANY
P.O. Box 241
Brunswick
Maine 04011
U.S.A.

Phone: 207 725 6534

The Zephyr Wind Dynamo is a sophisticated design with a low speed, high frequency, permanent magnet alternator direct coupled to a three bladed 20ft diameter (6m) downwind rotor. The reinforced plastic rotor blades are of light weight with tip spoilers for overspeed protection. A servo turns the rotor at right angles to the wind if overspeed or vibrations occur beyond the control of the spoilers.

Rated output of 15kW is achieved in a 30 m.p.h. (13.4m/s) wind. Cut-in is at 8 m.p.h. (3.5m/s) and 500W is produced at 10 m.p.h. Shut down takes place at 45 m.p.h. (20m/s). Weight on tower = 600lb (272kg).

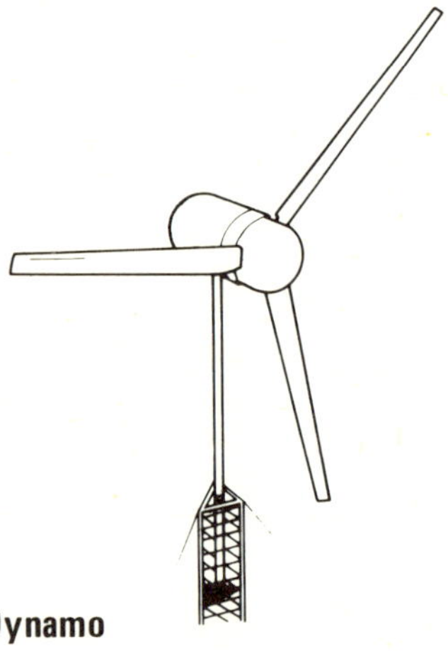

Zephyr Wind Dynamo

TRIMBLE WINDMILLS
Crimple Grange
Beckwithshaw
Harrogate
North Yorkshire HG3 1QU
U.K.

Phone: 0423 56006

Manufacturers of the 'Trimblemill' wind generator, a new windmill due to come onto the market at about the date of publication. Specification is given as:

Rotor diameter: 16ft (5m) 8 bladed (3 front, 5 rear).
Tower: 30ft (9.3m) Ductile Iron pipe.
Control: Electrical overload cut out, automatic furling in 53 m.p.h. winds, limited stress sail lashings, emergency mechanical brake.
Output: 22 m.p.h. (10m/sec) 5 kW 240V. 53 m.p.h. (23.6m/sec) 12.5kW 565V.
Generator: Designed specifically for windmill use by N.E.I. Clark, Chapman Engineering Ltd., (see generator section), it is integral in the hub and carries the blades. Power collection is via slip rings.
Extras: Optional control system for battery storage to give constant 240V 50Hz.

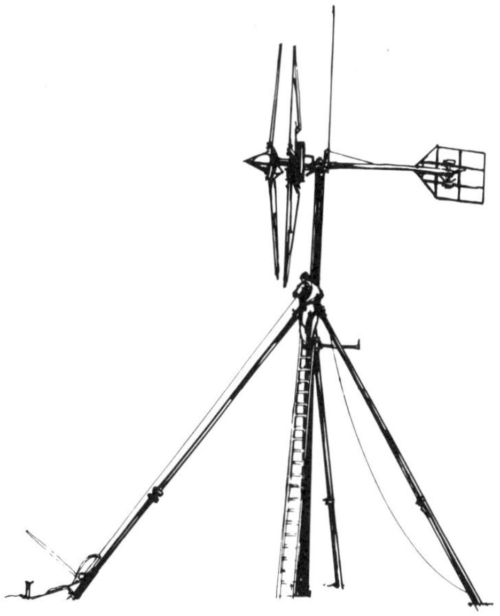

Trimblemill

Helion Windmill

TOPANGA POWER
Box 712
Topanga
California 90290
U.S.A.

Phone: 213 455 2458

This company manufactures a kit of parts to build the Helion windmill. The kit is to a high level of completeness and it is claimed that a competent builder can complete it in two days or less.

Alternatively, plans alone can be obtained from Helion, P.O. Box 4301, Sylmar, California 91342, U.S.A. (Phone 213 367 8291).

Specification as follows:

Rotor diameter: 12ft or 16ft (to choice) (3.6 or 4.9m).
Rated wind speed: 25 m.p.h. (11m/s).
Rated output: 1300W (12V system) or 1600W (24V system) at 18 m.p.h. (16ft) or at 25 m.p.h. (with smaller 12ft rotor).
Rotor type: Three aerofoils from aluminium sheet with centrifugal pitch change governing system. Rotor operates downwind of tower.
Transmission: Oil bath gears generator to 8.6 times rotor r.p.m.
Overall weight: 200lb (91kg).

Note that the illustration shows a prototype Helion windmill, similar in arrangement but differing in detail from the production model.

WINCO
Division of Dyna Technology, Inc.
East Seventh at Division Street
P.O. Box 3263
Sioux City
Iowa 51102
U.S.A.

Telex: 487110
Phone: 712 252 1821

This small 12* volt unit has been in production for some time and produces a maximum of 200 watts in a wind velocity of 23 m.p.h. Charging starts in a breeze of 7 m.p.h. When wind speeds exceed 23 m.p.h. the governor flaps deploy to safeguard the unit. Rotor diameter 6ft. Comes complete with 10ft stub tower.

Average usable energy per month

10 m.p.h. average – 20kWh
12 m.p.h. average – 26kWh
14 m.p.h. average – 30kWh

12ft Wincharger Unit

*24, 28, 32 or 36V available at extra cost.

Wesco 18m Prototype

THE WIND ENERGY SUPPLY CO. LTD (WESCO)
Iroko House
Bolney Avenue
Peacehaven
Sussex BN9 8HQ
U.K.

Telex: 877623 ContecG
Phone: Peacehaven (07914) 5051

Have so far mainly been involved in sophisticated windpower research and development, particularly in the field of combined heat and mechanical outputs from windpower. Have built an 18m (60ft) prototype (see photo) with a maximum output of around 190kW in a windspeed of 13.5m/s (30 m.p.h.) and are bringing the following series of machines onto the market:

- 5m (16ft) rotor suitable for a variety of applications.
- 5.5m (18ft) system for battery charging; designed for low wind operation and beacon/repeater station duty.
- 7m (23ft) system similar to the 5m one above but offering improved cost/performance (see detailed specification below).
- 18m (60ft) development of prototype mentioned above, with thermal/electric output via hydraulic transmission.

A more detailed specification of the 7m windmill follows:

Rotor diameter: 7m. Number of blades – 3 (downwind of tower).
Furling method: pitch change by torsion flexure/centrifugal weights.
Rated wind speed: 13.4m/s (30 m.p.h.).
Rated rotor speed: 311 r.p.m.
Output at rated wind speed: 15kW (415V 50Hz a.c. three-phase).
Tower: Square section, tubular steel lattice, 7.2m (24ft) high.

Optional extras include priority load controller, 380V three-phase alternator, economy and professional battery chargers, and low temperature spec. (–20°C).

Wind-Pumping Systems

AGRO-AIDS
27 Shrungar Shopping Centre
Mahatma Gandhi Road
Bangalore – 560001
India

Phone: 56149
Cables: INDEX BANGALORE

Agro-Aids are manufacturing the WP-2 windmill originally developed by the Indian National Aeronautical Laboratory. Its specification is as follows:

Rotor: 12 bladed, 16ft (4.88m) diameter.
Transmission: Direct acting, self-lubricating with 5in (125mm) stroke.
Governing: By moving tail vane.
Tower: Four post all-steel, 32ft (10m) high.
Output: 21m lift with 100mm pump in 20km/h (12 m.p.h.) winds = 2000 l/h. 4.6m lift with 200mm pump in 20km/h wind = 7500 l/h.

WP-2 Windmill

BOSMAN
Waterbeheersing En
Milieverbetering B.V.
Steggjesdijk 4
Postbus 3518
3364 Piershil (Z.-H.)
Netherlands

Telex: 24475
Phone: (01869) 13 16

Manufacturers of a small water pumping system to control water levels. The system, suitable for small pump drainage schemes, monitors the water level in the drainage ditch and, when the wind is available, will pump water from the ditch until it reaches the desired level, where the unit will automatically cut out.

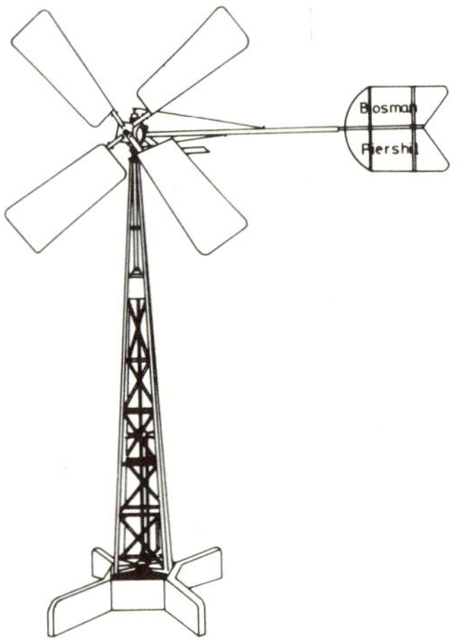

Bosman Water pumping system

S.A. BRUNO
Route du Mans
Bonchamps-les-Laval
53210 Argentre
France

Phone: (43) 53 65 90

also available from:

BRIAU S.A.
B.P. 43
37009 Tours Cedex
France

Telex: 750729F
Phone: (47) 61 38 17

Mistral Wind pumping unit

"Le Mistral" is a relatively simple wind pump intended for small water supplies. It requires virtually no maintenance. It is available with a choice of rotor sizes to cater for different applications as follows:

Rotor diameter: 2.00, 2.24, 2.50 metre.
Output in 24 hours: 3100, 4900, 8800 litre.
Maximum head: 30 to 40m.
Tower sizes: 4.5m, 6.0m, 8.0m, 10.0m.

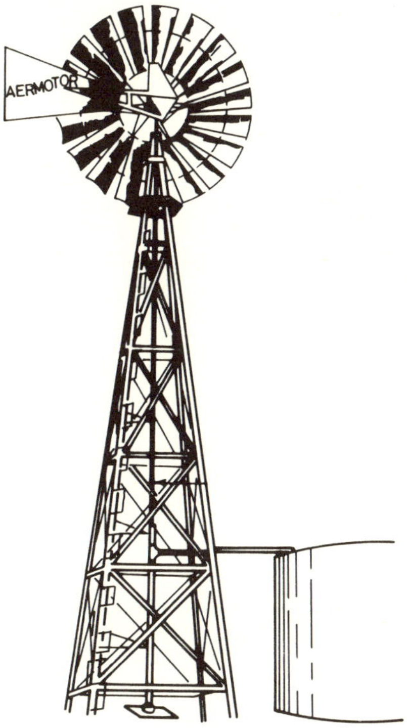

Aermotor Wind Pumping Unit

AERMOTOR
(Division of Braden Industries Inc.)
P.O. Box 1364
Conway
Arkansas 72032
U.S.A.

Phone: (501) 329 9811

Also:

W.D. Moore & Co
3 Keegan Street
O'Connor
West Australia 6163

Cables: Clipglaze
Phone: 374766

Aermotor have been in the water-lifting business since 1888 and apart from windmills, manufacture electric borehole pumps. The current windmill range, model 702, was first introduced in 1933. Versions are licence manufactured in Australia and Argentina as well as in the U.S. It is available with 6, 8, 10, 12, 14, and 16ft diameter rotors with towers in five heights from 21 through to 47ft. All models have oil filled gearboxes, automatic furling plus manual furling with a brake. Manufacturers figures for output are given below in U.S. gallons (1 U.S. gallon = 3.8 litre = 0.84 Imp.gall.).

AERMOTOR PUMPING CAPACITY

Cylinder dia:	*Capacity (g.p.h.)*		*SIZE OF AERMOTOR Total elevation in feet*					
(inches)	*6ft*	*8-16ft*	*6ft*	*8ft*	*10ft*	*12ft*	*14ft*	*16ft*
1¾	105	150	130	185	280	420	600	1,000
$1^7/_8$	125	180	120	175	260	390	560	920
2	130	190	95	140	215	320	460	750
2¼	180	260	77	112	170	250	360	590
2½	225	325	65	94	140	210	300	490
2¾	265	385	56	80	120	180	260	425
3	320	470	47	68	100	155	220	360
3¼	–	550	–	–	88	130	185	305
3½	440	640	35	50	76	115	160	265
3¾	–	730	–	–	65	98	143	230
4	570	830	27	39	58	86	125	200
4¼	–	940	–	–	51	76	110	180
4½	725	1,050	21	30	46	68	98	160
4¾	–	1,170	–	–	–	61	88	140
5	900	1,300	17	25	37	55	80	130
5¾	–	1,700	–	–	–	40	60	100
6	–	1,875	–	17	25	38	55	85
7	–	2,550	–	–	19	28	41	65
8	–	3,300	–	–	14	22	31	50

Manufacturers note: Capacities shown in the above table are approximate, based on the mill set on the long stroke, operating in a 15 to 20 mile-an-hour wind. The short stroke increases elevation by one-third and reduces pumping capacity one-fourth.

Dempster Wind Pump

DEMPSTER INDUSTRIES INC. Phone: (402) 223 4026
P.O. Box 848
Beatrice
Nebraska 68310
U.S.A.

Manufacture a range of windmills for water pumping with rotor diameter of 6ft to 10ft on towers of 22ft to 60ft.

Performance claimed is indicated as follows, in U.S. Gallons/hr. These capacities are based on a 15-mile per hour wind. Capacities are based on longest stroke of Dempster mills. If shorter stroke is used capacities will be reduced in proportion to length of stroke used.

Cylinder size	*6ft 5" Stroke* Elev.	*g.p.h.*	*8ft "A" 7½" Stroke* Elev.	*g.p.h.*	*10ft 7½" Stroke* Elev.	*g.p.h.*
1⅞	120	115	172	173	256	140
2	95	130	135	195	210	159
2¼	75	165	107	248	165	202
2½	62	206	89	304	137	248
2¾	54	248	77	370	119	300
3	45	294	65	440	102	357
3¼	39	346	55	565	86	418
3½	34	400	48	600	75	487
3¾	29	457	42	688	65	558
4	26	522	37	780	57	635

If the wind velocity be increased or decreased, the pumping capacity of the windmill will also be increased or decreased. Capacities will be reduced approximately as follows, if wind velocity is less than 15 miles per hour: 12 mile per hour wind, capacity reduced approximately 20%; 10 mile per hour wind, capacity reduced approximately 38%.

THE HELLER-ALLER COMPANY Cables: Helaler Phone: (419) 592 1856
P.O. Box 29
Corner Perry &
Oakwood Streets
Napoleon
Ohio 43545
U.S.A.

Baker Windmill

Heller-Aller have manufactured Baker Windmills, pumps, water systems, tanks and allied goods since 1886. Performance of the current range of models of windmill is indicated in the following table, (note that 1 U.S. gallon = 3.8 litre).

Pumping Capacities of Back-Geared "Baker" Windmills (in 15 mile-per-hour wind)

Total elevation (ft)	6 FOOT BAKER *Cylinder diameter (")*	*Capacity (U.S. g.p.h.)*	8 FOOT BAKER *Cylinder diameter (")*	*Capacity (U.S. g.p.h.)*	10 FOOT BAKER *Cylinder diameter (")*	*Capacity (U.S. g.p.h.)*	12 FOOT BAKER *Cylinder diameter (")*	*Capacity (U.S. g.p.h.)*
25	3	350	3½	900	4	1250	6	2400
35	2½	240	3	720	3½	925	5	1625
50	2¼	200	2½	450	3	700	4½	1425
75	2	160	2¼	350	2½	475	4	1125
100	2	150	2	250	2½	460	3	600
125	1⅝	120	1⅞	240	2	280	2½	525
150	–	–	1¾	220	2	280	2½	525
200	–	–	–	–	1⅞	260	2	325
250	–	–	–	–	1¾	215	2	325
300	–	–	–	–	–	–	1¾	200

Manufacturer's note: The above capacities are approximate. By the total elevation in feet we do not mean the depth of the well, but the distance to the cylinder.

M.B.P. (S.A.) PTY LTD
P.O. Box 2047
Adelaide
South Australia 5001

Cables: Modmaid
Phone: 43 6241

7-25 Manchester Street
Mile End South
South Australia 5031

M.B.P. manufacture windmills, syphon and borehole pumps, plus various tanks and other agricultural water supply equipment, until recently under the trade name of 'Metters'.

The windmills are conventional farm windmills with all normal facilities such as automatic and manual furling in wheel sizes from 6 to 14ft as indicated below.

Metters Windmill Installation

PUMP SIZE

Inches		2	2⅜	2½	2¾	3	3½	4
Millimetres		50.8	60.3	63.5	69.8	76.2	88.9	101.6
6 feet wheel 1.828 m								
Maximum head:	Feet	71	61	49	40	34	28	25
	Metres	21.64	18.59	14.94	12.19	10.37	8.53	7.62
Average daily output:	Galls	1,110	1,530	1.750	2,110	2,475	3,410	4,400
	Litres	5,001	6,955	7,956	9,592	11,251	15,502	20,002
8 feet wheel 2.438 m								
Maximum head:	Feet	127	102	86	72	62	44	35
	Metres	38.71	31.09	26.2	21.95	18.90	13.41	10.67
Average daily output:	Galls	1,320	1,860	2,090	2,540	2,970	4,125	5,280
	Litres	6,001	8,456	9,501	11,547	13,502	18,752	24,003
10 feet wheel 3.048 m								
Maximum head:	Feet	265	233	206	161	139	99	73
	Metres	80.77	71.02	62.79	49.07	42.37	30.18	22.25
Average daily output:	Galls	1,540	2,145	2,420	2,970	3,465	4,785	6,160
	Litres	7,001	9,751	11,001	13,502	15,752	21,753	28,003
12 feet wheel 3.658 m								
Maximum head:	Feet	292	255	233	199	169	139	112
	Metres	89.00	77.72	71.02	60.65	51.51	42.37	34.14
Average daily output:	Galls	1,650	2,287	2,585	3,180	3,700	5,115	6,500
	Litres	7,501	10,397	11,751	14,456	16,820	23,253	29,549
14 feet wheel 4,267 m								
Maximum head:	Feet	385	314	282	245	201	169	134
	Metres	117.35	95.71	85.95	74.68	61.26	51.51	40.84
Average daily output:	Galls	1,705	2,385	2,668	3,285	3,828	5,280	6,820
	Litres	7,751	10,842	12,129	14,934	17,379	24,003	31,004

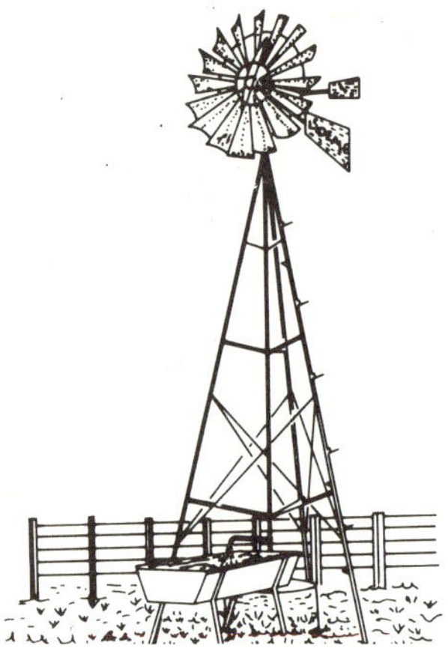

L'Idéale Wind pumping unit

ÉOLIENNES HUMBLOT Phone: (29) 94 09 09
8 rue d'Alger
Coussey
88300 Neufchateau
France

Manufacturers of direct drive wind pumping units intended for animal drinking trough supply and consequently rather smaller and less expensive than some of the American and Australian farm windmills. There are two main types and various permutations of rotors, head mechanisms and towers can be combined for different purposes. There is one large model listed last.

Model name	*Rotor sizes*	*Transmissions*	*Pumping head*	*Maximum output*	*Tower heights*
Cadeteol	1.75m	same for both	0 to 6m	500 l/h	4.4m
Supercadeteol	1.75m		0 to 12m	600 l/h	4.4m
Junioreol	1.75m		0 to 15m	600 to 2000 l/h	4.4m
	2.0m	Idée*	15 to 20m		5.25m
	2.25m	Idéale	20 to 30m		6.55m
	2.5m	Idéalo	30 to 40m		8.85m (choice)
	2.75m	Idéese (choice*)	40 to 55m		
Geanteol	3.50m	same (geared)	80m well	8000 l/h	9.30m
	4.0m		(100m total)		13.30m

*These transmissions are first two of fixed output, second two variable and numbers one and three are grease lubricated, the others oil bath. Spring shock absorbers are an optional extra to protect against iced up pumps damaging the drive train.

ETS PONCELET & CIE
B.P. No. 1
10380 Plancy L'Abbaye
France

Manufacturers of a range of small, relatively inexpensive, water-pumping windmills intended for filling cattle troughs, under the trade name "Oasis".

Model	*Rotor diameter*	*Tower height*	*Head range*	*Output*
1	1.60m	4.0m)		
1 "bis"	1.75m	4.0m)	0 to 30	200 to
1 "ter"	1.75m	6.0m)	(0 to 100ft)	1000
2	2.00m	6.0m)		l/h
2 "bis"	2.25m	6.0m)		(44 to
3	2.25m	7.5m)		220
4	2.50m	7.5m)		gall/hr)
5	2.50m	10.0m)		

Oasis Windmill

SOUTHERN STEEL WORKS LTD. Phone: 056 28633
Ballyhale
Co. Kilkenny
Ireland

The Ballyhale range of windpumps is available as follows:

Model	*Maximum hourly output**		*Rotor diameter*	
	(Galls)	*(Litres)*	*(ft)*	*(m)*
SS 1	300	1350	9	2.74
SS 2	800	3600	12	3.66
SS 3	1600	7200	15	4.57
SS 4	3000	13500	18	5.49

*The head at which this output occurs was not stated in the manufacturer's literature.

A water storage tank of 1100 gallons (4950 litres) can be supplied if required. These windmills furl automatically in a windspeed of 25 m.p.h. (40km/h).

REYMILL STEEL PRODUCTS Phone: 641
Sta. Rosa
Neuva Ecija
Philippines

Manufacturers of locally designed and developed steel wind-pumper with built-in governing and braking system. Three sizes available as follows:

Regular size:	tower height 40-50ft	windwheel diameter 10ft (3.05m)
Medium size:	tower height 40-50ft	windwheel diameter 12ft (3.6m)
Large size:	tower height 40-50ft	windwheel diameter 14ft (4.3m)

Pump stroke in all cases is 6 to 9 ins (150 to 225mm).

Reymill Windmill

S & L No.18 Windpump

STEWARTS & LLOYDS
P.O. Box 74
1930 Vereeniging
South Africa

Telex: 8-0480
Cables: Tubes Vereeniging
Phone: (016) 45133

This manufacturer offers a range of six sizes of windpump incorporating modern PTFE bearing bushes. The largest machine, which is direct drive, requires no maintenance as it has sealed for life ball-bearings and PTFE self-lubricating bushes. The smaller machines require occasional oil changes for their gearboxes. Automatic furling in storms plus manual furling from ground level is provided. The following models are available:

Model	*Rotor diameter*		*Pump stroke*	*Pump size*	*Head range*	*Flow range (max.)*
	(ft)	*(m)*	*(in)*	*(in)*	*(ft)**	*(Imp.gall/hr)*
6	6	1.83	5	1½ – 6	11 – 176	89 – 1430
8	8	2.44	5½	1½ – 6	13 – 209	83 – 1331
10	10	3.28	7	1½ – 8	12 – 396	75 – 2157
12	12	3.66	9¼	1¾ – 12	10 – 522	118 – 5549
14	14	4.27	9¼	2¼ – 12	18 – 554	161 – 4582
18	18	5.49	6	2 – 6	92 – 882	138 – 1828

*Note that the head range is for total head; the maximum depth of borehole is around two-thirds of the maximum total head given.

Basic 6m towers (20ft) can be extended to 9, 12 or 15m (30, 40 or 50ft).

THAI U Sa INDUSTRIAL FACTORY
No. 5g/15 M007
2 Pracharaj 2 Road
Dusit
Bangkok
Thailand

Phone: 585 2560
585 4815

Manufacturers of a range of direct drive, locally designed, multi-bladed water pumpers of relatively low cost.

Rotor dia.		*Tower height*		*Pump dia.*		*Output per hour**	
(ft)	*(m)*	*(ft)*	*(m)*	*(in)*	*(mm)*	*(Imp. galls)*	*(litres)*
6	1.82	40	12	2.5	63	267	1200
8	2.44	40	12	2.75	70	356	1600
10	3.05	40	12	3.0	76	444	2000
12	3.66	40	12	4.5	114	667	3000
14	4.27	40	12	5.0	127	888	4000

*Output in windspeed 10km/h (6 m.p.h.) for unspecified lift.

Thai Windpump Installation

Southern Cross Wind Pump

TOOWOOMBA FOUNDRY PTY Ltd
259 Ruthven Street
Toowoomba
Australia 4350

Telex: 40046
Cables: Foundfact
Phone: 32 3122

Toowoomba Foundry products under the "Southern Cross" tradename are widely known in Australia; other than windmills they include a very wide product range covering diesel engines, turbine and other pumps, irrigation machines, sprinklers, storage tanks, towers and aerial masts, etc. There are two ranges of windmills, the smaller, geared "IZ Pattern" and the larger, direct-drive "Seneschal". The following tables illustrate their performance according to the manufacturer in winds of 18 m.p.h. (29km/h). In winds of 12-16 m.p.h. (approx. 21km/h) the outputs would be reduced by 25% and in light winds of 8-10 m.p.h. (approx. 14km/h) the outputs would be reduced by about 50%.

Choice of pump cylinder diameter for various lifts and resulting flows

Size	*Mill*		*1¾in.*	*2in.*	*2¼in.*	*2½in.*	*2¾in.*	*3in.*	*3¼in.*	*3½in.*	*4in.*	*4¼in.*	*4½in.*	*5in.*	*6in.*	*8in.*
6ft.	"IZ"	Total lift in feet	73	60	51	43	37	32	27	24	19	17	15	12		
		Gallons per hour	90	120	155	180	230	275	320	370	485	545	610	755		
8ft.	"IZ"	Total lift in feet	132	109	92	77	66	57	50	44	34	31	28	23	16	
		Gallons per hour	100	120	170	210	250	300	350	405	530	600	670	830	1200	
10ft.	"IZ"	Total lift in feet	236	197	166	141	121	105	92	81	64	57	51	42	30	
		Gallons per hour	100	125	160	200	240	290	340	390	510	580	650	800	1150	
12ft.	"IZ"	Total lift in feet	315	263	222	189	162	140	123	108	85	76	68	56	40	23
		Gallons per hour	105	140	175	215	260	310	365	420	550	620	695	860	1240	2200
14ft.	"IZ"	Total lift in feet	443	370	312	265	228	197	172	151	119	107	96	79	56	32
		Gallons per hour	90	120	150	185	225	260	305	360	460	520	590	720	1040	1880

Size	*Mill*	*"Seneschal"*	*1¾"*	*2"*	*2¼"*	*2½"*	*2¾"*	*3"*	*3¼"*	*3½"*	*4"*	*4¼"*	*4½"*	*5"*	*6"*	*8"*	*10"*	*12"*	*14"*
17ft.	"R"	Total lift in feet	480	415	360	315	275	240	205	175	135	120	105	85	60	34	–	–	–
7in. Stroke		Galls. per hour	170	220	280	350	420	500	590	685	895	1010	1130	1400	2000	3600	–	–	–
17ft.	"R"	Total lift in feet	420	360	310	270	240	210	180	155	120	105	95	75	53	30	–	–	–
8in. Stroke		Galls. per hour	195	255	320	400	480	575	675	780	1020	1150	1290	1600	2300	4080	–	–	–
21ft.	"R"	Total lift in feet	–	–	575	500	440	390	345	310	240	210	190	155	105	60	38	27	–
8¼in. Stroke		Galls. per hour	–	–	270	330	400	480	560	650	850	960	1080	1330	1920	3400	5320	7660	–
21ft.	"R"	Total lift in feet	–	–	470	410	360	320	285	250	200	175	155	125	88	50	32	22	–
10in. Stroke		Galls. per hour	–	–	325	405	490	580	680	790	1030	1165	1310	1615	2320	4120	6440	9300	–
25ft.	"R"	Total lift in feet	–	–	–	710	635	575	515	460	350	310	280	225	155	88	55	39	28
9½in. Stroke		Galls. per hour	–	–	–	320	390	465	545	630	825	930	1045	1290	1860	3300	5160	7440	10200
25ft.	"R"	Total lift in feet	–	–	–	530	475	425	385	350	280	245	220	180	125	70	45	31	23
12in. Stroke		Galls. per hour	–	–	–	410	490	590	690	800	1045	1180	1320	1630	2350	4180	6520	9400	12900

WINDPUMPEN-ZENTRALE
Luetthoern 51
D 2330 Eckernfoerde
West Germany

Phone: (04351) 42024

"Pumpomat Windpumpen" is a series of unconventional paddle-bladed windpumps with an unusual governing system for water pumping windmills of variable pitch blades on smaller models and a vertical windvane which applies a brake on the large ones.

Model No.	*Rotor diameter*	*No. of blades*	*Water head*	*Nominal output*
1, 2	n/a	2	4m	200 l/hr
3	n/a	2	10m	180 l/hr
11, 22	n/a	4	4m	500 l/hr
33	n/a	4	10m	280 l/hr
44	n/a	4	16m	200 l/hr
55	n/a	4	22m	200 l/hr
111, 222	n/a	6	4m	750 l/hr
333	n/a	6	10m	300 l/hr
444	n/a	6	16m	300 l/hr
P360	2.5m	4	3.5m	10000 l/hr
P500	n/a	n/a	3.5m	30000 l/hr

Most of these pumps are intended for low lifts and there is a choice of diaphragm or piston pump with most models.

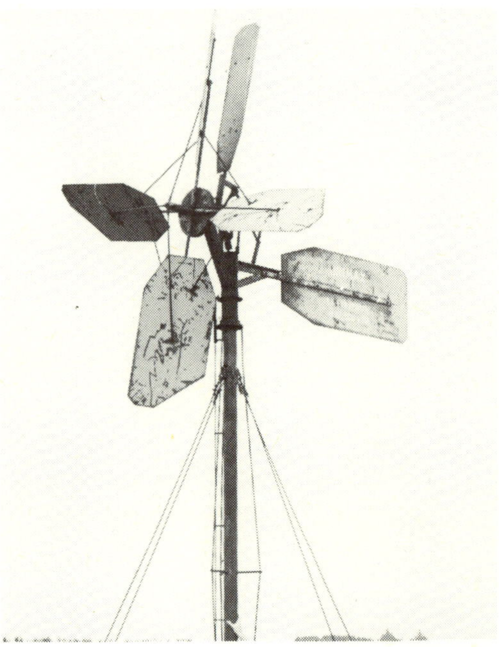

Pumpomat P360 Windmill

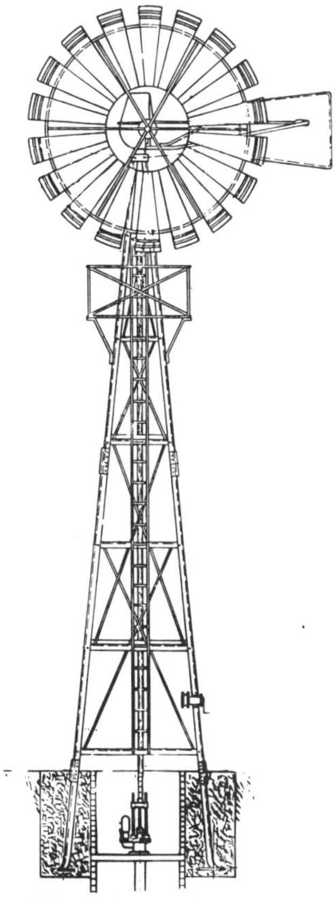

Climax Wind Pump

WYATT BROTHERS (WHITCHURCH) LTD
Wayland Works
Whitchurch
Salop ST13 1RS
U.K.

Phone: (0948) 2526-7

Wyatt Bros. manufacture the "Climax" range of windmills, pumps and well-heads. The windmills are conventional geared farm machines with automatic and manual furling and wheel sizes available from 6ft to 18ft (1.83 to 5.49m). The following table outlines their performance in a "good wind of 20-22 m.p.h." and in a 12 m.p.h. wind the manufacturer states that these figures will be reduced to about 65% of those indicated. A rule of thumb recommended for assessing daily output is to assume 65% of the figures indicated for 10 hours per day. The output is given in Imperial gallons = 4.5 litres or 1.2 U.S. gallons.

Size of mill	*Strokes per minute*	*Cylinder dia. (inches) and Actual gallons per hour*	*Total head in feet* 10	*50*	*75*	*100*	*125*	*150*	*175*	*200*	*250*	*300*	*350*	*400*
6ft	45	Cylinder	3¼	2½	2	1¾	–	–	–	–	–	–	–	–
		g.p.h. @ 5¾" stroke	420	250	155	120	–	–	–	–	–	–	–	–
8ft	42	Cylinder	3¾	3¼	2¾	2¼	2	–	–	–	–	–	–	–
		g.p.h. @ 5¾" stroke	500	380	275	182	145	–	–	–	–	–	–	–
		" 7¾" "	680	510	370	245	195	–	–	–	–	–	–	–
10ft	37	Cylinder	4¼	3½	2¾	2½	2¼	2	1¾	1¾	–	–	–	–
		g.p.h. @ 8" stroke	800	530	337	275	228	177	138	138	–	–	–	–
		" 9½" "	950	630	400	325	270	210	165	165	–	–	–	–
12ft	31	Cylinder	7	5	4	3¾	3½	3	2¾	2½	2¼	2	1¾	–
		g.p.h. @ 10" stroke	2260	1150	750	625	555	412	360	285	240	187	144	–
		" 11¼" "	2600	1325	860	720	640	475	415	330	275	215	165	–
14ft	29	Cylinder	8	6½	5	4½	4	3½	3¼	3	2¾	2¼	2	–
		g.p.h. @ 11¼" stroke	3115	1980	1210	980	780	600	510	410	372	250	200	–
		" 13" "	3600	2300	1400	1130	900	690	590	475	430	290	230	–
16ft	21	Cylinder	12	8	7½	6	5½	5	4¾	4	3¾	3¼	3	2¾
		g.p.h. @ 12" stroke	5400	2400	2125	1330	1120	930	855	600	525	392	336	284
		" 15" "	6750	3000	2650	1660	1400	1160	1070	750	655	490	420	355
18ft	17	Cylinder	15	12	10	8	7	5¾	5½	4¾	4¼	3¾	3½	3¼
		g.p.h. @ 12" stroke	6800	4360	3000	1920	1480	1000	920	665	544	425	368	320
		" 15" "	8500	5450	3750	2400	1875	1250	1150	830	680	530	460	400

Galvanized steel towers to carry "Climax" windmills are supplied in heights from 15 feet to 60 feet in multiples of 5 feet.

WAKES & LAMB LTD
Millgate Works
Newark-on-Trent
Notts NG24 4XB
U.K.

Phone: (0636) 4464

Manufacture the Newark series of conventional farm water pumping windmills with rotor sizes from 8ft (2.44m) up to 16ft (4.88m). Output range for the series is quoted as 475 gall/hr (2137 l/h) up to 2720 gall/hr (12,240 l/hr) at an unspecified wind.

Comet Windmill

SIDNEY WILLIAMS & CO. (PTY) LTD
P.O. Box 22
Dulwich Hill
New South Wales
Australia 2203

Phone: 56 2491

Manufacturers of the "Comet" range of windmills from 8ft to 35ft diameter mounted on towers of 20ft to 60ft height. The manufacturers say these units are of a well tried design and will pump water with a minimum of maintenance for many years.

The larger "Comet" sizes are believed to be the largest water-pumping windmills currently available anywhere.

MÁQUINAS AGRÍCOLAS FORTUNA LTDA
Divisão International
Rua João Adolfo, 118
conj. 710/711
C.E.P. 01050
São Paulo
Brasil

Phone: 36-5160/4497
Telex: OU 21724
Cables: BR-Fortuna

Manufacturers of the 'Aerogerador' wind generator. This machine is rated at 20 amps, 40 volts. No further details were available at press date.

Aerogerador Wind Generator

3

Water Power

HYDRO-ELECTRIC SYSTEMS
HYDRO-MECHANICAL AND PUMPING SYSTEMS

Background

Historically, water power was one of the first energy sources to be exploited by mankind. Water wheels and crude, wooden, vertical-axis Norse or bucket turbines have been in use in many parts of Europe and Asia since long before the Industrial Revolution, (and 2000 years or more in some places). By the time of the Industrial Revolution, water-wheel technology had been developed to a fine art and efficiencies approaching 70% were being achieved from the best machines.

Improved metallurgical skills, the need for higher-speed devices in order to generate electricity and a better understanding of fluid flow led to the development of turbines during the nineteenth century, which gradually supplanted the water wheel. Up until the 1930s, small water turbines were quite extensively used both in the developed and in the developing world to provide electrical power for places remote from a main electricity grid. However, the more developed industrial countries extended their electricity grids during the subsequent period and set an impossible example for larger, poorer and less densely populated countries to follow. As a result, low-cost mains electricity (usually heavily subsidised in the remoter extremities of the grid) pushed hydro-electric technology development towards large multi-megawatt systems designed to feed the new national grids. Most of the manufacturers of small turbines, faced with a contracting market for their wares, either went out of business or turned to pump manufacturing instead. So small turbine design remained substantially frozen at the 1930s level of advance until quite recently, when rapidly increasing fossil fuel costs created a new interest in such renewable energy sources as water power.

Current developments

Within the last few years, the few surviving manufacturers of small turbines have in many cases taken a new interest in what had generally become a minor sideline, while a number of new small turbine building enterprises are appearing, often utilising electronic controllers to obtain a great saving in cost combined with an improvement in reliability compared with the traditional hydro-mechanical governing of the 1930s and before.

The great attractions of hydro-power are:-

i. it is generally continuously available
ii. given a reasonable head, it is a concentrated energy source
iii. the available energy is readily predictable
iv. water-powered machines can be extremely cost-effective in the sense that they:-
    a. have a high power/weight ratio (i.e. power/cost ratio)
    b. have long working lives
    c. require little maintenance and have low running costs.

Against this are the disadvantages:-

i. suitable sites where a useful head of water can be engineered economically are not very common (although very many suitable sites are not being used).
ii. there is always a maximum power output available from a given site which limits further expansion of energy-consuming activities in the area.

The one example of a country which has set out to exploit as many as possible of its potential small-scale hydro-power sites in recent years is the People's Republic of China. Not only have the conventional large-scale grid-feeding hydro-projects common in other countries been implemented, but a very extensive small-scale hydro-electric programme has been set up. In 1949 there were only 50 small hydro-stations in China with a combined capacity of 5.6MW. Extensive river surveying was initiated in the mid 1950s, and the first major power station building programme began during the Great Leap Forward which fell short of its target but resulted in about 4500 small plants with a total capacity of the order of 200MW by 1959. This may seem a remarkable increase in ten years, but the period since is even more spectacular, as according to the New China News Agency there were some 60,000 small hydro-electric stations completed in China by 1975 (20,000 in the Yangtse basin alone). The average output of these stations is 36kW and total capacity of the small hydro-stations by 1973 was of the order of 1800MW. Small Chinese power plants are based on 1930s European designs, but have been refined to suit Chinese conditions. Chinese turbines do not as yet appear to be generally available on the international market (mainly because there has been little demand), but details will be given in future editions if this situation changes.

According to V. Smil writing in the journal *Water Power and Dam Construction* in March 1976 . . . "the well established guidelines for China's water power development — build large, medium and small plants simultaneously and most of them should be small and medium sized ones — will certainly continue to be implemented in the future. Small stations requiring limited investments operating at a low cost and yielding quick returns seem to be, when developed with a modicum of planning and staffed with trained personnel, a very sensible acquisition for a vast, overwhelmingly rural and relatively poor developing country."

Recognising the utility of small scale hydro-electric power in many of the hillier regions of the world (which are so often less developed than the more populated low lands), ITDG is working on a small turbine development programme in conjunction with Evans Engineering Land and Leisure Services Ltd., (see the following section), as described in more detail in Section 9.

Farmers and others in the more remote parts of the UK, USA and other industrial countries are coming to similar conclusions as a result of a great increase in diesel generation costs combined with the reduction or withdrawal of subsidies for mains connections coupled with greatly increased costs per kWh of electricity. As a result there is a great revival of interest in small-scale water power and many formerly disused water mill sites in the UK and other industrial countries are being investigated by their owners for possible small-turbine installations. Naturally, planners and others concerned with providing energy in the developing world are also interested in this possibility and it is hoped that the manufacturers listed in this section may be able to help satisfy this demand.

Water power sites and equipment

Virtually all water power applications depend on a head of water falling to a lower level. Figure 1 is a chart showing the power available (assuming a conversion efficiency of 70%) for a range of heads and flow rates. The greater the head the less the flow required for a given output, and vice-versa. The chart can be used to indicate the potential of a given site, although different efficiencies will effectively raise or lower the output proportionately (60% would yield 6/7 of the power indicated).

For various reasons, different types of equipment are needed to cope with different heads and end-uses of the mechanical power. Water wheels are obviously restricted to heads of half the diameter for undershot, or equal to the diameter for overshot while different types of turbine are best suited to different head ranges.

Low head turbines are generally of the propellor type and high head devices are usually impulse turbines such as the Pelton or similar Turgo wheel. The traditional medium head turbine, (still used on large-scale applications) is the Francis turbine, but its complex shape is very expensive to reproduce under modern conditions as a casting, so the Mitchell alias Crossflow alias Banki turbine (see the Ossberger-Turbinenfabrik turbine entry, which is slightly less efficient, is becoming popular for medium heads as it lends itself to fabrication from standard steel stock. However, propeller turbines and Pelton wheels can also be used for medium heads, although the former tend to run faster than is convenient while the latter can need multiple nozzles to make use of the full flow.

Water wheels turn relatively slowly and are therefore unsuitable for use with standard electricity generators requiring rotational speeds of 1500 r.p.m. or more. Gearing can be used (or belts or chains) to step the speed up, but the very high torque in the low speed end of the train requires large and expensive transmission components unless a rim drive is used. The water wheel is also extremely heavy and material-intensive in terms of its output, and requires a large and expensive shaft and bearings to cope with its high weight and torque. However it is still valuable for applications such as milling or driving low speed machinery via a belt and pulleys. Hydraulic transmission (using oil or water) may be another modern option in conjunction with a water wheel to achieve a higher speed step-up.

One of the main advantages of small-scale electricity generation is that it can involve the minimum of transmission and control equipment, by generating electricity at the voltage it is to be used at near to the point of use. Control may be by manual adjustment in response to the reading of a voltmeter or automatic control can be achieved by a mechanical/hydraulic governor or by an electronic "black box". The latter is a very cost-effective and reliable modern development that has only recently been possible with the advent of thyristors (see Evans Engineering's entry).

Another use for water power is to pump a small amount of water to a higher level, so that a hillside may be cultivated, or a village at a high level might have an on-the-spot piped water supply or a watering point might be provided for grazing animals. The traditional and simplest solution to this problem is to use a ram pump or hydram, several of which are included later in this section; a device such as this induces a water hammer effect caused by the sudden closure of an automatic clack valve and the sudden rise in pressure so caused forces a small proportion of the water flowing through it to a considerable height, (typically up to 40 times the supply head). However the greater the ratio of supply head to delivery head, the smaller the volume of water delivered. Hydrams need little attention and have extremely long useful lives, having only a single moving component, the clack valve. Turbines can also be used to pump water, and the Plata Pump (included) is a unique device specially developed for this purpose, which can operate at a very low head.

There are many places with fast flowing rivers but where no head can conveniently be engineered (since damming the river would be too great an expense or the hydraulic gradient is inadequate to allow a head to be contoured). As a result the Intermediate Technology Development Group is developing a turbine that is mounted on a floating pontoon and operates submerged in a stream at zero-head. This is a kinetic energy converter analagous to a submarine windmill (see also Section 9) for further details of the ITDG project). We expect our device and others will become available for widespread use within the next five years and that this will permit river energy to be tapped at very many more points than are possible at present. It may also become economic to apply this principle to tap the energy of marine coastal currents and tidal currents. At the moment the only commercial system of this kind is the Aquadyn (see the Natural Energy Centre entry).

Figure 1 Flow and head requirements for various power outputs at a conversion efficiency of 70%

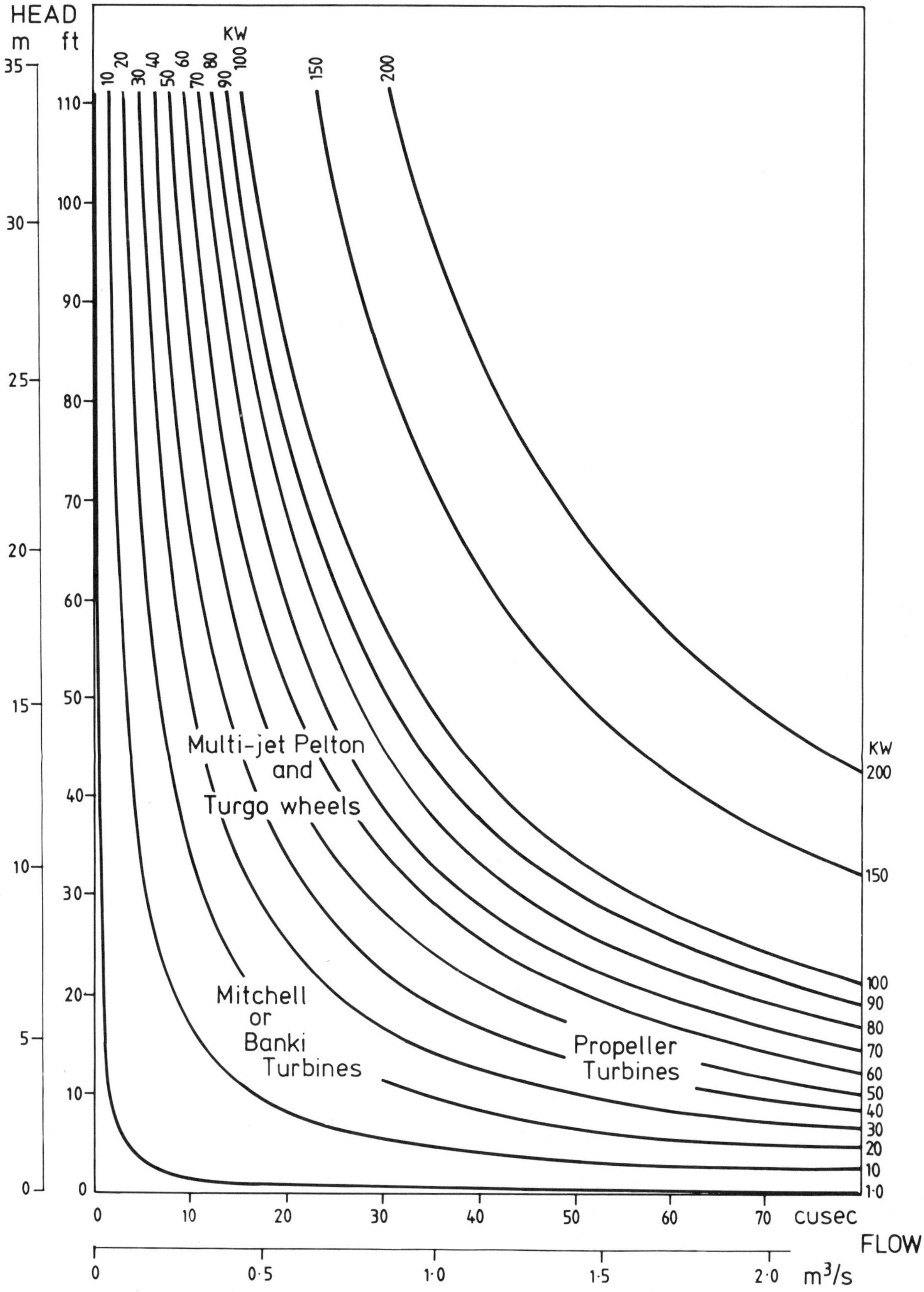

ALLIS-CHALMERS
Hydro-Turbine Division
Box 712
York
Pennsylvania 17405
U.S.A.
Phone: 717 792 3511

Allis-Chalmers Generating Unit

Manufacture a range of standardised hydroelectric generating units, all intended for use with heads of up to 15m (50ft). All units are horizontal shaft propeller turbines and outputs of from 50 to 50,000kW can be catered for. Obviously all but the smaller systems in this range are outside the general power range considered by this Guide, so only the smallest models in this range of ten sizes are included (the remaining sizes go up to 3000mm Ø).

For these models the following parameters apply (approx. figures).

Runner diameter	*Head range*	*Flow range*	*Power range*
750mm	2-15m	2-3.5m^3/s	30- 400kW
1000mm	2-15m	3-6 m^3/s	50- 700kW
1250mm	2-15m	5-9 m^3/s	80-1000kW

AB BOFORS-NOHAB
S-46101 Trollhättan
Sweden
Telex: 42084
Cables: Nohab
Phone: 0520 18000

A range of horizontal and vertical turbines with power outputs ranging from 100kW to 2000kW. Runner diameters vary between 7000mm and 2000mm for heads from 5m to 25m. Units are supplied complete with alternator and switch gear.

Bofors Vertical Shaft Turbine

BRIAU S.A.
B.P. 43
37009 Tours Cedex
France
Telex: 750729 F
Phone: (47) 61 38 17

Manufacture a packaged low head turbine/generator system with outputs in the range 4 to 50kVA. Either Francis or Propeller type runners are used, depending on head.

Performance envelope as follows:
Heads: from 2 to 20m (6 to 60ft)
Flows: from 200 to 4000 l/s (7 to 140ft^3/s)

They quote a rule of thumb for estimating the power of their system as follows:

$$P(kw) = \frac{5.5}{1000} Q\ (l/s).\ H\ (m)$$

A self-regulating brushless alternator is used, three-phase with output at 500V at a choice of 50 or 60Hz.

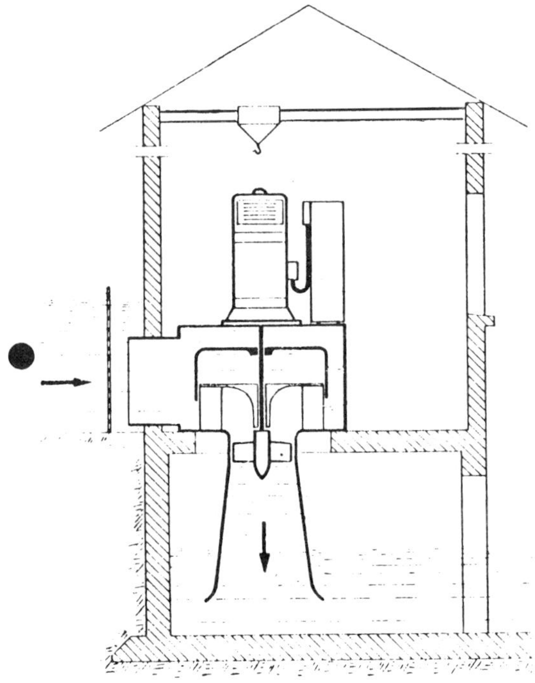

Briau Hydro Electric Unit

CANYON INDUSTRIES
5346 Mosquito Lake Road
Deming
Washington 98244
U.S.A.

Manufacture a miniature low head package turbine system called "Hydromite". The turbine and generator are integrated and designed for easy connection to a 4in (100mm) PVC/ABS standard plastic pipeline.

The output is at low voltage and can be used in conjunction with a storage battery bank. The performance envelope lies within the following limits:

Head: 8 to 34ft (2.4 to 10.4m)
Flow: 22 to 34ft^3/min (0.6 to 1m^3/min)
Power: 50 to 750 watts

The system is so small that its shipping weight is only 48lb (22kg).

Hydromite W/55 Amp Alternator

DREES & CO. GmbH
4760 Werl/Westf.
Postfach 43
Germany

Telex: 8 421 404 dres d
Cables: Dreesco Werl
Phone: 02922 5071/72

Manufacturers of a range of hydro electric systems from small micro plants to full size hydro-electric stations. The range includes Kaplan turbines, Francis Spiral turbines, and Pelton turbines.

Pelton Wheel Installation

GILBERT GILKES & GORDON LTD
Kendal
Cumbria LA9 7BZ
U.K.

Telex: 65125
Cables: Gilkes Kendal
Phone: 0539 20028

Long established manufacturers (over 100 years) of a wide range of different types of turbine.

'Hydec' Range

Small relatively low-cost, one or two jet Turgo impulse wheel with belt driven governor:

Net head: 30 to 300ft (10 to 100m)
Output (shaft): 10 to 200 b.h.p.
Nominal electrical outputs: 10, 25 and 50kW

Spiral cased Francis turbines for small outputs

This series are usually fitted with oil pressure governors to adjust the ring of gates built into the turbine casing.

Power range: 100 to 150 b.h.p.
Heads: 25 to 200ft (7.5 to 60m)

Open type Francis turbines

This series is usually fitted with oil pressure governors to vanes surrounding the runner. Runner sizes vary from 6in (150mm) up to 60in (1500mm).

Pelton Wheels and water motors

A wide selection of Pelton wheels is available, the smaller sizes being generally intended for use in water motors for use from ship's pressure mains (Butterworth line).

Water motors based on Pelton, Turgo, Francis and Propeller runners are available designed to operate at water pressures from 30lb/in^2 (2 bar) up to 150lb/in^2 (10 bar). The Axial Flow (propeller) motors operate at low pressures of 1lb/in^2 (0.1 bar) up to 13lb/in^2 (0.9 bar); all these pressures being "gauge" pressures.

The power range covered by the various water motors over their permissible pressure range is from 0.1 up to 42 b.h.p. (0.1 to 29kW). Obviously water motors could be applied for electricity generation.

Larger turbines

A comprehensive selection of large turbines with multi-megawatt capability is also available from this manufacturer.

Gilkes "Hydec" Impulse Water Wheel Installation

INDEPENDENT POWER DEVELOPERS INC.
Box 1467
Noxon
Montana 59853
U.S.A.

Phone: 406 847 2315

I.P.D. Impulse Turbine

Manufacturers of a range of small turbines and distributors of a variety of other small-scale power equipment including Dunlite wind generators storage batteries and inverters.

Two types of turbine are offered, a high head 3hp maximum Pelton wheel (min. head 60ft or 20m) and a low head 6hp maximum propeller turbine for heads from 5ft to 60ft (2m to 20m). Both are quite small units with direct coupled 32V generators. The system is intended for use in conjunction with a battery bank and inverter to provide considerably higher peak power surges than the generator can provide on its own; for example a peak output of 3kW, 6kW, 9kW, or 12kW can be provided by units rated for continuous outputs of 300W, 700W, 1700W and 8500W respectively. Monthly outputs from these systems would be 200kWh, 500kWh, 1200kWh, and 6000kWh respectively.

The output is converted from 32V d.c. to 115V, 60Hz a.c. by an inverter.

The system is made to high specifications and comes with a two year warranty.

I.P.D. also offer custom designed larger systems using Pelton wheels with outputs up to as much as 1000 b.h.p.

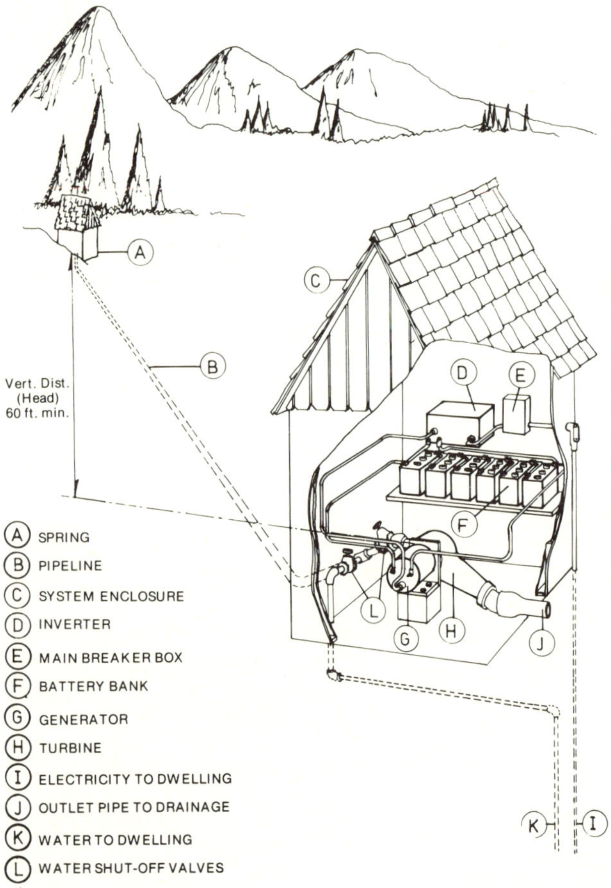

I.P.D. Hydroelectric Installation

Performance details
(Selected heads showing min. and max. flow)

Head ft	*m*	*Flow* gall/min	*m³/hr*	*Continuous power W*	*Monthly power kWh/ month*	*Model*
46	14	15	4.1	102	74	4P MA
80	24	10	2.7	119	86	..
80	24	25	6.8	301	216	..
103	31	11	3.0	168	120	..
103	31	27	7.4	417	300	..
161	49	34	1.1	100	72	..
161	49	34	9.3	826	594	..
184	56	4	1.1	109	79	4P SA
184	56	37	10.1	1015	730	..
230	70	5	1.4	172	123	..
230	70	43	11.7	1484	1168	..
288	88	6	1.6	258	185	..
288	88	46	12.5	1981	1426	..

ELEKTRO GMBH
St Gallerstrasse 27
8400 Winterthur
Switzerland

Telex: 76-299 Elektro

Model	*Type of Turbine*	*Head*	*Flow*	*Generator*
222	Pelton turbine	20-70m	0.5-2 /sec	W50 50 to 400 W
222	Pelton turbine	40-120m	1-3 /sec	W250 200 to 1500 W
Bu 242	Francis turbine	8-20m	10-30 l/sec	U286 500 to 2000 W
542,548	Pelton turbine	100-200m	15-30 l/sec	258 10 to 25W

Pelton Wheel and Electronic Controller

EVANS ENGINEERING
Land & Leisure (Services) Ltd
Priory Lane
St. Thomas
Launceston PL15 8DQ
Cornwall
U.K.

Phone: 0566 3982

This company acts as building contractors for micro-hydro stations in the U.K. and as consultants to the Intermediate Technology Development Group (see Section 9).

Products offered are:

1. Simple bare-shaft turbines of up to 5 b.h.p.
2. Packaged reaction and impulse turbines with brushless alternators and completely electronic controllers of up to 100kW.
3. Electronic control units applicable to any other turbine using the load control concept (involving a ballast circuit).
4. Instrument and control panels made to order.
5. Simple and duplex hydraulic rams (Hydrams) for pumping water.
6. Plywood/GRP tanks, sluice units, etc.
7. Reconditioned second-hand turbines of various sizes.

JYOTI LIMITED
Industrial Area
P.O. Chemical Industries
Baroda 390 003
India

Telex: 0175-214
Cables: Jyotipumps
Phone: 8352/8641

A range of turbines suitable for generating power from 25kW to 100kW where medium heads are available. The group also produces larger units for higher head installations.

5kW Micro-Hydel Units (Francis turbine vertical shaft)

Head	*Flow*
3m (10ft)	205 l/s (7.25ft^3/s)
6m (20ft)	150 l/s (5.28ft^3/s)
9m (30ft)	93 l/s (3.28ft^3/s)
12m (40ft)	73 l/s (2.6ft^3/s)

10kW Micro-Hydel Units (Francis turbine vertical shaft)

6m (20ft)	256 l/s (9.05ft^3/s)
9m (30ft)	170 l/s (6.0ft^3/s)
12m (40ft)	133 l/s (4.69ft^3/s)

25kW Micro-Hydel Units (Turgo type impulse wheel, horizontal shaft)

60m (197ft)	70 l/s (2.5ft^3/s)
130m (425ft)	31 l/s (1.1ft^3/s)

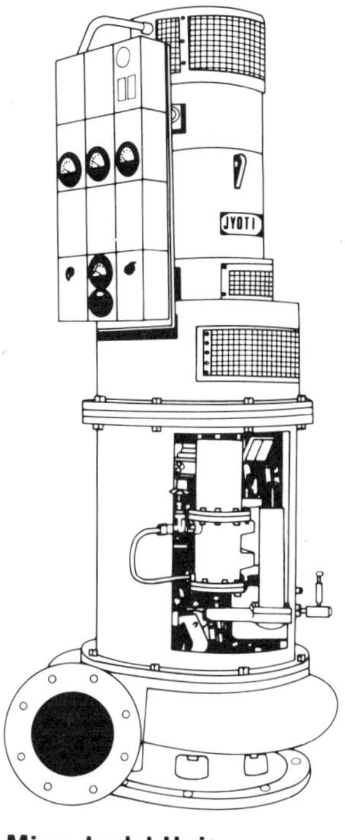

Micro-hydel Units

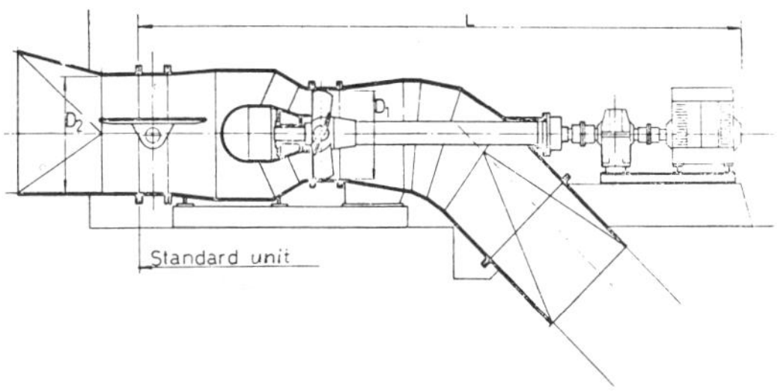

Type SH Water Turbine

KARLSTADS MEKANISKA WERKSTAD
Fack. S-681 01
Kristinehamn
Sweden

Telex: 66050 Kmwksn S
Cables: Verkstaden Kristinehamn
Phone: 0550 15200

Manufacturers of a range of turbines from micro to large multi-megawatt units. The small systems consist of propeller turbines available for either vertical or horizontal shaft installation as follows:

Runner diameters	=	0.7, 0.9, 1.15 and 1.5m
Heads	=	4 to 25m
Flow rates	=	1 to 15m^3/s
Outputs	=	50 to 1800kW

These units generally are supplied to feed a larger grid system through an a.c. asynchronous generator and therefore are non-adjustable, but a version with adjustable blades controlled by the upstream water level is also available.

JAMES LEFFEL & COMPANY
Springfield
Ohio 45501
U.S.A.

Cables: Leffel Springfield Ohio
Phone: 513 323 6431

Long-established manufacturers of water turbines (since 1862). Current production ranges from units having outputs of 5000hp down to quite small machines coming within the scope of this guide. The main small machines are the Hoppes hydro-electric system, the Samson turbine and a very small unit intended for laboratory use with either a miniature Francis or a propeller runner.

Hoppes Hydro-electric Units (integral turbine and generator sets)

Nominal elect. power output	*Minimum head & corres. flow*		*Maximum head & corres. flow*		*Penstock diameter for flow range*		
0.5kW (d.c. only)	8ft	104cfm	12ft	68cfm	12in	to	6in
1.0kW	8	190	25	68	12		6
2.0kW	8	330	25	110	16		6
3.0kW	8	470	25	158	18		10
5.0kW	8	760	25	260	20		12
7.5kW	11	800	25	380	20		16
10.0kW	12	980	25	480	20		16

Standard electrical rating is 3-phase 60Hz at 120, 240, or 480V but also available at 50Hz.

Improved Vertical Sampson Turbine (turbine supplied with vertical shaft)

Model	*Min. head & corres. power flow-rate & speed*				*Max. head & corres. power flow-rate & speed*				*Pen-stock dia.*
	ft	*hp*	*cfm*	*rpm*	*ft*	*hp*	*cfm*	*rpm*	
17E	3	1.1	252	161	26	29.5	741	474	24in
17D	3	1.5	328	161	26	38.5	967	474	26
17C	3	2.0	433	161	26	50.7	1273	474	28
17B	3	2.4	533	161	26	62.6	1571	474	30
17A	3	3.2	697	161	26	81.8	2053	474	36
20	3	4.2	914	140	26	107.0	2691	415	42
23	3	5.5	1209	127	26	141.0	3559	361	48
26	3	7.1	1545	108	26	180.0	4549	319	54
30	3	9.4	2057	94	26	241.0	6056	276	60
35	3	12.8	2789	81	26	327.0	8210	237	72
40	3	16.8	3657	70	26	428.0	10767	207	84
45	3	21.2	4629	63	26	542	13626	184	96
50	3	26.2	5714	56	26	669	16822	166	108
56	3	32.9	7168	50	26	839	21102	148	120
62	3	40.3	8787	45	26	1030	25868	134	
68	3	48.5	10570	41	26	1238	31115	122	
74	3	57.5	12517	38	26	1466	36848	112	

Laboratory Model Hydraulic Turbines

Turbine Model No 4W supplied with a 4in Francis and 4in propeller runner

Head (ft)	*hp*	*rpm*	*Discharge cu. ft. sec.*
5	.29	990	.685
7	.50	1170	.810
10	.83	1390	.972
12	1.10	1530	1.050
15	1.50	1712	1.160
17	1.81	1822	1.230
20	2.34	1977	1.325

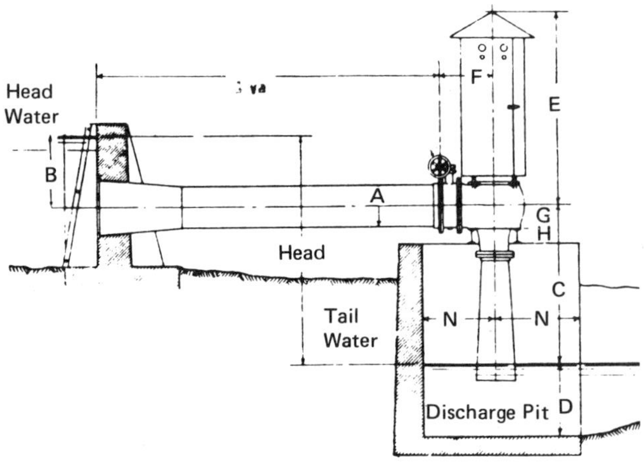

Leffel Turbine

MASCHINENFABRIK KÖSSLER GmbH
A-3151 St. Polten
St. Georgen
Austria

Telex: 015652
Cables: Turbinekossler
Phone: (0 27 46) 82 72

Manufacturers of two types of small scale hydro-electric sets, one with a horizontal shaft and the other with a vertical shaft configuration. Both systems are fully integrated generating systems. Outputs vary from 12kVA at 20m head and 70 l/s flow up to 1250kVA at 100m head and 1270 l/s flow. Outputs of 400/231V 50Hz are normally supplied, but can be varied to order.

The company also designs/manufactures/installs larger Impulse, Francis and Kaplan turbines of up to 50MW.

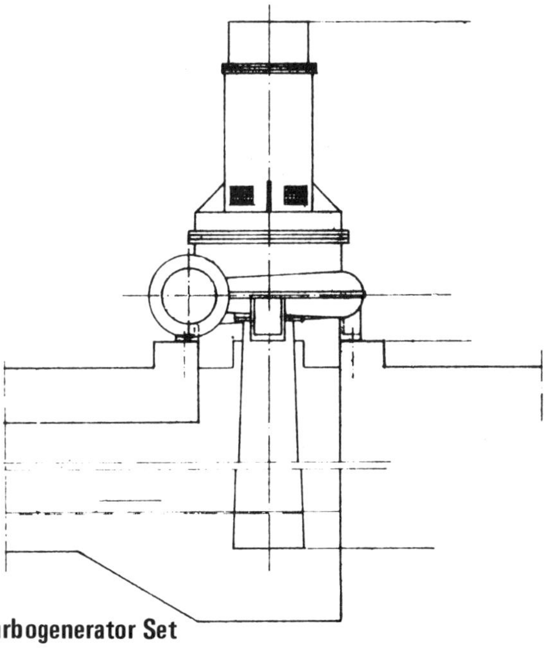

Turbogenerator Set

SMALL HYDROELECTRIC SYSTEMS
P.O. Box 124
Custer
Washington 98240
U.S.A.

Phone: 206 366 7203

Supply small impulse (Pelton) hydro-electric systems producing 5kW to 25kW from heads of 50ft to 350ft. The turbines are fitted with a Lima brushless alternator mounted on top of the turbine casing.

Peltex Unit

THE NATURAL ENERGY CENTRE
161 Clarence Street
Kingston upon Thames
Surrey KT1 1QT
U.K.

Telex: 21887 MONCO G
Phone: 01 549 5888/9

Distributors of a variety of power producing equipment from numerous sources. A unique item is the Aquadyn 12V charger. This is a small electric generator designed for charging the batteries and running electrical systems on yachts and small boats; it is capable of continuously running a small refrigerator, an automatic pilot and for trickle charging batteries, running general lighting, etc.

Power is produced by a propeller which is immersed under the boat and which drives a generator using the kinetic energy of the passing water. It could equally be immersed in the current of a river or stream as a free-stream turbine.

The only maintenance required is one oil change annually. Weight is 20kg and output is 15 amp at 12 volt with a 5kt current.

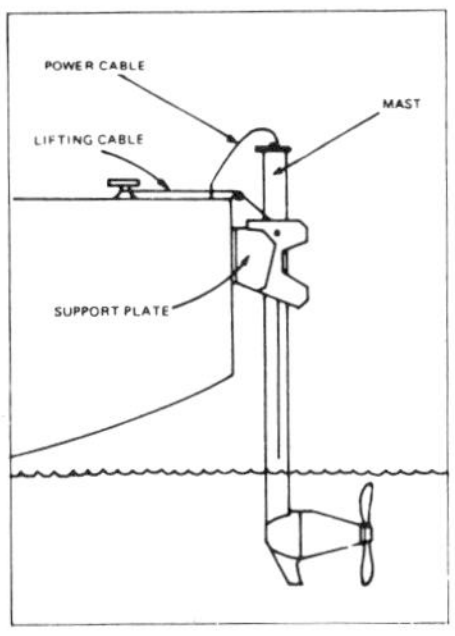

15 Amp/12V Aquadyn

OSSBERGER-TURBINENFABRIK
D-8832 Weissenberg
Postfach 425
Bayern
German Federal Republic

Telex: 06 24672
Phone: 09141 4091/2

Manufacturers of a range of Mitchell (Banki) cross-flow turbines. Units are self-contained, automatic, hydroelectric generating sets which only need be shut down once a year for an oil change and a grease. The units have automatically controlled output to cope with varying loads.

Power range = 1kW to 1000kW
Head range = 10ft (3m) to 800ft (240m)
Flow range = 1ft^3/s (0.03m^3/s to 250ft^3/s (7m^3/s)

A questionnaire is available for prospective customers.

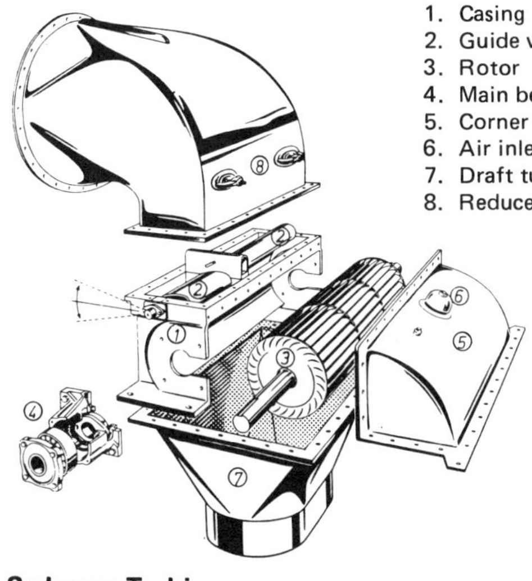

1. Casing
2. Guide vanes
3. Rotor
4. Main bearing
5. Corner casing
6. Air inlet valve
7. Draft tube
8. Reducer

Ossberger Turbine

NORTHERN WATERPOWER INC
P.O. Box 49
Harrisville
New Hampshire 03450
U.S.A.

Telex: 710 366 6762
Phone: 603 827 3367

NIAGARA WATERWHEELS LTD
706 East Main Street
Welland
Ontario L3B 3Y4
Canada

Phone: 416 734 7418

Northern Waterpower Turbine and Generator Set

These two related companies supply six models of horizontal-axis axial flow propeller turbine package, with outputs in the 20 to 300kW range.

The turbines incorporate a high specific speed runner with an inlet butterfly valve which is normally hand operated (motor operation optional). Flow control is by radial guide vanes actuated by servo motor controlled either by electro-hydraulic speed sensing or by tachometer relay governing, (the latter being for use when paralleled with larger systems).

Typical performance envelope for these models is:-

Head range:	10 to 60ft	3 to 20m
Flow range:	10 to 100ft^3/s	0.3 to 3m^3/s

Larger versions with outputs of up to 500kW at heads up to 30m are also available.

16ft Wheel Producing 20 b.h.p.

WESTWARD MOULDINGS LTD
Greenhill Works
Delaware Road
Gunnislake
Cornwall
U.K.

Phone: Gunnislake (0822) 832120

Manufacturers of fibreglass water wheels to order. Units can be overshot or breast wheels with diameters of 8ft, 16ft, and 20ft. Wheels can be built to any width and can be shipped in sections and erected on site.

Typical power outputs as follows:

Diameter	*Width*	*Nominal Power (max)*
8ft (2.4m)	2.5ft (0.76m)	6 b.h.p.
16ft (4.8m)	3.0ft (0.91m)	25 b.h.p.
20ft (6.0m)	4.5ft (1.4m)	35 b.h.p.

Hydro-Mechanical and Pumping Systems

CECOCO CHUO BOEKI GOSHI KAISHA
P.O. Box 8
Ibaraki
Osakafu
Japan 567

Cables: Cecoco Ibaraki
Phone: (0726) (22) 2442-3

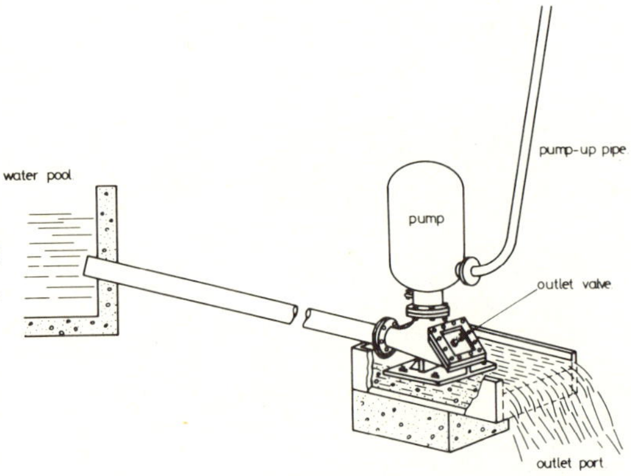

Ce-Co-Co Hydro-Hi-Lift Pump

The Ce-Co-Co Hydro-Hi-Lift series of hydraulic ram pumps can lift water up to fifty times the head available from the driving water source. Minimum driving head of 0.5m is necessary.

Pump Type No.	*No. 1*	*No. 2*	*No. 3*	*No. 4*	*No. 6*	*No. 8*
Outputs at:						
4X drive head	10	22	38	85	200	400 l/min
8X drive head	5	11	20	43	100	200 l/min
15X drive head	2	5	9	23	55	100 l/min
Driving flow	25	60	150	300	600	1200 l/min
Supply pipe dia.	40	50	75	100	150	200 mm

Blake's Hydram

JOHN BLAKE LTD
Royal Works
P.O. Box 43
Clayton-le-Moors
Accrington
Lancs. BB5 5LP
U.K.

Telex: 63242
Cables: Rams Accrington
Phone: (0254) 35441

A long established manufacturer of a range of hydraulic ram pumps, as follows.

Size of Hydram		1	2	3	3½	4	5X	6X	5Y	6Y
Driving flow	from	7	12	27	45	68	136	180	136	180
required (l/min)	to	16	25	55	96	137	270	410	270	410
Max delivery height poss.	m	150	150	120	120	120	105	105	105	105

Quantity of water raised in litres per 24 hours for each litre per minute of driving flow, for different delivery heads and driving water heads:

Delivery head (m)	10	30	60	100	125
Driving head (m)					
1	65	19.5			
2	156	53	19.5		
4		115	53	23	16
8			125	69	55
16			250	150	110

BRIAU S.A.
B.P. 43
37009 Tours Cedex
France

Telex: 750729 F
Phone: (47) 61 38 17

This company manufactures a range of 8 models of hydraulic ram in addition to small turbines, windmills, etc.

Output in % for water-fall height ⩾ 1,50 m
For lower height: consult BRIAU SA.

Type	*Absorbed flow in l/min*	*Ratio of delivery head to fall height* 4	6	8	10	12	15	20	*Pipe sizes* *Drive*	*Delivery*
A1	4 to 7	50	54	54	48	40	30	12	20/27	15/21
B2	8 to 15	52	56	56	50	44	33	16	26/34	20/27
C3	15 to 25	55	59	59	54	48	37	20	33/42	20/27
Do 3 A	25 to 45	55	60	60	55	49	38	22	40/49	26/34
D4	45 to 80	58	63	63	60	55	47	30	50/60	33/42
E6	80 to 160	60	65	65	64	60	55	42	66/76	33/42
G10	250 to 450	63	68	68	66	64	59	46	100	50/60
J12	750 to 1200	65	70	70	69	68	60	50	150	80

Calculation of Ram-flow

q: delivered flow in l/min
Q: absorbed flow by the ram in l/min
fh: fall height in m
DH: delivery height in m
n: output in % according to the above table

Example

Q: 50 l/min
fh: 4 m
DH: 32 m

Ratio, $\frac{DH}{fh} = \frac{32}{4} = 8$ output = 63%

Effective flow: $q = 50 \times \frac{4}{32} \times 63\% = 4$ l/min

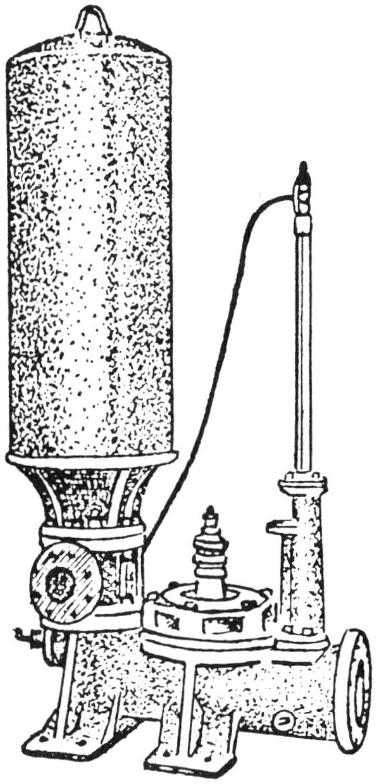

Briau Hydram

GREEN & CARTER LTD Phone: (0962) 880846
Vulcan Iron Works
Kingsworthy
Nr. Winchester
Hants S023 7QF
U.K.

Manufacturers of the Vulcan series of hydraulic ram pumps (Hydrams) for over 56 years. Capable of pumping water driven by as small a fall of water as 2½ft (0.75m) and of delivering water to heights of 400ft (122m). Special rams delivering up to 1000ft (300m) to special order.

Model No. (Drive pipe diam. (in.)	*Driving Flow (Imp. gall/min)*	*Diameter of delivery pipe (in)*	*(mm)*	*Delivery flow in 24 hours (Imp. gall)*
1¼	1 to 4	¾	22	100 to 600
1½	2 to 6	¾	22	400 to 1000
2	5 to 10	1	25	600 to 2000
2½	8 to 25	1	25	750 to 3000
3	12 to 30	1	25	1000 to 5000
4	20 to 45	1¼	30	2000 to 10000
6, 7	50 to 100	2	50	6000 to 30000

Vulcan Hydraulic Ram

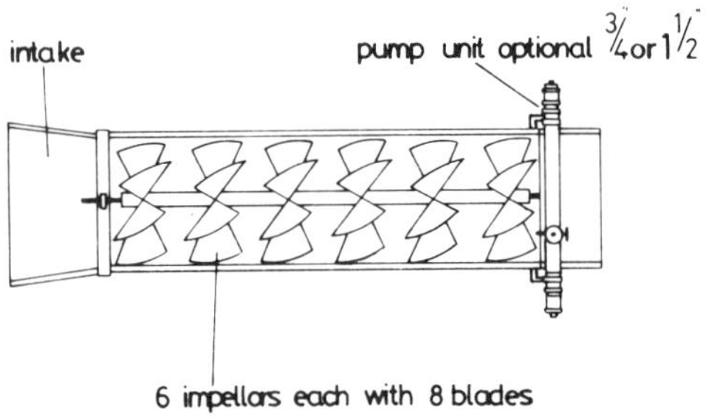

The Plata Pump

PLATA POWER LTD Cables: Nespat
P.O. Box 221 Phone: 78 694
Dunedin
New Zealand

This new water-powered pumping system, developed in New Zealand, consists of a plastic tube with a series of propellers mounted on a shaft inside it. It can operate on very low heads of water from 9in (0.23m) up to 2ft (0.6m) and operates best when the tube is running only half full with water. The shaft drives a choice of ¾in or 1½in pumps (20mm or 40mm pumps).

Performance envelope:

Driving flow required: 0.5ft^3/s (0.014m^3/s) up to 3ft^3/s (0.085m^3/s).

Head range for delivery: up to 125ft (38m) with 1½in pump; up to 300ft (90m) with 3in pump.

Outputs (typical): 68 gall/hr at 10ft head or 16 gall/hr at 300ft head, with 2ft^3/s driving flow and 15in driving fall (¾ pump). 316 gall/hr at 10ft head and 53 gall/hr at 100ft head with 2ft^3/s driving flow and 15in driving head (1½ pump).

Although the Plata Pump performs a similar function to a Hydram, it can operate with a much greater driving to delivery head ratio and it can also usefully handle much larger driving volumes of water than most ram pumps.

4

Bio-Mass and Thermal Energy

STOVES AND WATER HEATERS
HEAT PUMPS AND REGENERATORS
METHANE DIGESTERS
BURNERS & STEAM RAISING PLANT
GASIFICATION PLANT

"Bio-mass" is modern jargon for the oldest human energy resource. It means biologically derived material of any kind, all of which is potentially useful as a source of heat energy, since life on Earth is based on combustible compounds of carbon. The chain of life begins with photosynthesis, or the use of solar energy by plant life to synthesise organic matter from atmospheric carbon dioxide and water, and other life-forms in turn depend on this starting point in their food chain.

Dried organic matter is invariably capable of being burnt and the heat produced in this process represents released solar energy stored chemically. One of the main end-products of this combustion or oxidation is carbon dioxide, which is returned to the atmosphere. Fuels such as petroleum and coal consist of fossilised remains of ancient bio-mass. They are a more concentrated energy source, since the fossilisation process effectively compresses the combustible components of the original wood or marine organisms into a denser form.

The main forms of bio-mass in general use are of course wood, (and its derivatives such as charcoal) and dried animal dung. It is not generally realised by the minority of the human race within the world's moneyed economy (whether they are in so-called "developed" or "developing" countries) how important non-commercial or semi-commercial fuels are to the world's rural poor. Between 50 and 60 per cent of all people in the world spend a significant proportion of their time gathering twigs and firewood or drying the dung of their animals to provide for their basic energy needs such as cooking their food. Although the energy consumption of individual families in this way is small compared with the energy consumption of families in industrialised societies, the total firewood, charcoal and dung energy consumption for more than 2,000,000,000 people is of course substantial. It is hard for energy statisticians to obtain reliable figures, since gathered firewood is not registered by governments in the same way as commercial fuels, but a number of studies have been made. For example, using FAO figures, Earl[1] has shown that world energy consumption in 1970, given in millions of tonnes of coal equivalent (MTCE) was as follows:

World Energy Consumption in 1970 in MTCE

Coal	2419	*Geothermal*	1
Oil	2850	*Wood*	487
Gas	1418	*Dung*	90
Uranium	10	*Agricultural wastes*	10
Hydro	150		
		TOTAL	7435

In other words, wood and dung combustion accounted for an energy consumption equivalent to more than one fifth of the world's total oil consumption in that year and nearly four times the world's total hydro-electric output. Nuclear power, the greatest absorber of energy R&D expenditure in the industrialised world, is not in the running by comparison.

Earl also shows in the same reference that the populations of certain countries are heavily dependent on wood and dung for their fuel requirements. The table on the following page illustrates this by contrasting selected non-industrialised and industrialised countries.

As far as dung consumption is concerned, India is almost certainly the primary consumer. Serious studies[2] have shown that dung consumption in India in 1972 was approximately 100,000,000 tonnes which is the equivalent of 58MTCE in heating value.

Consumption of Fuel Wood Energy in Selected Countries

Country	*Energy consumption per capita from all sources (1973) (kg coal equivalent)*	*Fuel wood energy consumption per capita (1973) (kg coal equivalent)*	*Percentage of fuel wood based energy (%)*
Malawi	376	335	89
Nepal	259	248	96
Tanzania	1042	999	96
India	274	83	30
Kenya	447	299	67
Nigeria	480	435	91
Zambia	900	391	43
Brazil	1176	695	59
Libya	569	87	15
Algeria	479	9	2
USSR	4356	157	4
UK	5143	4	0
France	3570	52	1
USA	10817	43	0

Unfortunately, the main users of these energy resources are in the poorest section of the human community, precisely the section with the highest rate of population growth. Hence there is a severe crisis developing as demand for subsistence fuels exceeds supply in very many parts of the world. There is often a narrow gap between a sustainable rate of firewood collection and the catastrophic removal of vegetation followed by desertification which occurs when excessive demands are made. Dung is of course far better if used as a fertiliser rather than as a fuel, so that the continued use of dung as a fuel is causing severe deterioration of the quality of the soil for agriculture or replanting of trees and is therefore contributing to the problem. The so-called "oil crisis" which afflicts the wealthy is nothing by comparison with the "firewood crisis" that is beginning to afflict the poor.

Fortunately in the longer term there are good possibilities for alleviating the subsistence fuel crisis by a two-pronged attack; firstly the existing methods of combustion are invariably highly inefficient and low-cost more efficient burners could greatly reduce demand; secondly, the managed cultivation of fuel crops could add to the available supply, but this approach would only be economically viable given more efficient usage. In the short term there is little hope because of the magnitude of the problem and the severe neglect of the necessary technology. Here we have the world's fourth largest existing energy resource absorbing a negligible proportion of development effort compared with other much less significant energy conversion processes of value to the wealthier sections of the human race (whether they are in rich or poor countries). What R&D effort there is at present is mainly aimed at re-introducing wood fuel (or other solid fuels) into the wealthier sections of the community as a petroleum substitute, and is therefore motivated by the oil crisis of the "rich" rather than the wood crisis of the "poor".

Inevitably, most of the devices listed in this section, being commercial products, have been developed with other markets than the poorer half of humanity in mind, but nevertheless some may well find application in areas where they are not to be found at present. Other sections of this Guide, including Section 10 on information sources, consultants and R&D institutions will point to lower cost "non-commercial" devices (such as mud stoves) that could usefully be applied under various development and extension schemes in areas where subsistence fuels are proving to be in short supply. There are also other methods than direct burning by which bio-mass may be utilised, involving a variety of processes, and there are of course numerous alternative bio-mass fuels which can compete with the main fossil fuels of oil and coal.

As far as heat production is concerned, the relative merits of various fuels are perhaps best summarised in terms of the calorific values or total heat released when they are burnt. These can be expressed in terms of their weight or their volume, or for that matter, when sufficient data are available, in terms of their cost. The commonly used SI unit of heat is the MJ (Mega-joule) which is equal to 0.278kWh (kilowatt-hours), or 950 BTU (British Thermal Unit), or 0.0095 therms, or for that matter 0.373 hp-h (horsepower-hour), and the relative energy density of various fossil and bio-mass fuels is given in the table on the following page.

Also of interest is the productivity of various bio-mass fuels. Obviously this depends considerably on factors such as location, soil quality, etc. The most productive areas for photosynthetic material are the forested regions; agricultural land is generally less than half as productive in terms of bio-mass production. Earl *(ibid)* quotes the following perannum rates:

Photosynthetic Carbon Production Rates

Type of land	*Net primary production (tonne/ha)*	*Total world annual* natural production (giga-tonnes)*
A. Forest		
temperate deciduous	10	8.0
conifer and mixed	6	9.0
temperate rain forest	12	1.2
tropical rain forest	15	15.0
dry woodlands	2	2.8 = 36.0 GT
B. Non-forest		
agricultural	4	6.0 (man-made)
grasslands	3	7.8
tundra	1	1.2
deserts	1	3.2 = 18.2 GT
		Total: 54.2 x 10^9 tonnes

*total fixed carbon, including leaves, grass, roots, barks, etc.

Also given is the small proportion of this natural resource that is utilised (1970 figures); 6,700 million tonnes of coal equivalent just rots each year!

Clearly the firewood energy crisis is a crisis of geography; there is a considerable surplus of firewood available in regions other than where it is currently needed. Indeed the heat value of the unused fraction of the annual increment, at some 6,700 MTCE represents an energy surplus of around 90% of the world's total energy consumption in 1970. Which is not to say it could all be used, much of this wood would need more energy in its

Relative Heat Value of Various Fuels (Approximate Values)

	Calorific value/unit wt. (MJ/kg)	Calorific value/unit wt. (BTU/lb)	Calorific value/unit vol. (MJ/m^3)	Calorific value/unit vol. (BTU/ft^3)
A: Fossil fuels				
petrol/gasoline	44	19,000	32,000	860,000
fuel oil	44	19,000	39,000	1,050,000
paraffin/kerosine	45	19,500	36,000	970,000
diesel/gas oil	46	20,000	38,000	1,020,000
coal tar/asphalt	40	17,000	40,800	1,100,000
anthracite coal	35	15,000	56,000	1,500,000
bituminous coal	33	14,000	42,900	1,150,000
lignite (brown) coal	30	13,000	37,500	1,010,000
peat	20	9,000	18,200	490,000
coke	28	12,000	22,400	600,000
natural gas (methane)	56	24,000	40*	1,020
coal gas (methane)	9	4,000	20*	490
propane (cylinder gas)	48	21,000	90*	2,400
butane (cylinder gas)	47	20,000	120*	3,100
B: Bio-Mass fuels				
wood (oak)	18	8,000	14,400	390,000
wood (pine)	20	9,000	10,000	270,000
wood (acacia)	16	7,000	11,000	300,000
charcoal	28	13,000	11,000	300,000
sunflower stalks	20	9,000	10,000	270,000
wheat straw	18	8,000	–	–
beef cattle manure	14	6,000	–	–
methanol (methyl alcohol)	20	8,600	19,000	500,000
ethanol (ethyl alcohol)	28	12,000	28,000	700,000
bio-gas (65% methane)	20	8,600	23*	600
wood gas (typical)	–	–	11*	280

*Since these fuels are normally gaseous, the calorific value per unit volume is relatively low compared with liquid and solid fuels.

Use of Fuel Wood Stocks

Area	*Growing* stock* ($m^3 \times 10^9$)	*Annual increment* ($m^3 \times 10^9$)	*Used by industry* ($m^3 \times 10^9$)	*Used as fuel* ($m^3 \times 10^9$)	*Total used* ($m^3 \times 10^9$)	*Unused increment* ($m^3 \times 10^9$)	*Heat value of unused* ($TCE \times 10^9$)
developed countries	242	8.8	1.1	0.3	1.4	7.4	3.2
developing countries	382	9.0	0.2	0.8	1.0	8.0	3.5
total	642	17.8	1.3	1.1	2.4	15.4	6.7

*above ground wood in areas classified by the FAO as "forest".

Potential Bio-mass Values of Selected Crops

Species	*Location*	*Annual dry matter yield* (ton/acre)	*Annual dry matter yield* (tonne/ha)	*Heat value* (10^6 BTU/ acre/yr)	*Heat value* (GJ/ha/yr)	*Tonne oil equiv. per ha./yr.*
sunflower	USSR	13.5	30	200	530	12
forage sorghum	Puerto Rico	30.6	69	460	1210	28
hybrid corn	USA (Miss)	6	13	90	250	6
water hyacinth	USA (Fla)	16	36	240	630	14
sugar cane (average)	USA (Fla)	17	39	260	680	16
sugar cane (experiment)	USA (Cal)	32	72	480	1250	29
sudangrass	USA (Cal)	16	36	240	630	15
bamboo	S E Asia	5	11	70	210	5
eucalyptus	USA (Cal)	20	45	300	790	19
eucalyptus	India	17	39	260	678	16
eucalyptus	Ethiopia	21	48	320	834	19
american sycamore	USA (Ga)	3.7	8	60	160	4
algae (pond)	USA (Ca)	39	88	580	1520	36
tropical rainforest (typical)		18	41	270	710	17
subtropical deciduous forest (typical)		11	24	160	420	10

collection than it would be worth, but obviously there is a lot of potential that is not being applied because of a shortage of technology of the kind described in this section, to exploit it.

Since much of the non-exploited bio-mass is inconveniently located, the question arises of growing fuel crops where they are needed. Some figures for the potential value of a selection of crops that might be used as fuel on account of their high growth rates appear above.[3]

There are two primary mechanisms for oxidising bio-mass and releasing its energy content. The easiest and most widely used method is to burn it, or in some cases to apply heat, often by partial combustion, and thereby drive off gases (which might be burnt separately but usually are wasted) so as to leave charcoal which is a fuel that gives more consistent and controllable heat production. Another method is to use bacteria to carry out a process to yield combustible products. The various methods are summarised below:

Direct heat process	*Bacterial process*
complete combustion	methane (bio-gas) fermentation
pyrolisis (gasification)	alcohol by fermentation
carbonisation (charcoal)	
destructive distillation	

Burning processes are generally wasteful, in the sense that the only product is heat which is often used inefficiently, or even in the case of charcoal production, the wood volatiles are deliberately driven off and lost as the objective is to produce a more concentrated and smokeless fuel. Also, wet wood absorbs a large proportion of the available energy in producing steam. Pyrolisis plants or distillation plants can fractionate the products at the cost of more complication, while bacterial fermentation processes, although relatively slow, can yield a "digested" sludge of value as a fertiliser for reapplication to the land in addition to a fuel output as gas or alcohol.

It is perhaps of interest to note the products present in a typical sample of tropical hardwood;[4] all of which have potential uses:

In 1000kg of dry wood:	
charcoal	200kg (= 8400MJ)
gas (calorific value 10.5MJ/m^3)	140m^3 (= 1470MJ)
methyl alcohol	14 l (= 260MJ)
acetic acid	53 l
esters (methyl acetate, etc.)	8 l
acetone	3 l
wood oil and light tar	76 l
creosote oil	12 l
pitch	30 kg

The general lack of commercial technology for exploiting bio-mass is reflected by the make up of the following section of this Guide — clearly there is much potential for new products and it is known that there has been some increase in R&D in this area, although mainly orientated towards finding petroleum substitutes for industrialised societies. The most readily available technology consists of wood and other solid fuel burning stoves and heaters and a number of methane digesters (bio-gas) plants and wood gasification plants have appeared within recent years. Also, a few steam raising plants for use with steam engines are included. Some further comments on the type of equipment included follows.

Stoves and Water Heaters

All the stoves included are completely enclosed systems with full control of the air supply in order to achieve efficient and controlled combustion.

Open fires, traditionally used in many parts of the world for cooking and heating are notoriously inefficient — often needing ten times or more fuel than a good enclosed stove to perform the same function.

Benjamin Franklin is often credited with having developed the first efficient stove. In fact he reinvented what was already widely used in the Orient for many centuries. In 1740 he developed his so-called 'Pennsylvanian Fire-Place' to try and combat a serious firewood shortage by improving the efficiency of combustion — in his own words[5] "by the help of this saving invention our wood may grow as fast as we consume it . . ." Today the so-called Franklin Stove, widely used in the USA, bears little resemblance to the original invention it developed from, but at least Franklin stated the important objective that led Americans to give up the wasteful open fireplace they and their ancestors had brought with them from Europe.

A number of heat reclaimers are also included, which allow surplus heat in the flue gases to be used for heating either convected air or water, and thereby improve the efficiency of fuel combustion still further.

As explained in Section 9, the Intermediate Technology Development Group is investigating improved cooking stove designs as part of its Power Project activities.

Heat pumps and regenerators

These devices are generally used to move heat around, and since the energy required to move heat is often considerably less than the heat that is actually transferred, many of these devices have the unusual capability of appearing to deliver more heat than they use.

This is not the place to describe the principle of the heat pump in detail (almost any textbook on thermodynamics will do this). The principle of the heat pump is the same as that of a refrigerator, although the object is to take heat at low temperature from the outside air, the ground or a river and to raise the same heat to a higher temperature via a compression-heat exchange-expansion process; a refrigerator uses the same process to take heat out of food at a low temperature and to transfer it to the outside air at slightly above ambient temperature (hence the heat that is emitted from the back of a fridge).

The attractive feature of a heat pump is that the only power required is to drive a compressor which pumps the refrigerant fluid used for the heat transfer. Commonly the actual heat transferred can be over twice the energy needed to drive the compressor, so that a heat pump delivering say 6kW of heat can often require less than 3kW to drive it. Another feature of heat pumps is that sometimes they are reversible so that they can be used as air-conditioners in summer and heaters in winter.

Although the principle of the heat pump has been known for many years, it is only in recent years that they have begun to gain ground as viable heating equipment as a result of climbing electricity prices. While electricity was cheap in most countries it was always much simpler to heat using normal electrical resistance heaters.

Regenerators are systems that recover heat that would otherwise be wasted, through the use of a small energy input. Typical sources of waste heat are the flue gases of a stove or boiler or even the heat emitted in the stale air expelled by a ventilation system. Although some heat must usually be lost with flue gases (it is this heat which is needed to create a draught), many systems lose more heat than necessary, so the provision of a heat exchanger in the flue allows some surplus heat to be extracted either for space heating or to provide hot water; examples of such devices are included in this section and they are generally extremely cost effective. Also included is a heat regenerator which extracts heat from stale air being expelled from a building and transfers it to the fresh incoming air. An interesting application of this may be to transfer "cold" in the case of air conditioned buildings, where warm incoming air could be cooled from the stale outgoing flow.

This section includes only a limited selection of equipment in this category since heat reclamation is not strictly power generation. However it is generally cheaper to save energy than to generate power and it is hoped to expand this section in future editions of this Guide.

Methane digesters

Many organic materials can be fermented in a sealed container (with air excluded). Under these conditions anaerobic bacteria can break down the bio-mass and yield in the process a fuel gas (similar to naturally occurring "Marsh gas") consisting of about 60% methane and the rest mainly carbon dioxide. This gas, often called "Bio-gas" or in India "Gobar gas"* has reasonably good qualities for cooking or for running internal combustion engines.

Other merits of this process are that the digested sludge makes a useful fertiliser, so that unlike when bio-mass is totally burnt, it is possible to return much of the original material to the land and preserve the soil quality. In fact, the digestion process makes the nitrogen and various other chemicals present in the original bio-mass more accessible for plant growth than the normal aerobic (in air) rotting process. Also, unlike artificial N fertilisers the sludge contains humus which can greatly improve soil structure.

In addition to its use for fertiliser production, this process can be used for treating sewage and other potentially dangerous wastes, because the anaerobic process kills most pathogens harmful to people. Hence it can be applied as a small scale sewage processing system. It also has the advantage of producing an effluent without a seriously offensive smell.

At present the anaerobic process finds its main application in large process plants for municipal sewage treatment. Often, the considerable volumes of gas produced by this process are not effectively used by the municipalities concerned and surplus gas is simply "flared" off.

Small scale methane production was practised by French farmers and others during the Second World War when they were totally without petroleum fuel, but the subsequent "cheap petrol era" caused the process to be abandoned in industrialised societies (except for large-scale sewage treatment) until quite recently. On the other hand, considerable efforts to popularise the use of small-scale anaerobic digesters have been made in Asia, particularly in India and China, since the 1960s and onwards. It was realised in those countries that the anaerobic process had considerable merits in the rural economy if all its three uses for fuel production, fertiliser production and waste treatment could be combined on a small-scale at sufficiently low cost. The main difference between typical Indian and Chinese methane plants is that the Indian system requires a steel gas holder which rises as its fills, while the Chinese system traps the gas under a fixed masonry or concrete dome. The Chinese use the system to process human sewage as well as any other suitable wastes, but the Indians generally use cow dung (or Gobar) as the main input. It is claimed[6] that over a million houses in the Chinese province of Szechuan alone use bio-gas and that in one year the county of Yunghsing in the same province saved the use of 1,100 tons of coal, 15 tons of kerosene and over 800,000 work days formerly spent getting wood and coal. Experience in India has shown that the larger Gobar plants, generally attached to institutions such as Ashrams, are more successful than small "family-size" units and there have been difficulties in popularising the smaller plants.

There is a technical reason for the difficulty in obtaining such good results with small plants, as the process depends on steady temperatures for good gas production and the optimum rate of gas production occurs as a temperature of around 35°C. Larger plants have a greater thermal inertia, giving steadier temperatures in the digester and they also have a smaller surface area relative to their volume (since area goes up with L^2 and volume with L^3) so that heat losses are lower and a higher temperature is more easily maintained in a large plant. To be effective smaller plants need to be insulated, except where they are used in hot tropical regions having night temperatures not much lower than day ones. Obviously the same constraints apply to the commercially manufactured plants described in the following section.

This is not the place to give a detailed account of this complex biological process but it is worth giving a brief outline of the characteristics of the process. Figure 1 shows a schematic view of the digester system illustrating a number of end uses and optional additional processes. The following table[7] indicates typical gas yields from various input materials, (the gas being in all cases from 60 to 70 per cent methane).

Material	*Gas yield per unit dry matter*	
	(m^3/kg)	*(ft^3/lb)*
cow dung	.1 – .3	1 – 5
chicken droppings	.3	.5
pig dung	.4 – .5	6 – 8
farm wastes	.3 – .4	5 – 7
elephant grass	.4 – .6	7 – 9
chicken droppings and paper pulp	.4 – .5	7 – 8
chicken droppings and grass clippings	.4	6
sewage sludge	.6	10

In most cases the process requires from 20 to 30 days for completion (depending on temperature and other factors) – hence the digester capacity will require to be such that it can hold the requisite number of day's input. The input generally requires to be around 5-10% solids, the rest water; since many inputs are already wet, this does not necessarily mean that nine parts of water per part of input are added, just that in many cases some water must be added to achieve the optimum solids to water ratio. A critical factor is the ratio of carbon to nitrogen in the feed material, which ideally should be in the range 10 to 30 for good operation. In many cases with excessively carbon-rich inputs (e.g. vegetation or paper pulp), the process benefits if nitrogen is added (this can be in the form of ammonia, urea, urine, etc). Similarly, nitrogen-rich inputs (like chicken droppings or urine-rich stable or pig-sty wastes) can benefit from extra carbon in the form of grass clippings or paper (this being evident from the above table).

Burners and steam-raising plant

Unfortunately the limited number of entries in this section reflects the decline of steam power during the last few decades of increasing internal combustion engine usage, to the extent that small steam engines and steam-raising plants are today very rare. Also, many of the more recently developed types tend to be fuelled with oil rather than solid wastes, for convenience. There is a marked revival of interest in steam power, particularly to exploit non-petroleum fuels, and it is hoped that future editions of this Guide will include a wider choice of equipment in this section as work appears to be in hand in a number of countries. New technology such as fluidised bed combustors and heat pipes promise improved steam raising plant in future, with potentially high efficiency at relatively low cost.

An important element in the use of steam raising plant is safety. Virtually all countries have regulations governing the use and inspection of boilers and pressure vessels and it is as well to ensure that any equipment that you plan to use will comply (although most commercially available

*"Gobar" means cow dung.

products are almost bound to comply since the countries of manufacture impose their own regulations on the manufacturers). Obviously no attempt should be made to modify commercially bought equipment by tampering with the design as such actions could make a boiler dangerous. Modern mono-tube boilers are particularly safe, however, as the steam is raised in a single coil of tube and any accident, (and designs are such that they are usually fail-safe), will merely cause minor internal damage to the boiler. Mono tube boilers have the additional advantage of being quick to reach working pressure since the water capacity of the heated volume is small.

Another factor to bear in mind with steam raising plant is the need for water softening in many locations to prevent excessive "furring" of the boiler tubes, although in installations involving the condensing and recycling of exhaust steam (essential for reasonable efficiency), this problem only arises for small quantities of make-up water that may occasionally be needed.

Wood Gasification

This is another old technology, popular at the beginning of this century, and then later in times of emergency in some industrialised countries (e.g. during the Second World War) when petroleum was not available, but subsequently forgotten. The few commercial systems included in this issue stem from war-time experience in their respective countries of manufacture, but have been brought up-to-date in the face of the "oil crisis".

The attraction of these devices is that they enable internal combustion engines, particularly spark ignition (petrol) engines, to be run on wood and even on saw mill wastes, coconut shells or agricultural wastes.

The main problems are the accumulation of tars from certain wastes, and the need to scrub the gases to remove corrosive components. However these problems have been tackled in commercially available equipment, and again it is to be hoped and expected that wider variety of gasification plants will appear on the market as petroleum prices increase.

Footnotes

1. *Forest Energy & Economic Development,* D.E. Earl, Clarendon Press, Oxford, 1975.
2. Earl quoting from National Commission on Agriculture, India.
3. *Energy for Rural Development,* National Academy of Sciencies, Washington, 1976.
4. Earl — *Forest Energy & Economic Development.*
5. *Heating with Wood,* Larry Gay, Garden Way Publishing, Charlotte, Vermont, U.S.A.
6. *China Reconstructs,* May 1977, "Home-Made Gas for China's Countryside", article by Tai Mei-Tien.
7. *Methane Generation by Anaerobic Fermentation — an Annotated Bibliography,* Leo Pyle & Christina Freeman, Intermediate Technology Publications.

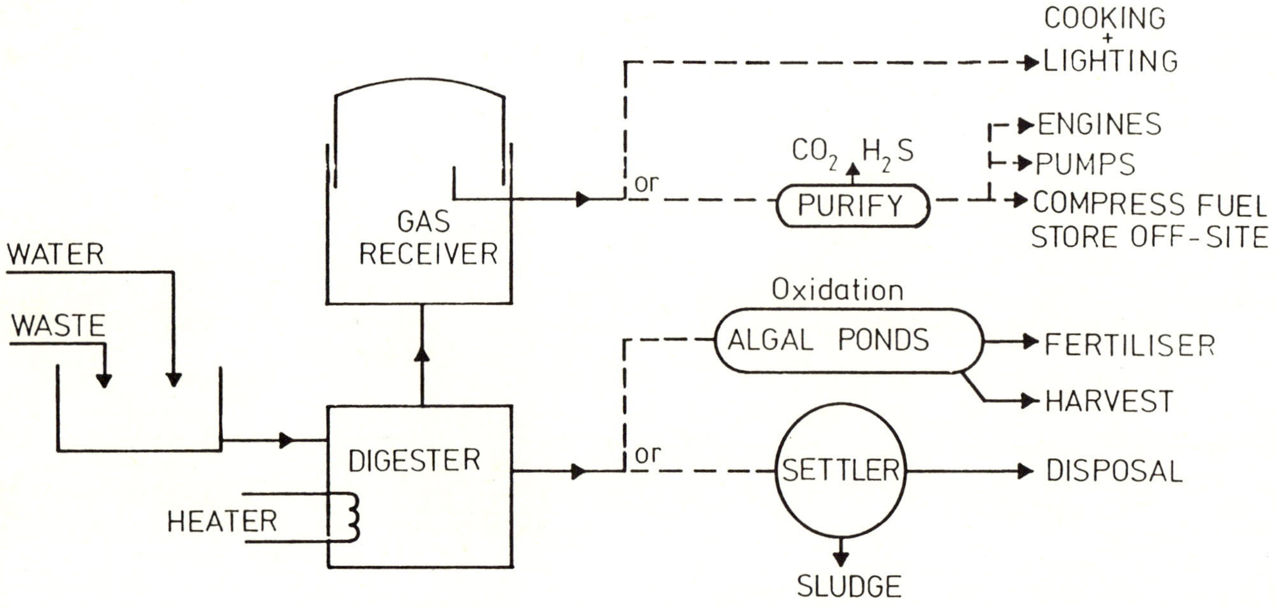

Schematic Diagram of Biogas System

Stoves & Water Heaters

AGAHEAT APPLIANCES Phone: (0952) 3973
Glynwed Domestic & Heating
Appliances Ltd
Aga Works
P.O. Box No.30
Ketley
Telford
Salop TF1 1BR
U.K.

Manufacturers of a range of domestic cooking and heating boilers using solid fuel. Some of these incorporate heat storage to give consistent cooking conditions. The model CB illustrated, has two hot plates, one fast and one slow and two ovens, one for roasting and baking and the other for slow cooking or warming.

This group also offers the well known "Rayburn" series of solid fuel cookers and boilers.

Aga Kitchen Range

BIRMINGHAM STOVE & RANGE COMPANY Phone: 205 322 0371
Box 2647
Birmingham
Alabama
U.S.A.

Manufacturers of a selection of stoves and ranges, and fires for burning coal and wood for home heating and cooking. Cooking stoves are in cast iron and models are available for wood and coal burning. A number of designs is also available for room heating.

Bonanza – Model 8316 Wood Range

THE ENTERPRISE FOUNDRY CO LTD Telex: 014 2288 Phone: 506 536 1160
Lorna Street
Sackville
New Brunswick
Canada

Long established manufacturers (since 1872) of a range of wood or coal burning and oil fired stoves and cooking ranges.

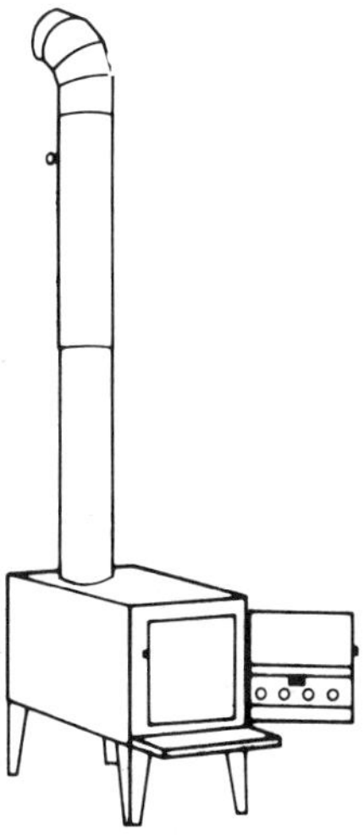

Enterprise BH-25 Wood Heater

FATSCO Phone: 616 926 7795
251-3 N Fair Avenue
Benton Harbor
Michigan 49022
U.S.A.

Long-established manufacturers of a range of small solid fuel stoves for heating and cooking purposes.

These stoves can burn almost any solid fuel and are quite small and relatively inexpensive. The smaller ones can be despatched from the manufacturers by ordinary postal delivery. All stove parts are individually replaceable and some stoves are available with a choice of black or stainless steel bodies. Most stoves have heat shields and are suitable for use in small boats as well as for land-based applications.

Fatsco Small Solid Fuel Stoves

JØTUL WOODSTOVES Phone: 0239 710100
Simon Thorpe
New Road
Newcastle Emlyn
Dyfed SA38 9BA
U.K.

Agents and suppliers of a range of Norwegian-built wood burning stoves for space heating and cooking purposes. Some models can be provided with a water heating kit for limited supplies of hot water. High efficiency is claimed for this range.

Jøtul 602

MALLEABLE IRON RANGE COMPANY Phone: 414 887 8131
715 North Spring Street
Beaver Dam
Wisconsin 53916
U.S.A.

Manufacturers of a range of cooking stoves for burning wood, coal or oil.

Monarch Wood-Coal Kitchen Range Model R9CW

ETS LARROQUE Phone: 40 22 47
401 Route de Seysses
31300 Toulouse
France

Manufacturers of wood burning stoves in a range of sizes. Unit is suitable for workshop or outhouse situation and due to simplicity is relatively cheap. It can be used for space heating or water heating.

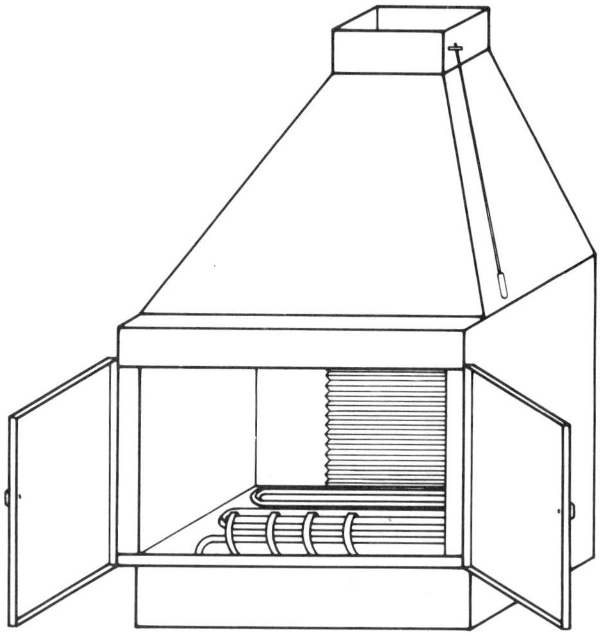

Larroque Wood Burning Heater

SCANDINAVIAN STOVES INCORPORATED Phone: 603 835 6029
Box 72
Route 12-A
Alstead
New Hampshire 03602
U.S.A.

Suppliers of a range of solid fuel (wood or coal) stoves, some of traditional Scandinavian cast iron design and some of modern design.

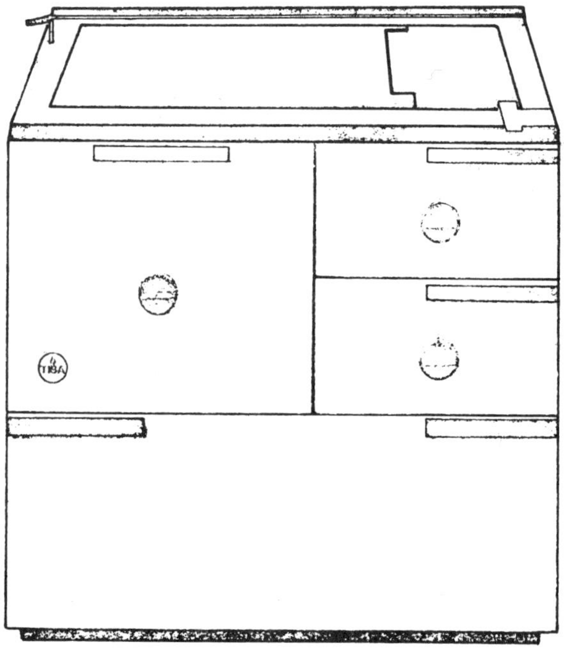

Tiba

Riteway Wood and Coal Heater

RITEWAY MANUFACTURING COMPANY
P.O. Box 6
Harrisonburg
Virginia 22801
U.S.A.
Phone: (703) 434 7090

Manufacturers of a range of efficient wood burning and dual fuel stoves and furnaces for space and water heating.

Model	*Fuel Capacity*	*Rating*	*Shipping Weight*	*Comments*
2000	4 cu.ft.	50 000 BTU/hr	200lb	Burns logs up to 24″ long
37	7½ cu.ft.	73 000 BTU/hr	400lb	Burns 24″ logs or coal

Plus a range of furnaces and boilers from 160 000 to 350 000 BTU/hr ratings.

TORRID MANUFACTURING CO. INC
1248 Poplar Place South
Seattle
Washington 98144
U.S.A.
Phone: 206 324 2754

Manufacturers of a wood burning stove for cooking and water heating. Unit is a traditional style and can be incorporated with a flue oven for baking or flue space heater for room heating.

Triumph Wood Stove

SHENANDOAH MANUFACTURING CO INC.
Box 386
Harrisonburg
Virginia 22801
U.S.A.
Phone: 703 434 3979

Manufacturers of various solid fuel stoves and heaters with thermostatic control to minimise fuel consumption. Models for mixtures of fuels or for wood and for coal/ coke are available.

Model R75 (illustrated) has a fire-brick lining and will burn for 12 hours on a single charge (when damped down). Shipping weight is 178lb (81kg).

Model R75

UNITED STATES STOVE COMPANY
P.O. Box 151
South Pittsburg
Tennessee 37380
U.S.A.

Phone: (615) 837 8631

Manufacturers of a large selection of stoves for heating and cooking purposes.

Models can be run on wood, coal, charcoal, oil, kerosene.

Wonderwood Stove

WATER MASTER SYSTEMS LTD
Coroglen Via Thames
New Zealand

Phone: Whenuakite 858

Manufacturers of a water heater using any fuel, be they waste oil, wood or any combustible waste. Claimed to heat 40-160 gallons per hour and a half. Also supply solar heating systems.

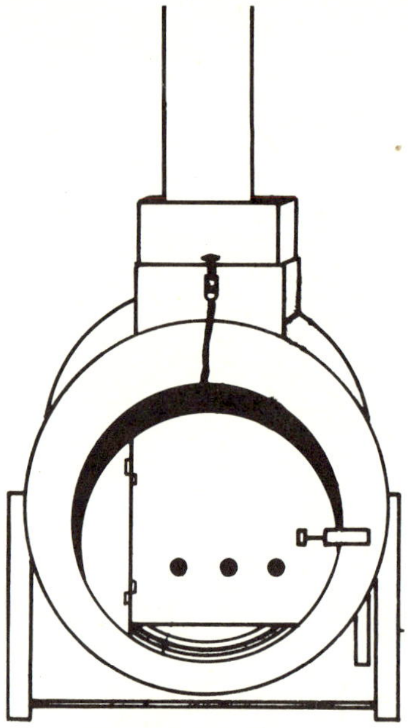

Water Master Heater

VALLEY COMFORT
Winlaw B.C.
Canada VOG 2JO

Phone: 604 226 7221

Manufacturers of cased wood burning stoves for home heating by radiation, natural convection or forced convection.

Model No.	*Fuel*	*Log Size (ft)*	*Chamber Lining*	*Max heat output BTU*	*Weight (lb)*
C26	Wood	1.5	*Stainless steel*	–	170
C31	"	2	" "		195
F51	"	2½	*Fire brick*	80-100 000	850
F71	"	3	" "	100-130 000	1050
RB3D	"	2½	" "	90 000	475
FB4D	"	2½	" "	120 000	550

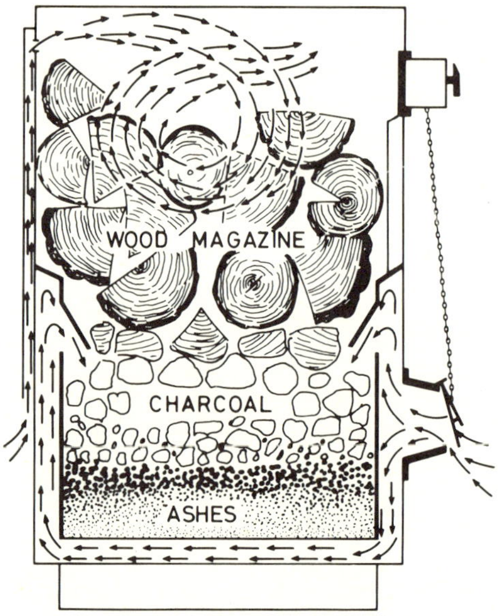

Valley Comfort Wood Stove

WAVERLY HEATING SUPPLY COMPANY
117 Elliott Street
Beverly
Massachusetts 01915
U.S.A.

Phone: 617 922 0581

Manufacturers of a wood burning, totally enclosed stove, of a modern design. Units will accept 20″ logs. The steel combustion chamber is encased in cabinet to provide convectional movement of heating air.

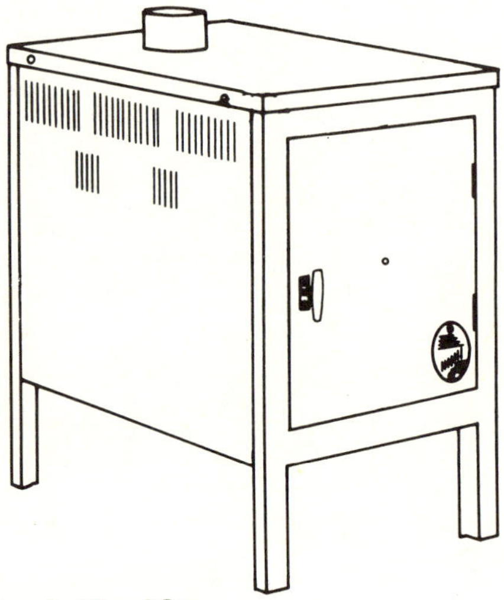

Waverly Wood Stove

Heat Pumps and Regenerators

BLAZING SHOWERS Phone: 707 882 9956
P.O. Box 327
Point Arena
California 95468
U.S.A.

Manufacturers of a water heater utilising the waste heat in the flue of a stove.

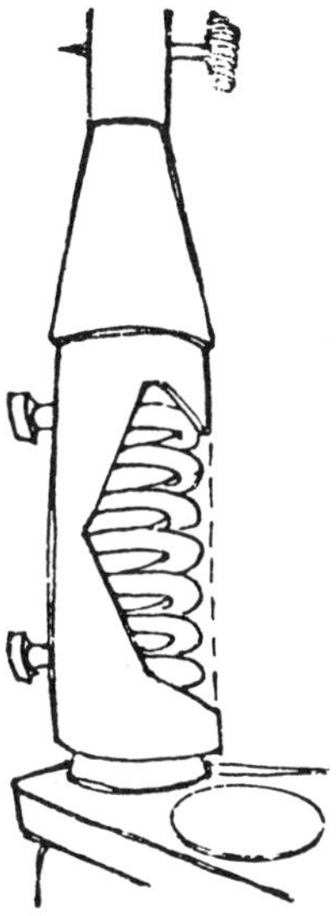

Flue Pipe Water Heater

DOLIN METAL PRODUCTS INC. Phone: 212 596 1400
475 President Street
Brooklyn
New York 112115
U.S.A.

Manufacturers of a flue heat reclaimer for recovery of heat that otherwise is lost up the chimney. Fan blown air is blown out of the unit and can be ducted to where required.

The fan requires a 110V, 60Hz electricity supply (a transformer can allow use on 240V systems or motor can be changed to one suitable for local conditions). Power consumption for the fan is 80W. Heat output is 11 000 to 16 000 BTU/hr (3.2 to 4.7kW).

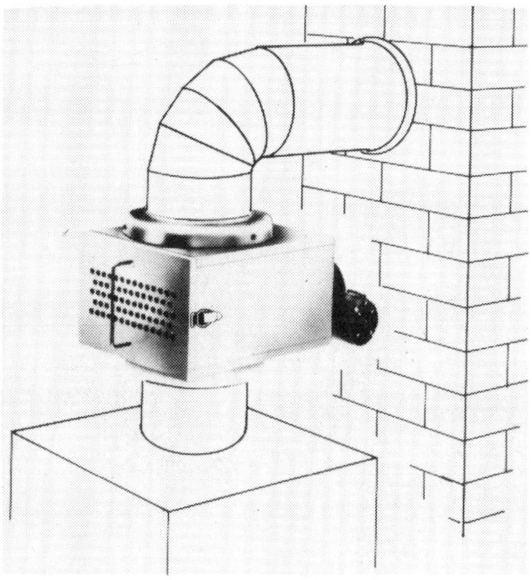

Dolin Flue Heat Reclaimer

CHIMNEY HEAT RECLAIMER CORPORATION Phone: 203 628 4738
53 Railroad Avenue
Southington
Connecticut 06489
U.S.A.

Manufacturers of a flue heat reclaimer to utilise heat normally lost up the chimney and also a wood fired boiler that incorporates the same system. The heat reclaimer incorporates an 80W electric motor to drive its fan for air circulation.

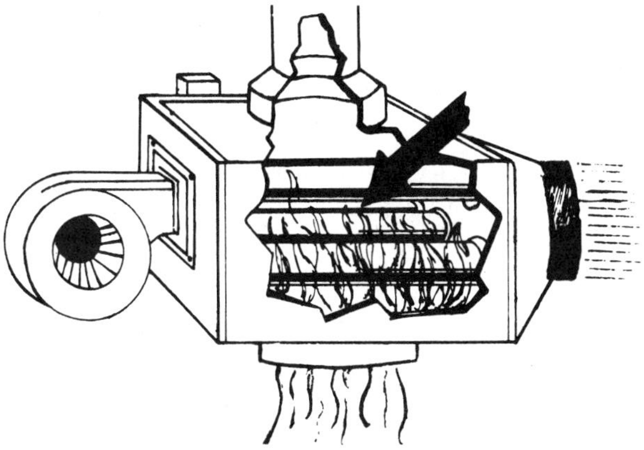

Chimney Heat Reclaimer

CURWEN & NEWBURY LTD Phone: (0373) 823646
Westcroft Works
Alfred Street
Westbury
Wiltshire BA13 3DZ
U.K.

Manufacturers of an air to air heat recovery system to reduce the heat lost from buildings due to forced ventilation. The unit extracts the heat from the outgoing air and warms the incoming fresh air. The unit can also be used in the reverse direction to minimise the loss of cool air from air conditioned buildings.

(Models available for high temperatures)

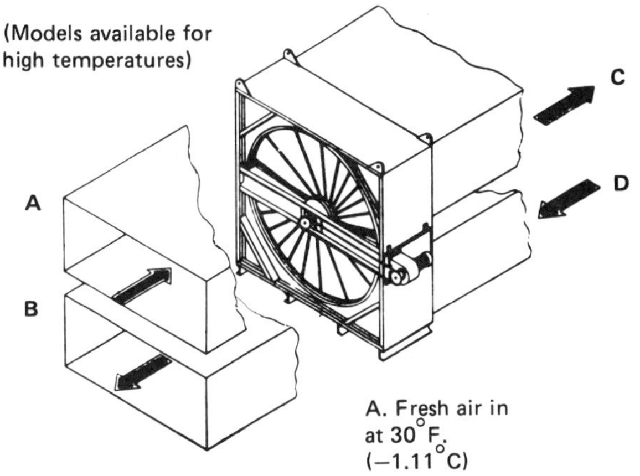

A. Fresh air in at 30°F. (−1.11°C)

B. To outside at 39°F. (3.89°C)

C. Fresh air with regenerated heat at 66°F (18.9°)

D. Contaminated exhaust air at 75°F. (23.9°C)

C.N. Heat Regenerator

ISOTHERMICS INC
P.O. Box 86
Augusta
New Jersey 07822
U.S.A.

Telex: 710 988 2292
Phone: 201 383 3500

Air to air heat recovery system for exhaust flows up to 600°F and ventilation rates of 750 cfm to 12500 cfm. Some units are produced for liquid to air heating. Air conditioning units and stove flue pipe heat recovery systems are also produced.

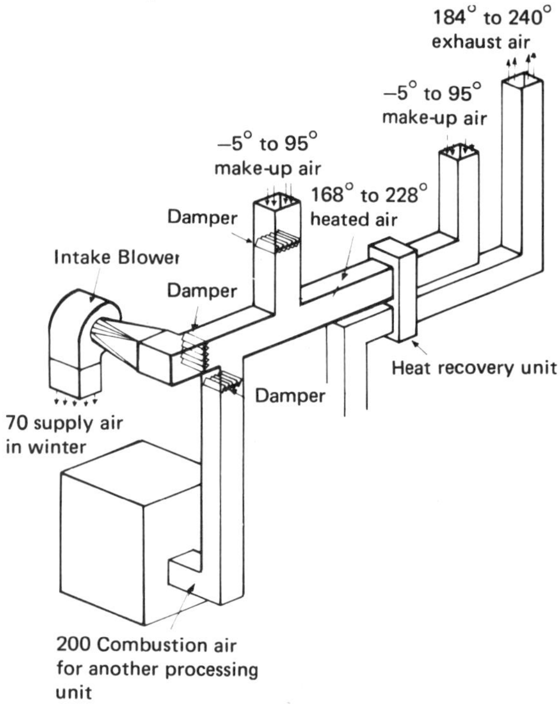

Iso-Vent

LENNOX INDUSTRIES LTD
P.O. Box 43
Lister Road
Basingstoke
Hants. RG22 4AR
U.K.

Telex: 858675
Phone: 0256 61261

This company is a major manufacturer of heat pumps (with additional branches in Germany, the Netherlands, France, the U.S.A. and Canada).

Heat pumps transfer heat from one temperature to another and therefore can be used either for heating or for cooling. The heating and cooling effect in terms of energy transfer is invariably considerably greater than the driving power needed since the heat source/sink is normally the atmosphere. Systems are available for the efficient heating and/or cooling of a variety of spaces ranging from private homes and small offices to large commercial/industrial premises. This is a much more efficient method of heating than the direct conversion of energy into heat.

Eleven heat pump models are available with nominal cooling capacities ranging from 7 to 39kW and nominal heating capacities from 6 to 32kW. The actual performance depends on a wide variety of factors, not least being the ambient conditions. The power requirement is generally between 0.25 and 0.5 of the actual output, again depending on specification and ambient conditions.

Lennox Unit

Methane Digesters

BIO GAS OF COLORADO INC.
5620 Kendall Court
Unit G
Arvada
Colorado 80002
U.S.A.

Phone: 303 422 4354

Designers of bio-gas systems for a range of installations. This organisation offers consultancy and design services in the field of bio-gas systems. These are developed for the American farmer and they include larger systems – for example a set of workshop plans are available for four digestion units sized from 2,250 to 18,000ft^3 capacity.

Colorado Demonstration Bio-Gas Unit

FARM GAS LTD
Heath Workshop
Lydham
Bishops Castle
Salop SY9 5HB
U.K.

Phone: (05883) 348

Manufacturer of two main types of methane digester, a 2500 gallon (11.25m^3) capacity system for farms and a small 400 gallon (1.8m^3) system for domestic wastes, garden wastes or for experimentation.

Specification: Model FG25, capacity 2500 gallons. Can handle slurry from 125 pigs, 16 cows or 2000 laying hens (slurry contains washing water as well as wastes). Gas production from 600ft^3 (16.8m^3) down to 400ft^3 (11.2m^3) per day depending on input and conditions. 400ft^3 output represents a 3kW gas heater running 24 hours/day, running a generator producing 1kW of electricity 24 hours/day or producing 200 gallons (900 litres) of hot water per day.

Farm Gas Ltd Digester System

SANAMATIC TANKS PTY LTD
P.O. Box 119
Bendigo
Australia 3550

Cables: Sanamatic Bendigo
Phone: 267 1464 (Melbourne)

Manufacturers of a range of bio-gas plants developed over a thirty year period by John Coulthard, for utilizing 500lb to 120 tons of effluent per day. The smaller unit comes complete with all necessary fittings and instructions for use for domestic purposes. This manufacturer is prepared to negotiate license manufacturing arrangements overseas.

The Sanamatic Tank is claimed to be of sufficiently low cost to pay for itself in one to two years when used in Australia. It consists of a specially developed butyl rubber liner suspended in a metal supporting frame, and is relatively portable and easily erected.

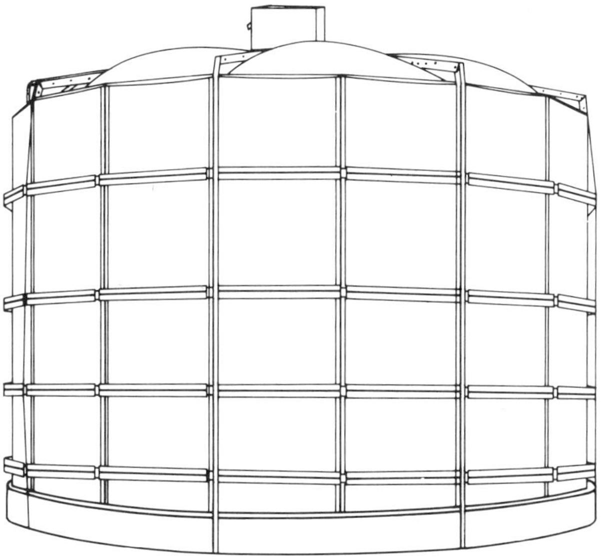

Sanamatic tank

Sectional View of Khadi and Village Industries Commission Model Gas Plant

KHADI & VILLAGE INDUSTRIES COMMISSION
Gramodaya
3 Irla Road
Vile Parle (West)
Bombay 400 056
India

Cables: Khadigram
Phone: 571323-9

Designers of bio-gas plants for construction by approved makers. These plants consist generally of a brick-lined pit with a fabricated steel gas holder suspended over it. Detailed drawings are also provided for people or groups who wish to construct their own units. Technical personnel to aid in such ventures can also be provided on a consultancy basis.

The following sizes of plant are available:

Capacity (m^3)	*2*	*3*	*4*	*6*	*8*	*10*	*15*	*20*
Approx. No. of cows required	2-3	3-4	4-6	6-10	12-15	16-20	25-30	35-40

Capacity (m^3)	*25*	*35*	*45*	*60*	*80*	*140*
Approx. No. of cows required	40-45	45-55	60-70	85-100	100-140	400-450

TUNNEL CO LTD Phone: Fort Ternan 7Y7
Tunnel Estate
P.O. Fort Ternan
Kenya

'Hutchinson' Methane Plant

Manufacturers of the 'Hutchinson' methane plant which has been developed over a period of twenty years. This manufacturer is interested in licensing arrangements to allow its products to be manufactured in other countries, since local manufacture is generally likely to be most economic for large and bulky systems such as methane digesters.

The following systems are available:

A. Continuously fed systems

Livestock requirements	*Plant*	*Bio-gas output per day (ft^3)*	*Size of gas holder (ft^3)*	*Comments*
5 cows	MK III No.1	25	50	Supplied complete plus gas light.
10 "	Mk III No.2	50	50	Complete plus cooker and light.
15 "	Mk III No.3	75	100	Complete plus cooker and light.
20 "	Mk III No.4	100	100	Complete plus cooker and light.
40 "	2x Mk III No.4	200	200	Complete plus cooker and light.
50 "	Mk II Domestic	450	300	Kit and plans supplied.
80 "	Mk II Large	750	400	Kit and plans supplied.
150 "	Mk II 2 Lid	1500	400	Kit and plans supplied.
200 "	Mk II 3 Lid	2000	400	Kit and plans supplied.

It is stressed that gas is a by-product and that the sludge from 5 cows will, in addition to producing the equivalent of 2 x 30lb propane cylinders per year, fertilise one acre (= 0.4 ha).

B. Batch fed plants

Each plant consists of a number of compartments each of which holds 300ft^3 of organic wastes and yields on average 75-100ft^3 of bio-gas per day. Each compartment is recharged every eight weeks. The residues from each compartment will treat 5 acres per year (= 2 ha). It is suggested that batch plants be used in conjunction with one of the continuous ones listed above in order to provide a supply of the active innoculum (active bacteria) to start the batch when a fresh load is introduced.

Mk I Plant 3 compartment 300ft^3 gas holder, supplied as kit plus plans.

Mk I Plant 4 compartment 300ft^3 gas holder, supplied as kit plus plans.

Mk IV Plant is a small batch plant, supplied ready built with an output of 30ft^3 gas per day.

This company also produces solar water heaters (see Section 1).
section).

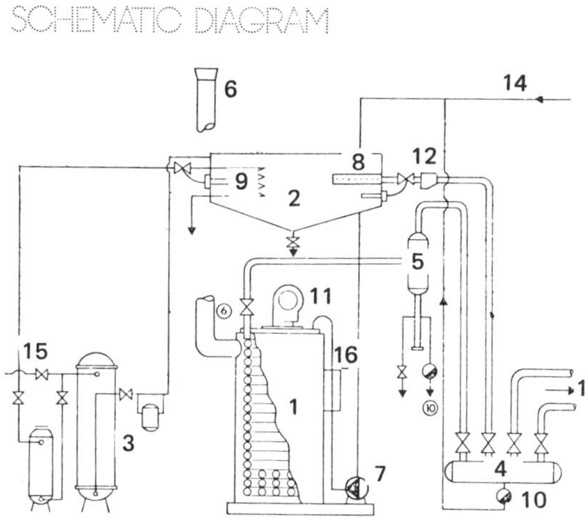

Dunlop Rapid Steam Generator

DUNLOP LTD
Energy Equipment Division
Holbrook Lane
Foleshill
Coventry CV6 4AA
U.K.

Telex: 31677
Phone: 0203 88733

Manufacturers of a range of steam generators to produce dry saturated steam at up to 120 atmospheric pressure. Units are gas, oil or dual fuel fired. Air requirements are less than $0.6m^3$/hr and the unit runs at 80% thermal efficiency.

1. Steam Generator
2. Feed Water Tank
3. Water Treatment Plant
4. Flow Distributor
5. Water Separator
6. Chimney
7. Feed Water Pump
8. Steam Lance
9. Cooling Coil
10. Steam Traps
11. Burner
12. Filter
13. Steam for use
14. Condense Return
15. Water Supply
16. Control Box

STONE-PLATT CRAWLEY LIMITED
Boiler Products
Gatwick Road
P.O. Box 5
Crawley
West Sussex RH10 2RN
U.K.

Telex: 877481
Cables: Stonelec crly
Phone: 0293 27711

A series of modern gas or oil-fired coil tube steam raising boilers giving 5 minute start up and quick response to load variations, complete with automatic control system. The manufacturers say the modular construction allows easy replacement of boiler tubes and other components, and that the various units are compact in arrangement.

Stone Vapor Boiler

Model	*Output nominal hp*	*Steam production at 100°C (212°F)*	*Fluid capacity*	*Heating surface*	*Dimensions L W & H*	*Weight (kg)*
7211	35	545 Kg/hr	36.8 l	$8.0m^2$	1686mm 1110mm 1550mm	952
7217	60	948 Kg/hr	43.2 l	$9.7m^2$	2100mm 1356mm 1940mm	1520
7227	100	1490 Kg/hr	75.0 l	$14.4m^2$	2266mm 1220mm 2010mm	1950
7245	150	2440 Kg/hr	126 l	$23m^2$	2860mm 1415mm 2120mm	3100
7260	200	3130 Kg/hr	207 1	$29m^2$	2910mm 1650mm 2330mm	3774

Design pressures in the range 13 to 79 bars (188 to 1150lb/in^2).

SEMPLE ENGINE CO. INC Phone: 314 961 6244
Box 6805
St. Louis
Missouri 63144
U.S.A.

Vertical tubular boilers burning coal or wood intended for use with steam engines.

Model	FT-40	FT-80	FT-160	FT-240	FT-480
Horsepower	5	10	20	30	60
Height (cm)	107	127	152	163	183
Shell diam. (cm)	46	63	81	99	120
Heating surface (m^2)	3.7	7.9	15.9	22.8	45.6
Operating (psig)	200	185	185	185	185
Pressure (bar)	13.8	13	13	13	13
Weight (kg)	304	508	980	1501	2485

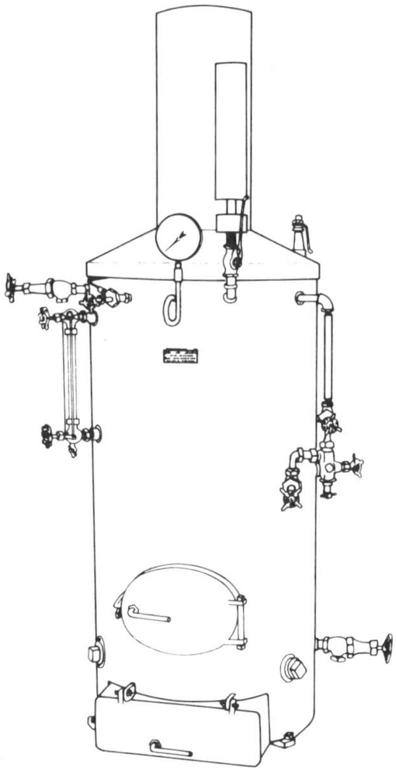

Semple Boiler

FREEHEET BURNERS
c/o Steam Power
106a Derby Road
Loughborough LE11 OAG

Manufacturers of an oil burner suitable for any oil from waste oil, such as old engine oil, to light fuel oils. Unit uses compressed air atomisation of the fuel (or steam atomisation if neither electricity or compressed air are available). There are five sizes of burner providing 7 000 to 1 925 000 BTU per hour. The company also produces warm air convectors for space heating and grain drying.

Freeheet Oil Burner

O'CONNOR ENGINEERING LABORATORIES INC
100 Kalmus Drive
Costa Mesa
California 92627
U.S.A.
Telex: 685 641
Phone: 714 979 3993

Watertube boilers with outputs of 300-10,000 pounds of steam per hour. The 1,000 pound per hour unit operates at 350 psig and 550°F with an overall thermal efficiency of 80%. Units are suitable for marine or industrial use.

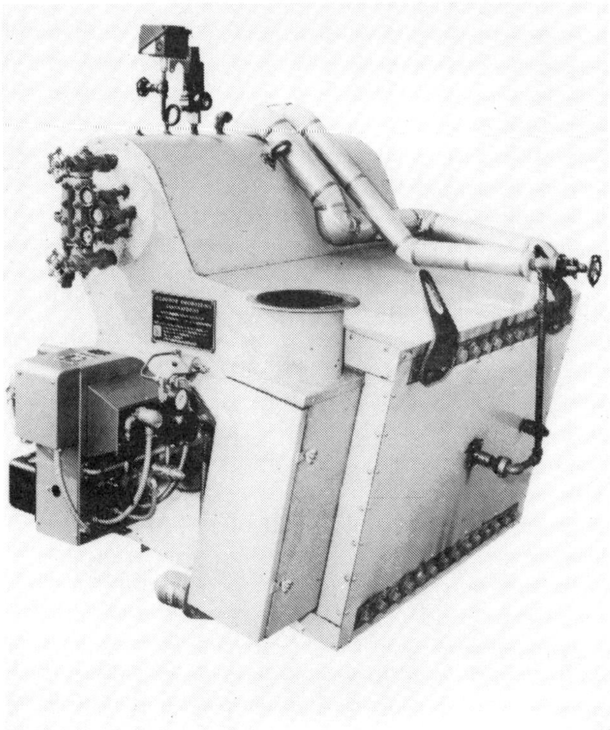

O'Connor Boiler

CECOCO AGRICULTURAL & SMALL INDUSTRIAL CENTER
Chuo Boeki Goshi Kaisha
P.O. Box 8
Ibaraki City
Osaka Pref
Japan 567

Cables: CECOCO Ibaraki

Suppliers of two gas producer plants developed in the early 1950's.

These units can be used to power a specially adapted car engine. Stationary units have an output of up to 100hp while the smaller units have an output of about 10hp and are suitable for motor transport.

These systems may be fuelled with wood chips, charcoal, coconut shells or other similar materials.

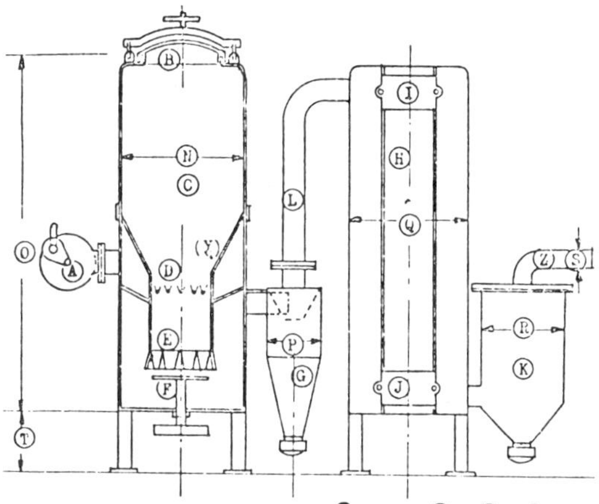

Cecoco Gas Producer

Dimension of Gas Producer (mm)

Type	*N*	*O*	*P*	*Q*	*R*	*S*	*T*	*Avail. Engine*
1	3500	900	1500	350	2000	40A	150	18-20HP
2	3000	900	1500	300	2000	40A	150	12-15HP
3	3000	700	1200	300	1800	32A	100	10-12HP
4	2500	650	1200	250	1800	32A	100	4- 8HP

Thalopat Wood Gas Producing Plant

THALOPAT A.G.
51 Feldstrasse
CH — 8400 Winterthur
Switzerland

Cables: Thalopat
Phone: 052 22 32 22

Manufacturers of a range of wood gas producing systems with outputs of 50,000 to 1,500,000 kcal/h using waste wood or other volatile wastes as fuel. Furnaces using oil firing can be changed to wood gas fired operation simply by changing the burner.

1kg of fuel oil is approximately 10,000 kcal which is equivalent to 3kg of waste wood. The larger systems have automatic or semi-automatic fuel feed.

Applications of systems include electricity generation using specially adapted engines, saw mill operation (using wastes as fuel), ploughing (with special cable-drawn plough), etc.

5

Internal Combustion Engines

DIESEL (COMPRESSION IGNITION) ENGINES
SPARK IGNITION ENGINES (PETROL AND GAS FUELLED)
GAS TURBINES

By far the most common power source in the 5 to 500 hp range is the internal combustion (i.c.) piston engine in either of its two main forms, the diesel (compression ignition) (c.i.) engine or the petrol (spark ignition) (s.i.) machine. Various rotary internal combustion engines (without pistons) and gas turbines have appeared in recent years, but remain relatively rare in this power range. And only one — a gas turbine — is included in this edition.

The main reasons for the success of the i.c. engine are its relatively high power/weight ratio, compact size and its "instant" startup capability, which led to its universal adoption for powering road vehicles, smaller ships and boats and small and mobile agricultural and industrial machinery. Mass production for these purposes made i.c. engines cheap, so they also became the primary power source for stationary applications such as electricity generating or pumping, where compactness and light weight are less important.

The internal combustion process, in which the fuel is burned inside the engine's power cylinder(s) was responsible for allowing compactness and light weight, as it eliminated the large boiler, with extensive heat transfer surface built into it, that is needed for the i.c. engine's primary predecessor, the external combustion (e.c.) steam engine. The other main external combustion engine, known as the Stirling engine after its inventor, uses air rather than steam as its working medium, but also requires sizeable heat transfer surfaces. However the internal combustion engine's main virtue is also its weakness, as in order to arrange easy combustion within the working cylinder it is necessary to use a clean and readily ignited fuel such as oil or gas; therefore i.c. engines are invariably dependent on petroleum-based fuels which are becoming increasingly expensive and scarce in many parts of the world. Therefore, hitherto almost extinct external combustion rivals described in the next chapter may be at the beginning of a period of revival for certain applications due to their ability to run on anything which burns such as wood, coal or agricultural wastes.

Some general characteristics of i.c. engines

This is not the book to describe the technicalities of engines in any detail. However, when comparing different commercially-available engines it soon becomes apparent that there is a lot of variation in types available even of one particular power output.

The two main categories which are differentiated in separate parts of the following section are diesels and petrol/gas engines. The former rely on the temperature increase caused by the sudden compression of air in the cylinder(s) to ignite a jet of fuel droplets sprayed into the cylinder at the appropriate instant (compression ignition) while the latter generally have the fuel mixed with the air prior to entering the cylinder and the mixture is ignited in the cylinder by an electrical discharge at a spark plug (spark ignition). Diesel engines therefore tend to be relatively heavier and more robust in construction in order to allow the high pressures required for compression ignition to be sustained and they also require a high-precision injection pump capable of metering and injecting the fuel into the cylinder at high pressure. Petrol engines tend to be lighter and more compact for their power output and they are also somewhat cheaper to manufacture. However the diesel is inherently more efficient and often has a longer life and better reliability than a spark ignition engine of comparable power. Their greater efficiency is because the ignition method allows the engine to operate reliably with a weaker mixture (a spark ignition system requires a minimum concentration of fuel to allow the flame to spread from the spark plug) and in addition, the higher compression ratio of the diesel effectively allows it to "breathe more deeply". The compression ratio for a petrol or s.i. engine must be limited to prevent premature ignition caused by the heat of compression, so it cannot breathe so "deeply", to use a valid human analogy.

Therefore, the petrol engine tends to be cheaper, lighter, more compact but thirstier for fuel than a diesel of the same power. As a result, private cars have petrol engines which allow a sportier performance and lower first cost, while commercial vehicles usually are diesel powered to give them better fuel economy, a longer working life and higher reliability. For stationary applications, the main virtue of petrol engines is where light weight is important to allow easy portability, or low initial cost and simplicity are important.

A general characteristic following from this for both types of piston i.c. engine is that the smaller and lighter it is for a given power output the cheaper will be its first cost and the shorter will be its life. Obviously long-life equipment generally requires better-quality materials and manufacturing, which invariably cost more, but the simplest way to get more power from a smaller engine is to design it to run at a higher speed. The faster a given size of engine runs, the more air it can breathe and the more fuel it can burn – consequently as the power is closely related to fuel consumption, the more power it will produce. However, the life of an engine both between major overhauls and *in toto* is related to the total number of revolutions – so a fast revving engine will obviously wear itself out faster than a larger, slower revving one of the same power.

The consequence of this is that it is best to use a large, heavy, slow-running engine for any applications requiring continuous reliable operation for long periods, while a cheap, compact high speed machine may be appropriate for intermittent applications or where portability is important. It will be shown later that it is a mistake to consider first or capital cost as a primary choice criterion in many cases, as the actual running costs depend equally or even more on such aspects as total engine life (ammortization period), maintenance costs and fuel costs.

It follows from this that it is also better to run an engine at a lower power than its "maximum rated power" to prolong its life – indeed in many cases such as high altitude or hot locations, it is essential to "de-rate" the engine to a lower power by using an engine of greater maximum rated power than the load requires. A typical rating factor might be 0.8, implying that to obtain a horsepower of P we must use an engine of $\frac{0}{0.8}$ P rated output, = 1.125P. Another reason to derate the engine is that the maximum efficiency (that is the minimum fuel consumption relative to the power output) is usually obtained at a power output and speed two thirds approximately of that corresponding with the peak power output – so as well as extending the life of the engine, much better fuel economies may be obtained. Manufacturers will generally advise on optimum rating factors to suit your requirements, but if in doubt, it is wiser to derate too much rather than too little.

Other factors that become apparent when studying the various engine options available are as follows:-

a. Four-stroke or two-stroke

Both s.i. and c.i. engines can be designed to run in such a way that ignition takes place either every other revolution (four-stroke or four-cycle) or once per revolution (two-stroke cycle). Four stroke engines tend to be more efficient and slower revving while two-strokes are simpler to build and have a high power/weight ratio that makes them suitable for small-scale low cost applications. Many s.i. two-strokes are lubricated by mixing oil with the petrol supply – this removes the need for oil changes but is rather wasteful, tends to cause a smoky exhaust and there is a risk of an inexperienced operator trying to run it on neat petrol and thereby damaging it.

b. Air cooled or liquid cooled

All types of internal combustion engines require some form of cylinder cooling to prevent overheating. This can be arranged either via a water jacket and radiator (as used for most cars) or by blowing air past a series of cooling fins (as on most motor-cycles). Each method has its pros and cons. Water cooling is more compact in many cases for higher powered engines and the water jacket tends to deaden the engine noise and allow slightly quieter running – also the running temperature is readily controllable by a thermostat which can regulate the water flow. However water can leak out or freeze (the latter can be prevented by using anti-freeze in winter) and it is obviously an important regular maintenance function to check coolant levels with a liquid cooled engine. As loss of coolant generally causes severe engine damage if the machine is allowed to continue running, various safety devices can be fitted to warn of overheating for that or any other reason; obviously they are worth fitting to unattended engines and are often supplied as standard. Air cooled engines cannot lose their coolant, but it is important to ensure that the cooling fins do not get clogged with dust or dirt and they generally require a large and powerful fan to drive sufficient cooling air. This fan is often belt-driven – so the belt needs to be kept correctly tensioned and in good condition. Small, light-weight portable engines tend to be air cooled as this allows a simpler more compact design. Therefore air cooling is more common with small engines. There is little to choose between air and liquid cooling for medium engines and most large machines are liquid cooled.

c. Cylinder arrangement

Most engines have vertical cylinders with the crankshaft below. For various reasons, generally dictated by the need for compactness when engines were developed for automotive use, certain multi-cylinder engines have horizontal cylinders opposing each other in pairs, or a vee arrangement. Multi-cylinder engines tend to run more smoothly and quietly than ones of the same power with fewer cylinders, but they also tend to be more expensive (having more components), to be harder to maintain but to be rather more compact.

d. Special features

Many of the engines described can be supplied just as a basic engine lacking essential accessories (which can then be chosen from other sources). More often they are supplied complete with a starter motor, electrical system for charging a starter battery (or this can be done via a mains voltage battery charger where the engine drives a mains voltage generator), cooling radiator (with liquid cooled machines), etc. They are often offered mounted on skids and in many cases as a complete generating or pumping set. A number of generating sets of this kind is described in Section 7. In such cases a choice of generator, electrical output, instrumentation and other ancillaries is commonly offered and described in detail in the manufacturers' brochures. A common optional extra with larger diesels in the range considered is supercharging. This artificially forces a greater volume of air and fuel through the engine to boost the power output of a given size of engine and it can be useful at high altitude locations.

e. Other important criteria to consider

Probably the most important point when choosing an engine is the availability of spare parts and servicing skills — the most efficient machine to an ideal specification is useless when you cannot get spares for it and very expensive if an enormous spares inventory is required due to its rarity locally. Hence the presence of an efficient and helpful local agent should usually be a vital first consideration. The section at the end of this book on agencies, while not totally comprehensive, will, it is hoped, give some guidance in this direction. If no agent is listed for a certain engine in your country, it may well still be worth enquiring of the manufacturer in case one exists that we were not aware of. It is worth seeking the opinion of other engine users in your country as agents have a vested interest in praising their wares — and many have agencies for several rival engines and may for good or bad reasons favour one or other type.

The economics of internal combustion engines

The i.c. engine has two major disadvantages in most non-industrial countries, namely (a) it and its spares usually have to be imported and (b) it uses petroleum-based fuel, so obviously it is a heavy consumer of hard currency which is often scarce. Consequently its widespread use does more for the economies of the industrial and the OPEC countries than for the economies of the poorer nations. This book indicates that in certain circumstances other alternatives which are less well-known might be worthy of consideration. However, for many years to come, in most parts of the world, it will be difficult to find other power sources in the range under consideration that are quite as convenient or well understood as i.c. engines. In the long run further depletion of oil reserves and increasing oil prices will make the alternatives increasingly attractive. Meanwhile, the main components of engine running costs are:-

- Fuel consumption — (related primarily to power output) and fuel purchase, delivery and storage costs.
- Capital cost which is related to size and power output and interest rates governing the annual charge on the investment.
- Lubricant consumption (generally from 1 to 5% of fuel consumption) and lubricant purchase, delivery and storage costs
- Operational life of engine — this varies from as little as 1000 hours for very small, high-speed, light weight two-stroke engines, to perhaps 2000 to 3000 hours for medium spark ignition and small diesels, to five or ten years for the heavier duty medium-sized petrol and diesels, to decades for some of the larger, slow-speed diesels if they are carefully maintained. Calculations can be badly upset if bad maintenance causes premature destruction of an engine.
- Repair and maintenance charges.

Vinod Mubayi and Tien Le* have shown in a detailed comparative analysis of the costs of various alternative energy sources when applied to pumping irrigation water, that an equation can be derived to estimate the approximate costs of operating small diesel engines in early 1977 as follows:-

$$e = \frac{16.275}{4h}\left(\frac{700}{P} + 110\right) + 0.322C_f + \frac{3}{P}\text{ US cents hp-h}$$

where

h = hours operation per year
P = maximum rated power output in hp
C_f = total delivered cost of fuel in US cents per litre

Obviously this is a gross approximation based on numerous assumptions described in the paper concerned, but it does allow certain interesting points about the relative effects of various cost elements to be seen. Notably, that even at low fuel prices, fuel costs become dominant compared with capital costs when engines are utilised for periods in excess of about 1500 to 2000 hours per year (there is a total of 8760 hours in a year); obviously this applies even more so with higher fuel costs more common outside the USA. Hence an apparently "cheap" engine in capital terms is often more expensive to run than a higher first cost alternative, if fairly heavy utilisation is envisaged, while a low cost machine may be best for rare operation such as for stand-by duty.

In order to give an initial indication of the character of the various engine models offered by each manufacturer listed in this guide, in addition to the major sub-division of engines into compression (diesel) and spark (petrol/gas engine) ignition categories, the reader will find a note of special features such as which are two-stroke (to distinguish them from the majority which are four-stroke). Also, the range of models from each manufacturer is tabulated to show the model number or name, cylinder arrangement, type of cooling, cylinder capacity in cubic cm (a measure of engine size), maximum power output, typical rated output at a speed appropriate for continuous operation, and lastly the weight of the engine which may give an indication both of the first cost and the life of the engine (both being generally related to weight, first cost being roughly proportional to weight and life being proportional to $\frac{\text{weight}}{\text{horsepower}}$).

Lastly, mention should be made of the sole internal combustion engine in this edition which is in the rotary class. This is a small gas turbine, having its primary value in its extremely high power to weight ratio, a factor common to engines of this kind (which makes them so suitable for aircraft propulsion). Small gas turbines have not been very successful in the past due to difficulties in scaling down larger units and the high cost of rather sophisticated components. However, new materials and manufacturing techniques and larger manufacturing volumes as more are produced for aircraft auxiliary power units and ground starters, may lead to further units of the kind described becoming available for small-scale power production. The unit described is in use in a number of places for ordinary electricity generation.

Pros and cons of small turbines are:-

Pro: high power, little vibration, light weight, small size, good reliability given correct maintenance.

Con: expensive first cost, poor fuel economy compared with diesel, rather unusual maintenance skills and spares needed.

Other rotary i.c. engines include the Wankel produced by NSU and Mazda. These are not in widespread use and like the turbine are not as economical as piston engines, but are lighter in relation to their power output.

**Irrigation in Less Developed Countries,* Vinod Mubayi & Tien Le, Policy Analysis Division, Brookhaven National Laboratory, Upton, NY 11973 USA, March 1977.

Diesel (compression ignition) Engines

2800 Cased Stationary Diesel Unit

ALLIS CHALMERS
P.O. Box 563
Harvey
Illinois 60426
U.S.A.

Telex: 910-257-2135
Cables: Founders – Milwaukee Wisconsin
Phone: (312) 339 3300

Water-cooled, turbocharged engines with power output from 60 b.h.p. to 450 b.h.p. Designed for industrial heavy plant irrigation pumping, electricity generation and marine use. Can be supplied as complete generating sets including 75 and 100kW continuous rating.

Smaller sizes of engine include:

Model	*Cylinder*	*Cooling*	*Gross Power (b.h.p.)* *Max.*	*Cont.*	*@ r.p.m.*	*Nett Weight (kg)*
D-262	4	water	77	60	2000	363
2800 Mk1	I	water	85	75	2200	476
2900 Mk2	I	water	135	100	2400	499
3500 Mk2	I	water	175	136	2200	646

Plus five further sizes up to 450 b.h.p.
Various three-phase generating sets available 94kVA/60Hz, 77kVA/50Hz up to 340kVA.

W44 Air-cooled V4 Diesel Engine

BERMOTOR LIMITED
21 London Road
Tunbridge Wells
Kent TN1 1DA
U.K.

Telex: 95446
Phone: (0892) 37588

This is a Renault Company offering the following range of air and water-cooled industrial engines.

Model	*Cylinder*	*Cooling*	*Cont.*	*@ r.p.m.*	*Nett Weight (kg)*
21	I	air	4	3000	51
51	I	air	6	3000	68
71	I	air	11	3000	100
42	2V	air	20	3000	142
44	2V	air	20	3000	197

Water-cooled equivalent industrial power units are also available with power outputs of 10 b.h.p. to 70 b.h.p. Units can be supplied as complete pumping sets.

CHRYSLER CORPORATION
P.O. Box 1
Marysville
Michigan 48040
U.S.A.

Cables: Chrymarine

Water-cooled diesel engines suitable for industrial and automative installations. Some units are specifically designed for marine power use or electricity generation.

Model	*Cylinder*	*Cooling*	*Cap. (cc)*	*Gross Power (b.h.p.)* *Max.*	*Cont.*	*@ r.p.m.*	*Nett Weight (kg)*
IN-433	4	water	2180	61	36	2600	215
IN-633	6	water	3270	92	53	2600	300
CI641-100	6	water	4030	110	83	2800	360
CI655-100	6	water	5460	130	95	2500	427

Model CI641-100 available with brushless generator 60kW @ 3000 r.p.m.

Model IN433

BRITISH LEYLAND U.K. LIMITED
Leyland
Preston PR5 1SN
U.K.

Telex: 67655
Cables: Leymotors Leyland
Phone: (07744) 21400

100 Series

A range of water-cooled engines designed for automotive, marine and construction machinery use. Can be supplied as complete engine gearbox and axle units. A variety of gearboxes and transmissions also available.

Model	*Cylinder*	*Cooling*	*Cap. (cc)*	*Gross Power (b.h.p.) Max.*	*Cont.*	*@ r.p.m.*	*Nett Weight (kg)*
100	4	water	1490	37	27	2750	186
120	4	water	2520	60	41	2250	268
4/98	4	water	3770	75	62	2200	391
6/98	6	water	5655	115	94	2200	490
401†	6	water	6540	138	122	2600	567
AV505†	6	water	8180	151	110	2000	760
410*	6	water	6540	155	136	2600	580
500	6	water	8200	170	150	2600	753
680†	6	water	11100	202	178	2200	953
AV760†	6	water	12470	212	170	2000	1156
510*	6	water	8200	220	194	2200	835
690*	6	water	11100	240	210	220	968

*turbocharged
†available in horizontal configuration (all the rest are vertical)

DAIMLER-BENZ A.G.
7000 Stuttgart 60
West Germany

Telex: 07-23901
Phone: 07-23901

Water-cooled units with power outputs ranging from 28 b.h.p. to 430 b.h.p. Units are designed for automotive, industrial and marine use.

Model	*Cylinder*	*Cooling*	*Cap. (cc)*	*Gross Power (b.h.p.) Max.*	*Cont.*	*@ r.p.m.*	*Nett Weight (kg)*
OM615	4	water	2200	60	38	2600	184
OM616	4	water	2400	65	43	2600	189
OM314	4	water	3780	85	56	2200	302
OM352	6	water	5680	130	68	1800	400
OM352A*	6	water	5680	168	83	1800	410
OM360	6	water	8720	192	119	1800	640
OM346	6	water	10810	210	93	1500	778
OM355	6	water	11580	240	147	1500	857
OM401	6V	water	9570	192	81	1500	616
OM402	8V	water	12760	256	148	1500	756
OM403	10V	water	15950	320	180	1500	915
OM404	12V	water	20910	430	230	1500	1064

*turbocharged

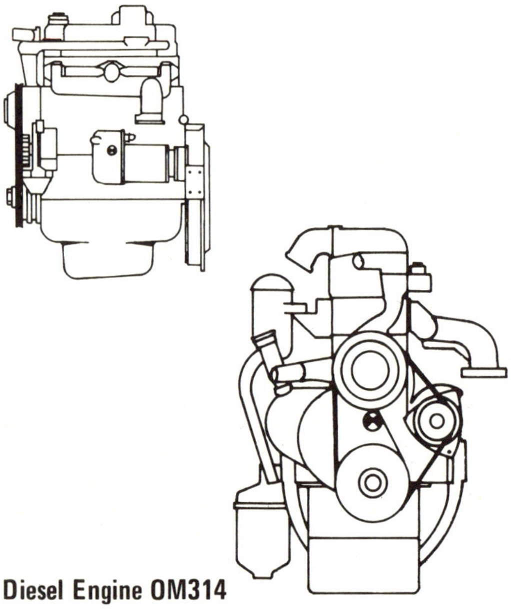

Diesel Engine OM314

CATERPILLAR OVERSEAS SA
118 Rue du Rhone
P.O. Box 408
1211 Geneva
Switzerland

Telex: 22706
Cables: Catoversea-Geneva
Phone: (022) 206222

Range of water-cooled engines 64kW to 970kW. Units can be supplied suitable for truck, marine, agricultural and pumping applications. Most of the range have higher outputs than covered by the scope of this guide.

Model	*Cylinder*	*Cooling*	*Cap. (cc)*	*Gross Power (b.h.p.) Max.*	*Cont.*	*@ r.p.m.*	*Nett Weight (kg)*
3304NA	4	water	7000	100	85	2000	718
*3304T	4	water	7000	165	125	2000	736

*Turbocharged

3304 Marine Propulsion Engine

GM Bedford 220 Fan to Flywheel Diesel Unit

DETROIT DIESEL ALLISON INTERNATIONAL-EUROPE
P.O. Box 6
London Road
Wellingborough
Northants NN8 2DL
U.K.

Telex: 31329
Cables: Genmopower Wellingborough
Phone: (0933) 71122

Water-cooled engines with power outputs from 40 b.h.p. to 1600 b.h.p. Units designed for automotive industrial or marine use and can be supplied as complete (brushless) generating sets, rated at 30, 35, 44 and 71kW, 50 or 60Hz.

Detroit-Diesel Allison fan to flywheel two-stroke diesels with integral injector/pump units

Model	*Cylinder*	*Cooling*	*Cap. (cc)*	*Gross Power (b.h.p.) Max.*	*Cont.*	*@ r.p.m.*	*Nett Weight (kg)*
2-71	2	water	2300	68	40	1500	436
3-53	3	water	2610	78	64	2200	438
3-53*	3	water	2610	97	64	2200	438
3-71	3	water	3496	106	75	1800	693
3-71*	3	water	3496	113	82	1800	693
4-53	4	water	3480	108	87	2200	623
4-53*	4	water	3480	123	87	2200	623
4-71	4	water	4660	160	117	1800	954

*Denotes 4-valve version. Numerous larger models available, plus versions with torque converter and 35, 50, 55, 75 105kVA and larger generator sets.

GM Bedford Diesels (four-stroke)

220F/F	4	water	3614	65	56	2500	356
330F/F	6	water	5420	100	82	2500	460
500F/F	6	water	8200	164	95	1600	600

These models available as fan to flywheel suffix (F/F), complete power unit (P) and marine (M).

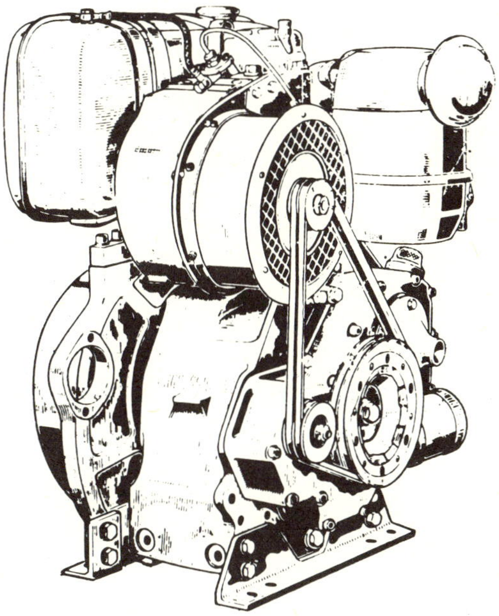

Hatz Diesel Engine

MOTORENFABRIK HATZ KG
P.O. Box 20
D-8399 Ruhstort/Rott
W. Germany

Telex: 057260
Cables: Hatzmotor
Phone: 08531-3022

Manufacturers of a range of four-stroke air-cooled diesel engines with power outputs of 1.5 hp to 80 hp. Units are suitable for automotive or stationary installations.

DA Series Air Cooled Engine

DORMAN DIESELS LIMITED
Tixall Road
Stafford ST16 3UB
U.K.

Telex: 36156
Cable: Dorman Stafford
Phone: (0785) 3141

A series of robust, medium speed, four-stroke diesel engines as follows:

Model	*Cylinder*	*Cooling*	*Cap. (cc)*	*Gross Power (b.h.p.) Max.*	*Cont.*	*@ r.p.m.*	*Nett Weight (kg)*
4DA	4	air	4150	75	46	1500	470
6DA*	6	air	6230	112	69	1500	572
8DA	8V	air	8310	144	86	1500	711
8F*	8V	water	9120	177	107	1500	685
6LD*	6	water	9882	145	124	1500	859
6LE*	6	water	11340	170	147	1500	859

*Turbocharged variants available.

Other models up to approximately 950 b.h.p. are available.

ELVE CORPORATION
Elve Chambers
Green Street (off Bank Street)
Fort
Bombay 400023
India

Telex: Elve OII-3856
Cables: Elbuscon
Phone: 263981/2/3

Water-cooled slow speed diesel engines with power outputs of 7.8 b.h.p. and 11 b.h.p. Units can be supplied as complete pumping units.

Model	*Cylinder*	*Cooling*	*Cap. (cc)*	*Gross Power (b.h.p.) Max.*	*Cont.*	*@ r.p.m.*
6/1	1	radiator or tank	553	7.4	5	600
8/1	1	radiator or tank	622	11	8	850

Manufacturers claim 5 year operational life.

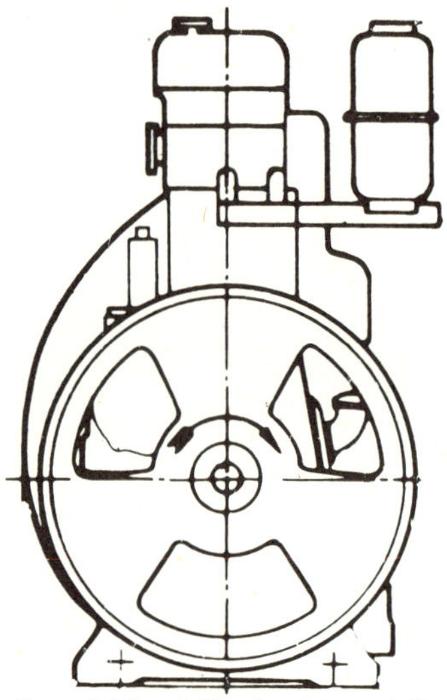

Brand Diesel Engine Type 8/1

K Series Single Cylinder Air-Cooled Diesel Engine

FARYMANN DIESEL
P.O. Box 100
684 Lampertheim
West Germany

Telex: 465710
Cables: Farymanndiesel
Phone: (6206) 2001

Air-cooled and water-cooled diesel engines with outputs ranging from 1hp to 26hp. The units are aimed at the small industrial and the marine auxiliary engine market and those at the lower end of the range are very small and compact. Model K50 is claimed to be the smallest industrial diesel in the world.

Model	*Cylinder*	*Cooling*	*Cap. (cc)*	*Gross Power (b.h.p.) Max.*	*Cont.*	*@ r.p.m.*	*Nett Weight (kg)*
K50	I	air	200	3.5	3	2500	33
K54	I	air	242	4.5	3.8	2500	33
K14	I	air	298	6.0	5.2	2500	51
L14	I	air	412	8.0	6.8	2500	60
A10	I	air	582	11.0	9.5	2500	68
A12	I	air	618	12.0	10.0	2500	69
L20	IH	air	412	8.0	6.8	2500	62
A20	IH	air	582	11.0	9.5	2500	70
A22	IH	air	618	12.0	10.0	2500	71
G20	IH	air	1625	14.0	14.0	2000	180
R10	2V	air	1160	18.0	14.5	2000	130
R12	2V	air	1236	21.0	15.5	2000	132
P10	2V	air	1276	22.0	17.0	2000	150
S10	2V	air	1558	26.0	21.0	2000	160
A40	IH	water	582	12.0	10.0	2500	75
A30	I	water	582	12.0	10.0	2500	73
R30	2V	water	1160	24.0	18.0	2500	130
S30	2V	water	1560	32.0	25.0	2500	170

ENFIELD INDUSTRIAL ENGINES LIMITED
Somerton Works
Cowes
Isle of Wight
U.K.

Cables: Diesels Cowes
Phone: Cowes 4711

Hand started units suitable for use in stationary, automotive and marine installations. Centrifugal governor speed control for factory setting between 1200 r.p.m. and 2000 r.p.m. An over-riding control can be fitted varying the speed from maximum to 800 r.p.m.

Model	*Cylinder*	*Cooling*	*Cap. (cc)*	*Gross Power (b.h.p.) Max.*	*Cont.*	*@ r.p.m.*	*Net Weight (kg)*
85 single	1	air	568	7.5	5.5	1500	121
85 twin	2	air	1135	14.9	11	1250	147

Marine versions available, all models with hand or electric start options.

Model '85' Single Cylinder Unit

FORD MOTOR COMPANY LIMITED
Royal Oak Way South
Daventry
Northants NN11 5NT
U.K.

Telex: 311552
Phone: (032 72) 71111
Cables: Fordparts, Daventry
Fordparts, Daventry

2401E Industrial Diesel Engine

Model	*Cylinder*	*Cooling*	*Cap. (cc)*	*Gross Power (b.h.p.) Max.*	*Cont.*	*@ r.p.m.*	*Nett Weight (kg)*
2504E	3	water	2868	41.5	30	1500	311
2512E	3	water	3294	57	39	1500	324
2514E	4	water	4196	69.5	53	1600	401
2711E	4	water	4150	71	58	2000	340
2712E	4	water	4150	80	63	2000	340
2713E	6	water	5950	100	78	2000	442
2714E	6	water	6220	108	85	2000	442
2715E	6	water	6220	120	93	2000	442
2704ET*	6	water	5945	150	116	2000	455
High speed range							
2401E	4	water	2360	58	42	2750	215
2402E	6	water	3540	87	63	2750	292

*available for marine use with or without intercooler.

KLOCKNER-HUMBOLDT-DEUTZ AG
P.O. Box 800 509
Deutz-Mulheimer Strasse III
5000 Cologne
W. Germany

Telex: 08 873 201
Cables: Deutzmotor Köln
Phone: (0221) 8221

Wide range of air cooled engines as indicated. Two stage combustion version offers low exhaust emissions to California 1977 standard.

Model	*Cylinder*	*Cooling*	*Cap. (cc)*	*Gross Power (b.h.p.) Max.*	*Cont.*	*@ r.p.m.*	*Nett Weight (kg)*
Air cooled direct injection engines.							
F1L208	1	air	410	8	7	3000	65
F1L210	1	air	670	14	12.5	3000	80
F1L411D*	1	air	690	14	8.75	2500	110
F2L411D*	2	air	1390	28	23	2500	146
F2L912*	2	air	1880	32	20	2300	235
F3L912*	3	air	2830	58	47	2300	270
F4L912*	4	air	3770	80	63	2300	300
F5L912*	5	air	4710	100	79	2300	380
F6L912*	6	air	5660	120	95	2300	410
BFL6L913	6	air	6128	160	123	2000	485
F5L413R*	5	air	7350	136	113	2300	623
F6L413R*	6	air	8820	163	136	2300	740
F6L413*	6V	air	8480	163	136	2300	595

Plus larger units up to 455 b.h.p.

*Denotes variants with two-stage combustion, offering reduced exhaust emissions.
Water cooled series from 200 to 9000 b.h.p. also available.

F2L411W

L. GARDNER & SONS LIMITED
Barton Hall Engine Works
Patricroft
Eccles
Manchester M30 7WA
U.K.

Telex: 668023
Cables: Gardworks Eccles Manchester
Phone: 061 789 2201

6LXB 10.45 Litre Diesel Engine

Water cooled four-stroke diesel engines for automotive, marine and industrial purposes.

Model	*Cylinder*	*Cooling*	*Cap. (cc)*	*Gross Power (b.h.p.) Max.*	*Cont.*	*@ r.p.m.*
6LXB	6	water	10450	180	120	1500
8LXB	8	water	13938	240	160	1500
8L3B	8	water	24138	260	214	1150

R.A. LISTER & COMPANY LIMITED
(Member of Hawker Siddeley Group)
Dursley
Gloucester GL11 4HS
U.K.

Telex: 43261
Cables: Machinery Dursley
Phone: Dursley 4141

Air and water cooled diesel engines with a power range from 2.5 b.h.p. to 250 b.h.p. Small units have hand starting, larger units are started electrically. Numerous accessories are available for most models for a wide variety of applications, plus a very comprehensive international distribution network. A wide selection of marine engines and auxilliary engines based on these power units is also available.

Model	*Cylinder*	*Cooling*	*Cap. (cc)*	*Gross Power (b.h.p.) Max.*	*Cont.*	*@ r.p.m.*	*Nett Weight (kg)*
8/1	1	water	–	8	6	650	–
16/2	2	water	–	16	12	650	–
LT1	1	air	408	5.25	4	1800	80
ST1	1	air	633	12.5	7.3	1800	107
ST2	2	air	1266	25.0	14.6	1800	170
ST3	3	air	1900	37.5	21.9	1800	215
HR2	2	air	2090	37	21.5	1500	280
HR3	3	air	3135	55	32	1500	370
HR4	4	air	4180	73	43	1500	432
HR6	6	air	6270	111	64	1500	560
HRS6	6	air	6270	–	80	1500	626
JA6	6	air	10620	173	109	1500	950
JAS6	6	air	10620	–	122	1500	975
HRW2	2	water	2090	37	21	1500	306
HRW3	3	water	3135	55	32	1500	419
HRW4	4	water	4180	73	43	1500	507
HRW6	6	water	6270	111	64	1500	652
HRW56	6	water	6270	–	80	1500	684
JW6	6	water	10620	173	109	1500	1180
JWS6	6	water	10620	–	140	1500	1200

Lister Diesel Engines

PETTERS LIMITED
Power Generation Division
Hamble Lane
Hamble
Southampton S03 5NJ
U.K.

Telex: 47626
Cables: Petter Hamble
Phone: (042) 122 2061

This manufacturer is part of the Hawker Siddeley Group and offers air and water cooled diesel engines with power outputs ranging from 2 b.h.p. to 50 b.h.p. Power off-take via a clutch at the machined flywheel housing end on the larger units, or by pulley attached to an extension of the crankshaft on the small units. Various units are available as a generating and refrigeration or pumping set.

Model	*Cylinder*	*Cooling*	*Gross Power (b.h.p.) Max.*	*Cont.*	*@ r.p.m.*	*Nett Weight (kg)*
Compact range						
AA1	1	air	3.8	2.8	2500	42.6
AB1	1	air	5.5	3.8	2500	44.5
AB1W	1	water	5.5	3.8	2500	46.8
AC1	1	air	7.2	5.0	2500	47
AC1W	1	water	7.2	5.0	2500	46
AC1Z	1	air	6.0	5.0	2500	49
BA1	1	air	11.0	7.5	2000	105
BA2	2	air	22.0	15.0	2000	147
AC2	1	air	13.2	11.0	3000	78
Standard range						
PAZ1	1	air	3.3	2.5	1500	108
PH1*	1	air	9.0	6.3	1500	185
PH2*	2	air	18.0	12.5	1500	252
PJ1*	1	air	12.4	8.5	1500	210
PJ2*	2	air	24.7	17.0	1500	276
PJ3*	3	air	37.1	25.5	1500	362
PJ4*	4	air	49.5	34.0	1500	435

*Water cooled variants available with suffix W to model number.

AC1Z Single Cylinder

PRAGOINVEST
Foreign Trade Corporation
P.O. Box 890
Praha 9
Czechoslovakia

Telex: 122 379
Cables: Pragoinvest Praha

A range of air cooled medium speed, four-stroke diesel engines with one, two, three or four cylinders. Units are suitable for various light industrial uses, generating electricity, etc.

Slavia S 95 Type Diesel Engine

MECCANICA S.p.A
CP 313
Via A.C. Ducati 3
Borgo Parigale
40100 Bologna
Italy

Telex: 51492
Phone: (051) 405049

A range of Italian built Ducati four-stroke diesel engines with forced air-cooling via a flywheel blower. Starting normally by a rope but units can be supplied for hand crank starting or electric dynastart.

Model	*Cylinder*	*Cooling*	*Cap. (cc)*	*Gross Power (b.h.p.) Max.*	*Cont.*	*@ r.p.m.*	*Nett Weight (kg)*
1S7	I	air	345	7	5.5	2500	48
1S9	I	air	345	9	6.2	2500	48
1S11	I	air	432	11	7.8	2500	50
1S16	I	air	673	18	13	2500	86
1S20	I	air	746	20	14.2	2500	88
2IS22	2	air	864	22	15.2	2500	86

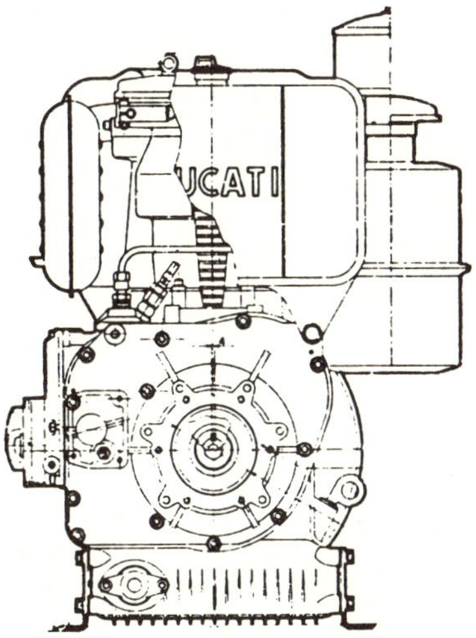

IS16 Four-stroke Diesel Unit

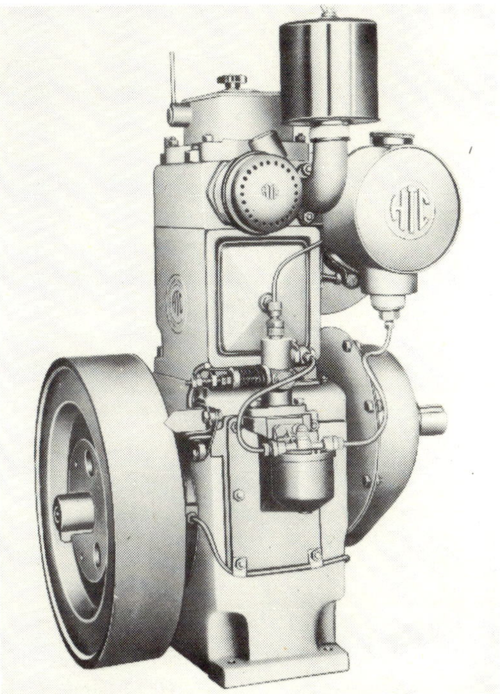

H.T.C. DIESEL ENGINES
PRIVATE LIMITED
Beaumon Chambers
27/33 Nagindas Master Road
Bombay 400 023
India

Telex: 011-3242
Cables: Hotraco
Phone: 272188

Single-cylinder, open-flywheeled, hand-cranked engines, which can be supplied with clock-wise or anti-clockwise rotation.

Model	*Cylinder*	*Cooling*	*Cap. (cc)*	*Gross Power (b.h.p.) Max.*	*Cont.*	*@ r.p.m.*	*Nett Weight (kg)*
AAVRS-5	I	air	553	–	5	1500	155
RVRS-5	I	water	553	–	5	1500	–
VRS-5	I	water	553	–	6.5	1500	–

Engines offered as pumping set on base-plate, trolley or skids.

Model VRS5

INDIAN NATIONAL DIESEL ENGINE CO. LIMITED
6 Little Russel Street
Calcutta 700071
Registered Office:
Hall & Anderson Buildings
Park Street
Calcutta 700016
India

Telex: 7211, 3283
Cables: Innatdec
Phone: 43-1938

Manufacturers of four-stroke diesel engines which can be supplied as pumping or generating sets.

Model	*Cylinder*	*Cooling*	*Cap. (cc)*	*Gross Power (b.h.p.) Max.*	*Cont.*	*@ r.p.m.*	*Nett Weight (kg)*
PH1	I	air	659	8.2	6.25	1500	188
PH1W	I	water	659	8.2	6.25	1500	178
PH2	2	air	1318	16.4	12.5	1500	255
PH2W	2	water	1318	16.4	12.5	1500	250

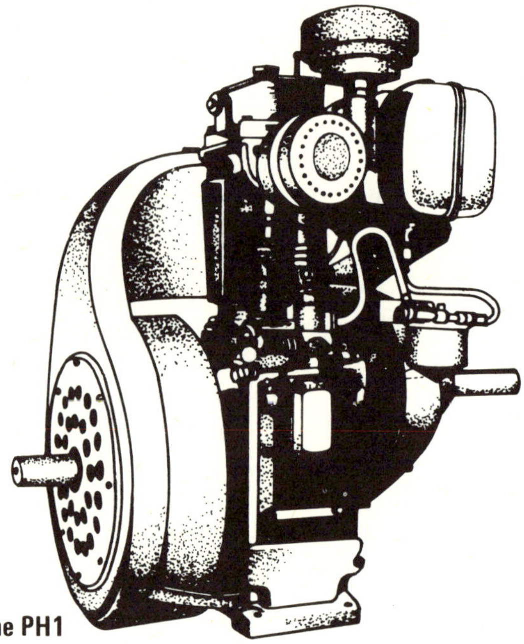

Air Cooled Diesel Engine PH1

Ruggerini P101/2 (40 h.p.)

RUGGERINI MOTORI S.p.A.
42040 Villa Bagno
Reggio Emilia
Italy

Telex: 51137
Cables: Ruggerini Motori Rubiera
Phone: (0522) 55221

All models can be equipped with cables and control panel for dynastart and electric start and fuel cut-off switch for remote control. Both horizontal and vertical shaft models are available.

Model	*Cylinder*	*Cooling*	*Cap. (cc)*	*Gross Power (b.h.p.) Max.*	*Cont.*	*@ r.p.m.*	*Nett Weight (kg)*
RD80	1	air	375	8.0	6.3	3000	42
RD850	1	air	482	10.3	9.0	3000	49
RD901	1	air	540	14.2	12.6	3000	60
RD920	1	air	565	16.0	13.5	3000	65
P101	1	air	746	20.6	17.7	3000	95
RD901/2	2	air	1080	26.0	23.0	3000	88
P101/2	2	air	1492	40.0	36.0	3000	143
P105/2	2	air	1750	45.0	40.0	3000	150
RD80V	1H	air	375	8.0	6.3	3000	42
RD850V	1H	air	485	10.3	9.0	3000	49
RD920V	1H	air	565	16.0	13.5	3000	65

Various clutches, reduction gears, starters, etc., are available.

NISSAN DIESEL MOTOR COMPANY LIMITED
Kowa Building
3-7-1 Kanda Nishiki-cho
Chiyoda-ku
Tokyo
Japan 101

Telex: J24905 Nideko
Cables: Nissandiesel Tokyo

Model	*Cylinder*	*Cooling*	*Cap. (cc)*	*Gross Power (b.h.p.) Max.*	*Cont.*	*@ r.p.m.*	*Nett Weight (kg)*
SD226							
marine	4	water	2164	46	40	2600	375
SD22	4	water	2164	66	36	2600	190
SD33	6	water	3246	99	53	2600	300
ED30	4	water	2956	91	39	2000	295
ED6	6	water	5654	152	85	2200	470
ND6	6	water	6842	140	87	2000	600
PD60	6	water	10308	190	122	1800	822

SD22 Diesel Unit

Various other models up to 350 b.h.p. are also available, and some units are supplied as an enclosed power-pack as an optional extra.

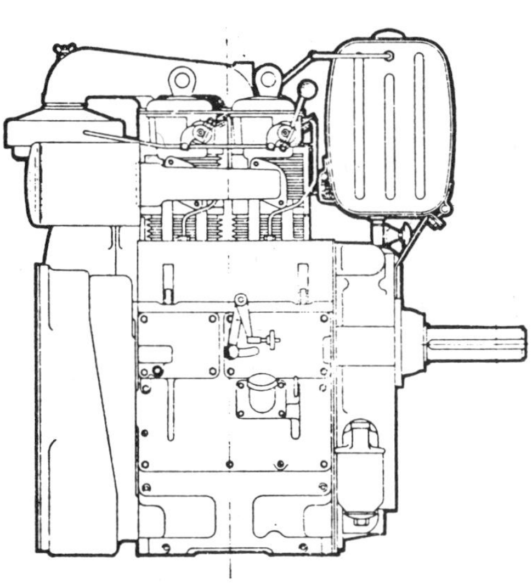

Ruston 'YWA' Air-Cooled Diesel

RUSTON & HORNSBY (INDIA) LIMITED
Chinchwad
Poona — 411 019
India

Telex: 303 Greaves (PN)
Cables: Rustonind
Phone: 82601/2/3/4/5

Medium speed engines of unitary construction to allow easy maintenance and repair of individual cylinders and pistons. Industrial and marine versions with the facility of power off-take being from either end of the crankshaft.

Model	*Cylinder*	*Cooling*	*Cap. (cc)*	*Gross Power (b.h.p.) Max.*	*Cont.*	*@ r.p.m.*	*Nett Weight (kg)*
2YDA	2	air	2463	42	22	1500	480
3YDA	3	air	3695	68	46	1600	557
4YDA	4	air	4928	92	63	1600	620
6YDA	6	air	7391	135	92	1600	793
1YWA	1	air	–	15.5	9.8	1500	245
2YWA	2	air	–	31.5	19.6	1500	308
3YWA	3	air	–	51	29.4	1500	372
2YD	2	water	2463	42	27.5	1500	524
3YD	3	water	3695	65	42	1500	584
4YD	4	water	4928	90	55	1500	673
6YD	6	water	7391	135	84	1500	876
6YDX	6	water	7391	182	–	–	–
Slow speed horizontal engines with exposed flywheel.							
1HR	1H	water	1900	6	5.5	550	367
1XHR	1H	water	1900	9	8	800	367
1YHR	1H	water	4300	11	10	450	635
1ZHR	1H	water	4300	12	11	475	635
2XHR	1H	water	6000	15.5	14	475	839
3HRO	1H	water	9300	20	18	400	1270
HRY	1H	water	9300	28.6	26	430	1283

MEDALIST UNIVERSAL MOTORS
1552 Harrison Street
P.O. Box 2508
Oshkosh
Wisconsin 54901
U.S.A.

Telex: 262-717
Cables: Unimot
Phone: (414) 231 4100

Manufacturers of a range of three water cooled, four-stroke diesel engines with rated power outputs of 12, 17 and 25 b.h.p. at 2800 r.p.m. Specifically designed for marine applications, but can be supplied with or without a final drive reduction. Closed system fresh water cooling can be supplied on some models for sea water use.

17 h.p. Universal Marine Diesel

V8.640 Diesel Engine

PERKINS ENGINES LIMITED
Eastfield
Peterborough PE1 5NA
U.K.

Telex: 32501
Cables: Perkoil Peterborough
Phone: (0733) 67474

Model	*Cylinder*	*Cooling*	*Cap. (cc)*	*Gross Power (b.h.p.) Max.*	*Cont.*	*@ r.p.m.*	*Nett Weight (kg)*
D3.152	3	water	2500	49	38	1800	210
4.108	4	water	1760	45	26	1800	150
4.203	4	water	3300	61	52	1800	236
4.236	4	water	3860	81	64	1800	249
4.248	4	water	4070	84	65	1800	249
6.354	6	water	5800	114	87	1800	390
T6.354*	6	water	5800	137	117	1800	417
V8.540	8V	water	8830	180	135	1800	623

*Turbocharged.

Two further V8's up to 250 b.h.p. are also available.

These engines are widely used in industrial, earth moving and agricultural machinery, and for powering generator sets.

KIRLOSKAR OIL ENGINES LIMITED
13 Laxmanrao Kirloskar Road
Kirkee
Pune – 411 003
India

Telex: 014-245-Koel Pn
Cables: Koel Pune (India)
Phone: 55346/7/8

Kirloskar RA3 Diesel Unit.

Air and water cooled engines with a power range of 5 to 370 b.h.p. Manufactured for automotive, marine, agricultural and industrial use. Can be supplied as complete generating or pumping sets.

Model	*Cylinder*	*Cooling*	*Cap. (cc)*	*Gross Max.*	*Power (b.h.p.) Cont.*	*@ r.p.m.*	*Nett Weight (kg)*
TA1	1	air	660	9.38	6.0	1500	164
TA2	2	air	1320	18.75	12.0	1500	250
TV1	1	water	660	10.88	7.0	1500	166
TV2	2	water	1320	21.76	14.0	1500	247
KA27	1	air	276	4.5	2.7	2000	91
CA1	1	air	580	6	5	1500	172
CA2	2	air	1160	12	10	1500	248
AV1	1	water	553	8.1	5	1500	166
AV2	2	water	1106	16.2	10	1500	247
RA2	2	air	1728	21	16.5	1500	323
RA3	3	air	2592	31.5	24.7	1500	433
RA4	4	air	3456	42	33	1500	475
RA6	6	air	5184	63	48.5	1500	595
RDA2	2	air	1930	26.5	19	1500	325
RDA3	3	air	2900	36	28.5	1500	435
RDA4	4	air	3870	47	37	1500	475
RDA6	6	air	5800	70	55	1500	595
RV2	2	water	1728	27.5	16.8	1500	328
RV3	3	water	2592	41.25	25.2	1500	438
RV4	4	water	3456	55	33.6	1500	480
RV6	6	water	5184	82.5	50.4	1500	600
RBV2	2	water	2090	30.8	25.8	1650	338
RBV3	3	water	3135	46.2	38.7	1650	448
RBV4	4	water	4180	61.5	51.6	1650	490
RBV6	6	water	6270	92.5	77.4	1650	610
RE4	4	water	2700	55	31.25	2000	235

Kirloskar also manufacture a selection of 1 to 45kVA generating sets and 16 to 61 b.h.p. pumping sets. Also, marine variants and larger units.

VOLVO PENTA
P.O. Box 392
401 26 Goteborg 1
Sweden

Telex: 20755
Cables: Penta
Phone: 23 54 60

Volvo Penta D70B

Range of units with power outputs ranging from 44kW to 201kW, designed for industrial machinery, marine auxiliary use, as power packs, and can be supplied complete with a large range of ancillary equipment.

Model	*Cylinder*	*Cooling*	*Cap. (cc)*	*Gross Max.*	*Power (b.h.p.) Cont.*	*@ r.p.m.*	*Nett Weight (kg)*
D70B	6	water	6730	165	100	1800	670
TD70B*	6	water	6730	215	139	1800	680

*Denotes turbocharged version – a range of larger units up to 329 b.h.p. is also available.

Inboard marine engines:

MD5A	1	water	440	7.5	–	–	111
MD7A	2	water	740	13	–	–	175
MD11C	2	water	1120	–	20	2300	230
MD17C	3	water	1680	–	30	2300	290
MD70B	6	water	6730	139	106	2000	800

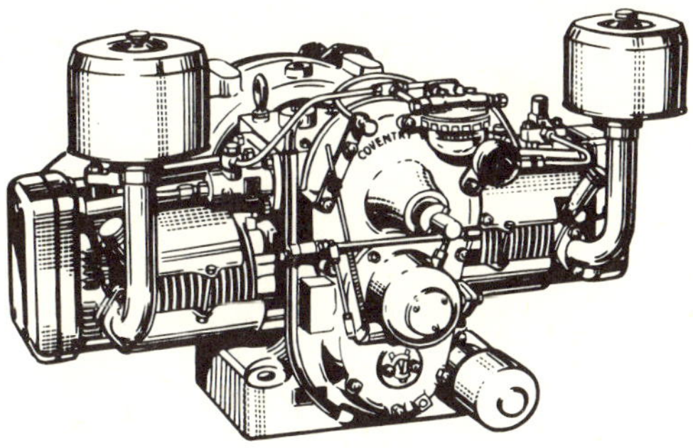

Model HDA

A.N. WEAVER (COVENTRY VICTOR) LIMITED
Smiths Industrial Estate
Humber Avenue
Coventry CV3 1JL
U.K.

Cables: Precision Coventry
Phone: 0203-452625

A range of air and water cooled general purpose engines as follows:

Model	*Cylinder*	*Cooling*	*Cap. (cc)*	*Gross Power (b.h.p.) Max.*	*Cont.*	*@ r.p.m.*	*Nett Weight (kg)*
WD3	1	water	636	11	7.3	1500	125
AD3	1	air	636	9	6	1500	154
HDA	2H	air	1448	23	15.4	1500	207
HDW	2H	water	1448	30	17.8	1500	197

Some of these units are available adapted for marine and other specialised uses.

VOLKSWAGENWERK AG
Heinrich Nordoff Strasse
D-3180 Wolfsburg 1
Germany

Telex: 09586-0 vwwd
Cables: Volkswagenwerk/Wolfsburg
Phone: 22-3535

1500 cc water cooled diesel engine producing 33 kW at 4000 r.p.m. Unit based on engine used for Volkswagen Golf car, and is intended for industrial applications.

Model 068.2, 1471cc, water-cooled, 4 cylinder in-line. Maximum power 50 b.h.p. Maximum continuous rating (typical) 25 b.h.p. @ 2000 r.p.m. Weight (dry) 129 kg. 12V electrical system complete with starter.

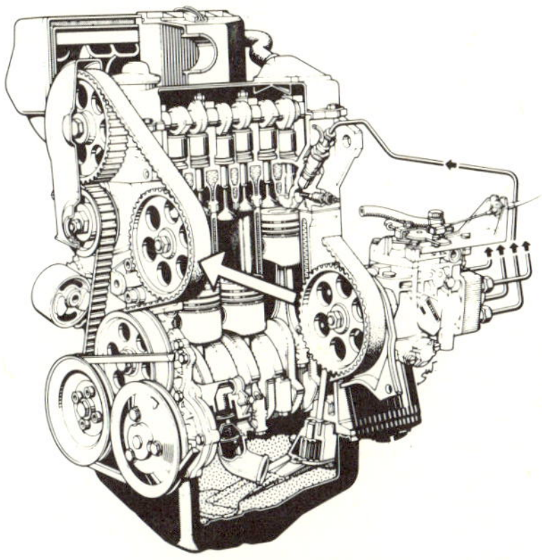

Volkswagen 1500cc Diesel Engine

SPILLING CONSULT AG
Sonnenweg 4
CH-5610 Wohlen
Switzerland

Telex: 57 939
Cables: Spill Ch
Phone: 057-6 73 57

Manufacturers of modular diesel engine with multi-fuel capability consisting of a single cylinder module which can be assembled in any combination from 1 to 8 units in line.

	Cylinder	*Cooling*	*Capacity*	*Gross Power (b.h.p.) Cont.*	*@ r.p.m.*
each module	I	water	13200cc	100	1000
supercharged				180	1000

Spilling Six-cylinder Diesel Engine

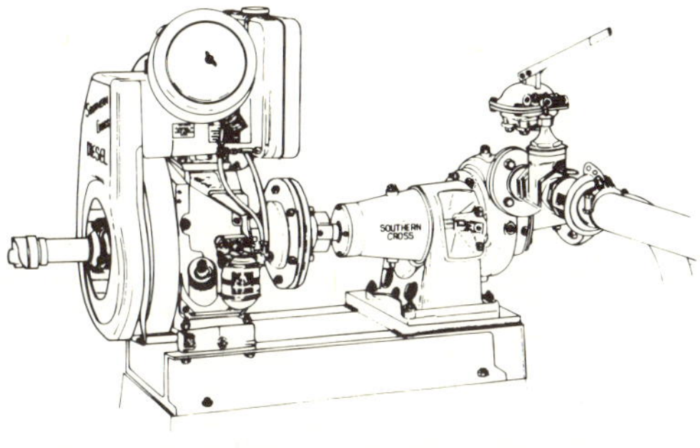

Southern Cross 6 h.p. Air-cooled Diesel Engine

TOOWOOMBA FOUNDRY PTY LTD
P.O. Box 109
Toowoomba
Queensland 4350
Australia

Telex: Socross 40046
Cables: Foundfact Toowoomba
Phone: 076 32 3122

Manufacturers of "Southern Cross" slow to medium speed diesel engines. A series of air cooled, four-stroke cycle engines as follows:

Model	*Cylinder*	*Cooling*	*Cap. (cc)*	*Gross Power (b.h.p.) Max.*	*Cont.*	*@ r.p.m.*	*Nett Weight (kg)*
EF-D	1	air	316	3.5	2.5	1500	85
EF-E	1	air	591	6.0	5	1500	139
EF-H	2	air	1182	12	10	1500	236
ED-C	1	air	1104	12	11	1500	326
ED-E	2	air	2092	24	20	1500	388
ED-G	4	air	4184	50	44	1500	572

GRAHAME PUTTICK LIMITED
Sandwich
Kent
U.K.

Telex: 96336
Cables: Puma Sandwich
Phone: (03046) 2901

Puma 10 b.h.p. Diesel Engine

A range of exposed flywheel water cooled single cylinder engines designed for slow speed operation and simplicity of maintenance.

Model	*Cylinder*	*Cooling*	*Capacity (cc)*	*Gross Power (b.h.p.) Cont.*	*@ r.p.m.*	*Nett Weight (kg)*
PZ3	1	water	1950	7	600	420
PZ4	1	water	1950	8	700	420
PZ10	1	water	1950	10	800	425

These engines are available as complete pumping and generating sets.

Robson RC Diesel Engine

JOHN ROBSON (SHIPLEY) LIMITED
P.O. Box 31
Ives Street
Shipley
Yorkshire
U.K.

Telex: 517386
Cables: Robson Shipley
Phone: Shipley 52041

Single cylinder, water cooled, low speed, totally enclosed, horizontal diesel engines. Starting is by hand for which purpose an automatic decompressor is fitted and a starting handle is provided. Drive may be taken from either side from a pulley that bolts to the flywheel, and the units can be supplied as complete pumping sets.

Model	*Cylinder*	*Cooling*	*Cap. (cc)*	*Gross Power (b.h.p) Max.*	*Cont.*	*@ r.p.m.*	*Net Weight (kg)*
RB	1H	water	1800	14	13	1100	457
RC	1H	water	3850	24	22	880	1016
R16	1H	water	2150	18	16	1100	476
R26	1H	water	4600	29	26	880	1057

Total operational life of up to 20 years claimed.

ROLLS ROYCE MOTORS LIMITED
Diesel Division
Shrewsbury
Salop SY1 4DP
U.K.

Telex: 35171/2
Cables: Roycar Shrewsbury
Phone: (0743) 52262

Rolls Royce C8T Diesel Engine

A range of water cooled diesel engines with outputs from 100kW to 690kW. Units are supplied on skids or mounting feet and as generating sets.

Model	*Cylinder*	*Cooling*	*Cap. (cc)*	*Gross Power (b.h.p.) Max.*	*Cont.*	*@ r.p.m.*	*Nett Weight (kg)*
C6N	6	water	12170	180	150	1500	1133
C8T	8	water	16200	340	294	1500	1470

Plus a variety of other models up to 730 b.h.p. nett.

Saab-Scania DS8

SAAB-SCANIA
S-151 87 Södertälje
Sweden

Telex: 10200 Scania s
Cables: Scania Södertälje
Phone: 0755-341 40

Prime Mover Engines

Model	*Cylinder*	*Cooling*	*Cap. (cc)*	*Gross Power (b.h.p.) Max.*	*Cont.*	*@ r.p.m.*	*Nett Weight (kg)*
D8	–	water	–	167	99	1500	–
DS8	6I	water	–	210	130	1500	–
D11	6I	water	–	215	144	1500	–

D8 engine available as 73kW generating set, 131 hp marine engine or 99 hp (continuous) at 1500 r.p.m. prime mover.

Other engines also available up to 400 hp.

YANMAR DIESEL ENGINE
COMPANY LIMITED
1-11-1 Marunouchi
Chiyoda-ku
Tokyo
Japan 100

Telex: 0222-2310, 0222-4733
Cables: Yanmar Tokyo
Phone: 03-213-8111

TS (C) Series Engine

Water cooled engines designed for use in marine, industrial and agricultural applications.

Model	*Cylinder*	*Cooling*	*Cap. (cc)*	*Gross Power (b.h.p.) Max.*	*Cont.*	*@ r.p.m.*	*Nett Weight (kg)*
TS50C	1H	water	270	5	4	2000	62
TS60C	1H	water	335	6	5	2000	70
TS70C	1H	water	380	7	6	2200	82
TS130C	1H	water	635	13	11	2200	130
TS105C	1H	water	515	10.5	9	2200	96
TS115C	1H	water	750	15.5	13	2200	150
TS180C	1H	water	870	18	15	2200	170
TS80C	1H	water	435	8	7	2200	88

This series available in generating sets of 1.5, 2, 3, 5, 7.5, 10, and 12kVA (various voltages) and as marine auxillary engines.

Marine engines (fitted with clutch and various reduction gear arrangements).

YSB8	1H	water	330	8	5	2200	107
YSB12	1H	water	510	12	8	2200	135
SB8	1	water	330	8	5	2200	114
SB12	1	water	510	12	8	2200	137
SKE	1	water	664	13	12	2400	170
2SE	2	water	1134	18	16	2200	250
2QM	2	water	1100	22	20	2600	225
3QM	3	water	1650	33	30	2600	280
2TE	2	water	1650	25	22	2000	380
3TE	3	water	2475	37	33	2000	490
1SME	1	water	1050	15.5	14	1800	240
2SME	2	water	2100	33	30	2200	430
3SME	3	water	3150	50	45	2200	530
3ESDE	3	water	4600	63	56	1800	680
4ESDE	4	water	6130	84	74	1800	800
3BNE	3	water	10200	132	115	1600	1480
4BNE	4	water	14960	175	150	1600	1730
6BNE	6	water	20400	260	240	1600	2550
3KDE	3	water	8500	90	82	1450	1240
4KDE	4	water	11350	121	110	1450	1701

Plus a range of larger engines up to 1980 b.h.p.

XDP 4.88 Diesel

SOCIETE COMMERCIALE DE MOTEURS – C.L.M.
B.P. 420
49 rue Noël Pons
92000 Nanterre
France

Telex: 620162
Cables: Cogemot-Nanterre
Phone: 780 7211

This manufacturer is a subsidiary of the PSA-Peugeot-Citroën group producing a range of water cooled four-stroke diesel engines (as used in Peugeot and Citroën vehicles), with outputs from 68 b.h.p. to 106 b.h.p.

Model	*Cylinder*	*Cooling*	*Cap. (cc)*	*Gross Power (b.h.p.) Max.*	*Cont.*	*@ r.p.m.*	*Nett Weight (kg)*
XDP 4.88	4	water	1946	55	40	3000	153
XDP 6.90	6	water	3168	90	61	2500	246
XDP 4.90	4	water	2112	61.5	40	2500	172
CRD 90B	4	water	2175	61.5	41	2500	205

WITTE ENGINE CORPORATION
(A Hawker Siddeley Company)
P.O. Box 386
555 East 56th Highway
Olathe
Kansas 66061
U.S.A.

Telex: 042-6262
Phone: (913) 764 3512

A range of diesel engines, the 120/HS being a hand cranked, slow speed unit, with an output of 26 b.h.p. Speed regulation is to ± 5% no load to full load and it is designed for continuous 24 hour running conditions. Units are also available for marine applications for boats up to 45ft.

Model	*Cylinder*	*Cooling*	*Cap. (cc)*	*Gross Power (b.h.p.) Max.*	*Cont.*	*@ r.p.m.*	*Nett Weight (kg)*
120M	2H	water	1966	30	22	1500	398
120	2H	water	1966	33	26	1800	545

This heavy duty four-stroke diesel engine is available as a complete generating set and there is also the Model G-260 natural gas spark ignition engines with many common components.

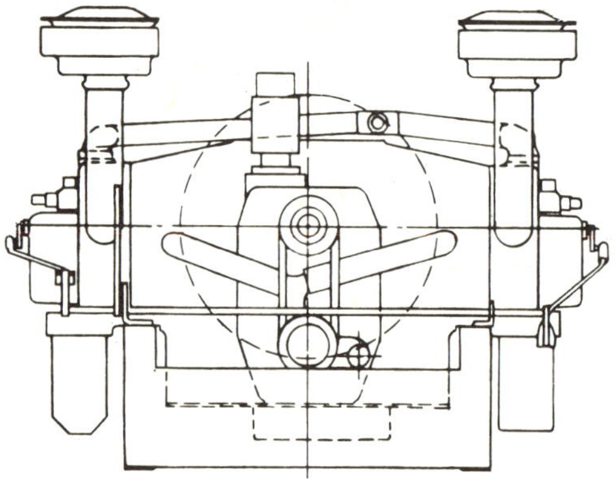

120/HS Unit

SDMTW 6, 7 Litre Unit

MOTORENFABRIK ANTON SCHLÜTER
Münchner Strasse 32
D-8050 Freising
Germany

Telex: 0526515
Cables: Schlüterwerke Freising
Phone: (08161) 13051/55

Water cooled diesel engines with power outputs from 30 hp to 200 hp, suitable for vehicle power, construction machines and stationary installations. Units can be supplied with a variety of power take-off assemblies on the flywheel casing.

Model	*Cylinder*	*Cooling*	*Cap. (cc)*	*Gross Power (b.h.p.) Max.*	*Cont.*	*@ r.p.m.*	*Nett Weight (kg)*
SDMW2	2	water	2376	38	34	1800	227
SDMW3	3	water	3564	57	50	1800	325
SDMW4	4	water	4752	76	68	1800	420
SDMW6	6	water	7128	114	100	1800	595
SDMW8	8	water	9504	152	136	1800	775

WHITE ENGINES INCORPORATED
101 Eleventh Street S.E.
Canton
Ohio 44707
U.S.A.

Telex: 98-3439
Cables: Hercano
Phone: 74 979 3993

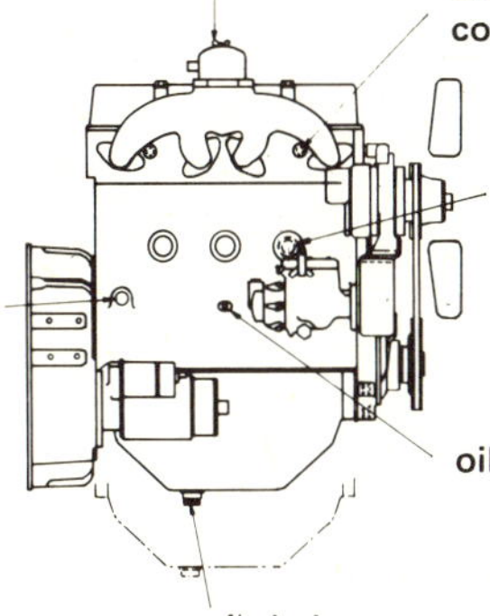

Model D-2000/D-2300

Water cooled engines designed for automotive and industrial use. Can be supplied as enclosed generators or open power units.

Model	*Cylinder*	*Cooling*	*Cap. (cc)*	*Gross Power (b.h.p.) Max.*	*Cont.*	*@ r.p.m.*	*Nett Weight (kg)*
D-2000	4	water	3250	68	53.5	1800	435
D-2300	4	water	3710	72	61	1800	500
D-3000	6	water	4880	114	80	1800	500
D-3000T*	6	water	4880	117	105	1800	510
D-3000T*	6	water	4880	131	96	1800	510
D-3400	6	water	5560	115	92	1800	438
D-4800	6	water	7850	150	127	1800	870
D-4800T*	6	water	7850	180	180	1800	870

*T denotes turbocharged version.

Spark Ignition Engines (petrol and gas fuelled)

DP-60 Engine

COOPER ENERGY SERVICES
19 North Center Street
Corry
Pennsylvania 16407
U.S.A.

Telex: 91-4554
Phone: (814) 665 8281

A range of heavy duty, low-speed, horizontal, two-stroke engines, developed for pumping oil in oilfields but well-suited for driving reciprocating water pumps. Very little maintenance required — engines have particularly long operational life.

Model	*Cylinder*	*Cooling*	*Cap. (cc)*	*Gross Power (b.h.p.) Max.*	*Cont.*	*@ r.p.m.*	*Nett Weight (kg)*
EA-22	1H	water	4340	22	17	450	1317
EA-30	1H	water	5410	30	22	450	1340
E-42	1H	water	9290	42	35	400	1961
DP-60	1H	water	13930	60	40	350	2813
DP-80A	1H	water	21800	80	60	300	3691
DP-125	1H	water	36150	125	100	280	6143
DP-165	1H	water	46330	165	125	260	6347
DP-250	2H	water	72300	250	175	250	8111
DP-325	2H	water	92660	325	230	250	8467

JLO MOTORENWERK GMBH
P.O. Box 1620
D-2080 Pinneberg
W. Germany

Telex: 02 189 113
Cables: Ilo Pinneberg
Phone: (04101) 2141

Two stroke engines.

Model	*Cylinder*	*Cooling*	*Cap. (cc)*	*Gross Power (b.h.p.) Max.*	*Cont.*	*@ r.p.m.*	*Nett Weight (kg)*
L35	1	air	35	1.7	1.1	3600	4.5
L77	1	air	73	2.6	2.3	3600	7.5
L97	1	air	98	4.0	3.0	3600	8.1
L101	1	air	98	4.3	3.8	3600	11.8
L152	1	air	148	6.0	5.4	3600	14.2
L197	1	air	198	7.3	6.7	3600	16.6
L252	1	air	247	8.0	7.9	3600	27.2
L253	1	air	247	9.5	9.1	3600	24.1
L372	1	air	372	14.3	13.5	3600	32.6

There is also a series of specialised models specifically designed for various applications including spraying and lawnmowers.

L152 Engine

BRIGGS AND STRATTON CORPORATION
P.O. Box 702
Milwaukee
Wisconsin 53201
U.S.A.

Telex: 02-6776
Cables: Basco
Phone: (414) 461 1212

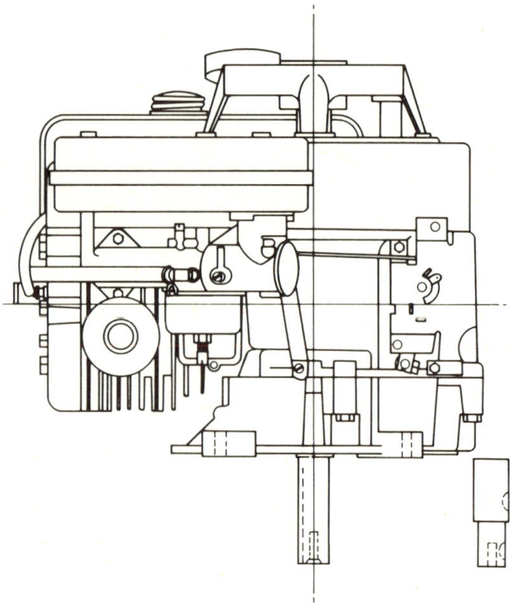

Series 170700

Range of four-stroke air cooled engines available with wide choice of accessories, mountings and shaft sizes.

Model	*Cylinder*	*Cooling*	*Cap. (cc)*	*Gross Power (b.h.p.) Max.*	*Cont.*	*@ r.p.m.*	*Nett Weight (kg)*
92500	1H	air	148	3	2.3	3000	8.8
92900	1H	air	148	3	2.4	3000	8.9
110900	1H	air	187	4	3.0	3000	11.0
100900	1H	air	170	4	2.9	3000	13.8
130900	1H	air	205	5	3.7	3000	13.9
170700	1H	air	274	7	5.3	3000	19.7
191700	1H	air	318	8	6.0	3000	22.2
251700	1H	air	399	10	7.9	3000	26.9
60100	1	air	109	2	1.5	3000	10.1
80100	1	air	127	2.5	2.0	3000	10.1
80200	1	air	127	3	2.3	3000	10.9
80300	1	air	127	3	2.3	3000	11.5
100200	1	air	170	4	2.9	3000	13.4
130200	1	air	205	5	3.7	3000	13.9
170400	1	air	274	7	5.3	3000	20.0
190400	1	air	318	8	6.0	3000	20.4
251400	1	air	399	10	7.9	3000	28.7
233400	1	air	376	9	7.3	3000	41.4
243430	1	air	392	10	7.9	3000	43.5
302430	1	air	494	13	10.2	3000	48.1
326430	1	air	531	16	12.2	3000	48.3

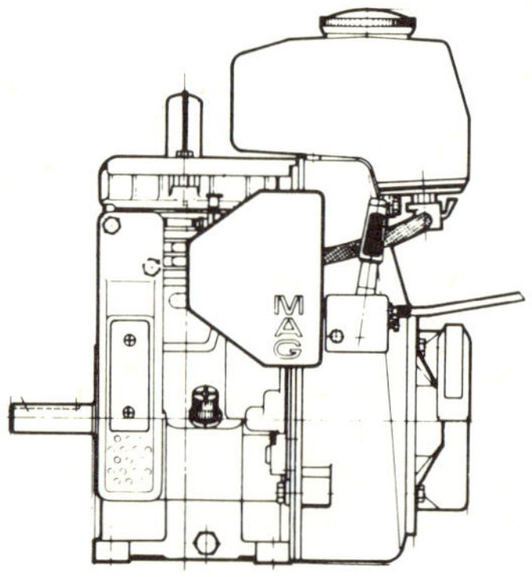

MAG1014 SRL

VILLIERS MAG ENGINES LTD
Factory 47
Pensnett Trading Estate
Brierley Hill DY6 7NA
U.K.

Telex: 338558
Cables: Vilmag
Phone: (038 44) 6553

MOTOSACOCHE S.A.
CH-1211 Geneva 26
Switzerland

Telex: 23139
Phone: (022) 42 01 60

Model	*Cylinder*	*Cooling*	*Cap. (cc)*	*Gross Power (b.h.p.) Max.*	*Cont.*	*@ r.p.m.*	*Nett Weight (kg)*
MAG1014 SRL	1	air	141	3.5	2.8	3000	17
MAG1026 SRL	1	air	258	7.0	5.5	2600	28

036.1 Water-cooled 1100cc Industrial Engine

VOLKSWAGENWERK AG
Heinrich Nordhoff Strasse
D-3180
Wolfsburg 1
Germany

Telex: 09586-0 vww d
Cables: Volkswagenwerk/ Wolfsburg
Phone: 22 3535

Four-stroke industrial engines

Model	*Cylinder*	*Cooling*	*Cap. (cc)*	*Gross Power (b.h.p.) Max.*	*Cont.*	*@ r.p.m.*	*Nett Weight (kg)*
036.1	4	water	1093	49	50	3600	107
049.6	4	water	1588	61	46	3000	111
122	4H	air	1192	34	28	3000	94
126A	4H	air	1584	46	36	2800	100

K181 (8hp) Unit

KOHLER INTERNATIONAL LIMITED
High Street
Kohler
Wisconsin 53044
U.S.A.

Telex: 2 688
Cables: Kohlerint
Phone: (414) 457 4441

Range of four-stroke, single and twin cylinder, air cooled petrol engines with power take-off.

Model	*Cylinder*	*Cooling*	*Cap. (cc)*	*Gross Power (b.h.p.) Max.*	*Cont.*	*@ r.p.m.*	*Nett Weight (kg)*
K91	1	air	145	4.0	3.6	3400	19.5
K141	1	air	278	6.25	5.6	3000	29.5
K161	1	air	278	7.0	6.2	3000	29.5
K181	1	air	305	8.0	7.4	3000	29.5
K241	1	air	391	10.0	8.6	2800	53.5
K301	1	air	476	12.0	11.2	2800	53.5
K321	1	air	512	14.0	12.3	2800	53.5
K341	1	air	588	16.0	14.9	2800	55.3
K532	2	air	880	20.0	18.5	2600	85.0
K582	2	air	946	23.0	19.8	2600	85.0

Purpose-built generating sets are also available.

KIRLOSKAR OIL ENGINES LIMITED
Laxmanrao Kirloskar Road
Poona 411 003
India

Telex: 014-245
Cables: Koel Poona
Phone: 55341

Small air cooled two-stroke engines.

Model	*Cylinder*	*Cooling*	*Cap. (cc)*	*Gross Power (b.h.p.) Max.*	*Cont.*	*@ r.p.m.*	*Nett Weight (kg)*
KP70	1	air	72	3.0	3.0	5000	4.4
KP75	1	air	35	1.7	1.5	4800	3.75

Engines available with optional recoil starter.

KP75/1 Petrol Unit

Ford 1599cc Natural Gas Engine

FORD MOTOR COMPANY LIMITED
Royal Oak Way South
Daventry
Northants NN11 5NT
U.K.

Telex: 311552
Phone: (032 72) 71111
Cables: Fordparts, Daventry

Range of water cooled petrol engines with power outputs as indicated below.

Model	*Cylinder*	*Cooling*	*Cap. (cc)*	*Gross Max.*	*Power (b.h.p.) Cont.*	*@ r.p.m.*	*Nett Weight (kg)*
2261E	4	water	1093	47	30	3000	110
2262E	4	water	1298	55.5	38	3000	115
2263E	4	water	1298	68	37	3000	115
2264E	4	water	1599	65.5	47	3000	120
2265E	4	water	1599	84.5	51	3000	120
BDA	4	water	1600	114	93	4500	114
2503E	3	water	2590	46	31	1500	310
2511E	3	water	3294	57	41	1500	323
2513E	4	water	4196	74	53	1500	402
2606E	4V	water	1996	80	63	3000	150
2613E	6V	water	2495	117	77	3000	181
2614E	6V	water	2994	136	99	3000	181
2652E	4V	water	1498	60	45	3000	128
2653E	4V	water	1699	74	55	3000	130
2655E	6V	water	2293	108	76	3000	160
2656E	6V	water	2551	124	85	3000	163

KAWASAKI HEAVY INDUSTRIES LIMITED
World Trade Centre Building
4-1 Hamamatsu-cho
2-chome
Minato Ku
Tokyo
Japan

Telex: J22672/J26888
Cables: Kawasakiheavy Tokyo
Phone: Tokyo (03) 4352516

KF24

Air cooled four-stroke and two-stroke engines with power outputs ranging from 12 b.h.p. to 28 b.h.p. Typical applications include stationary power units and vehicle motor units.

Four-stroke cycle models

Model	*Cylinder*	*Cooling*	*Cap. (cc)*	*Gross Power (b.h.p.) Max.*	*Cont.*	*@ r.p.m.*	*Nett Weight (kg)*
KF24*	1	air	98	2.3	1.6	1800	9.6
KF34*	1	air	132	3.4	2.3	1800	13
KF53*	1	air	181	5.0	3.5	1800	16
KF64*	1	air	244	6.5	5.0	1800	20.5
KF81*	1	air	324	8.2	6.0	1700	32
KF100*	1	air	397	10.0	7.5	1600	40
KF140*	1	air	552	14.0	10.5	1600	60
KF160*	2H	air	680	16.0	12.0	1600	85
KF200*	2H	air	794	20.0	15.0	1600	85

*two versions available — low speed, rated output quoted

Two-stroke cycle models

KT12	1	air	23	1.2	0.8	7000	2.2
KT15	1	air	27	1.5	1.0	7000	2.5
KT18	1	air	35	1.5	1.0	5000	3.7
KT30	1	air	50	2.4	1.7	6000	6.5
KT33	1	air	80	3.3	2.2	1750	12
KT43	1†	air	110	4.3	3.0	1600	13
KT60	1	air	169	6.0	4.5	1750	27
KT300	2†	air	554	28.0	28.0	4500	48

†this model is available in a vertical shaft option, intended for grass cutting machines.

Main uses for these engines are for small pumps and generators. See also the range of Kawasaki purpose-built generating sets in Section 7.

Rotax 125cc Engine

BOMBARDIER-ROTAX GMBH
P.O. Box 5
A-4623 Gunskirchen
Austria

Telex: 25546 bomrot a
Cables: Bombrotax Gunskirchen
Phone: (07246) 271

Magneto ignition industrial engines. 2 stroke cycle.

Model	*Cylinder*	*Cooling*	*Cap. (cc)*	*Gross Power (b.h.p.) Max.*	*Cont.*	*@ r.p.m.*	*Nett Weight (kg)*
75	1	air	75	3	2.54	4500	9
95	1	air	95	4.75	4.25	4500	9.5
125	1	air	125	4.5	3.3	3000	16
150	1	air	150	5.8	4.2	3000	16
151	1	air	151	7.4	6.2	4000	16
161	1	air	165	7.6	4.5	3000	15
185	1	air	185	9.2	5.2	3000	15
200	1	air	199	7.5	6.0	3000	23
250	1	air	247	9.7	7.2	3000	23
250AS	1	air	247	14.3	11.0	4000	23
300	1	air	299	12.0	9.0	3000	23
500	2	air	494	16.2	14.2	3000	38
635	2	air	635	38.5	38	5000	40
640	2	air	635	38.5	38	5000	40
373	2	air	368	24	23.5	5250	33
637	2	air	635	40	39	5250	45

W217 Air-cooled 4 b.h.p. Unit

BERMOTOR LIMITED
21 London Road
Tunbridge Wells
Kent TN1 1SY
U.K.

Telex: 95446
Cables: Bermotor Tunbridge Wells
Phone: (0892) 37588

Multi-fuel spark ignition engines, four-stroke, fuelled with petrol (gasoline), paraffin (kerosene) or propane (l.p.g.). Power figures given for petrol-fuelled versions.

Model	*Cylinder*	*Cooling*	*Cap. (cc)*	*Gross Power (b.h.p.) Max.*	*Cont.*	*@ r.p.m.*	*Nett Weight (kg)*
18A	1	air	–	1.5	1.0	2500	11.3
318A	1*	air	–	2.0	1.3	2500	11.3
117B	1*	air	–	3.3	2.2	2500	14
217	1*	air	–	4.0	2.8	2500	14
239A	1*	air	–	4.5	3.3	2500	20
610A	1	air	–	6.0	4.0	2000	36
810	1	air	–	8.0	5.0	2000	36
112	1	air	–	10.0	8.0	1600	70

*vertical shaft variants available in these sizes

800-45	4	water	845	24.0	15.0	2500	98
688-45	4	water	1108	37.0	22.0	2500	108
810-45	4	water	1289	42.5	27.0	2500	108
821-45	4	water	1565	70.0	60.0	4200	–

Various engines available as generating sets, battery chargers, site welders, self-priming irrigation and drainage pumps, etc.

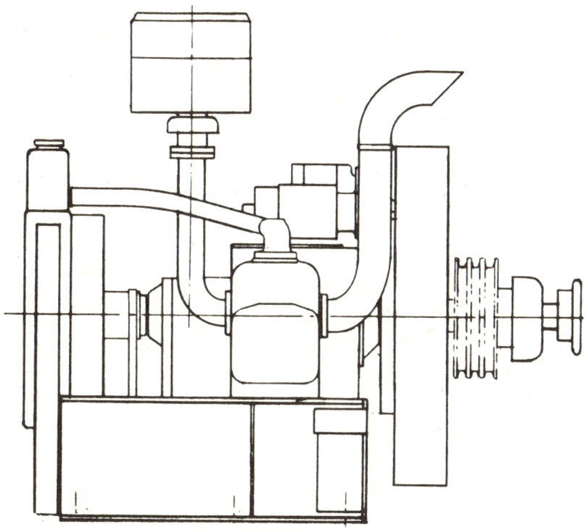

G-260 Engine (Oilfield) Unit

WITTE ENGINE CORPORATION
(A Hawker Siddeley Company)
P.O. Box 386
555 East 56th Highway
Olathe
Kansas 66061
U.S.A.

Telex: 042-6262
Phone: (913) 764 3512

Model G-260 natural gas or propane engine:

Model	*Cylinder*	*Cooling*	*Cap. (cc)*	*Gross Power (b.h.p.) Max.*	*Cont.*	*@ r.p.m.*	*Nett Weight (kg)*
G-260	2H	water	1966	17	8	600	545

This engine is available as a single or three-phase generating set and there is also the Model 120 diesel engine (see page 26) with many common components.

WHITE ENGINES INCORPORATED
101 Eleventh Street S.E.
Canton
Ohio 44707
U.S.A.

Telex: 98-3439
Cables: Hercano
Phone: 216 454 5631

Water cooled engines with power outputs 35 b.h.p. to 140 b.h.p. Can be supplied as closed generator power units, with housing, instrument panel, governor, etc.

Model	*Cylinder*	*Cooling*	*Cap. (cc)*	*Gross Power (b.h.p.) Max.*	*Cont.*	*@ r.p.m.*	*Nett Weight (kg)*
G1600	4	water	2670	77	55	2000	190
G2000	4	water	3240	84	69	2000	270
G2300	4	water	3700	89	76	2000	270
G3000	6	water	4880	130	100	2000	360
G3400	6	water	5550	144	115	2000	360

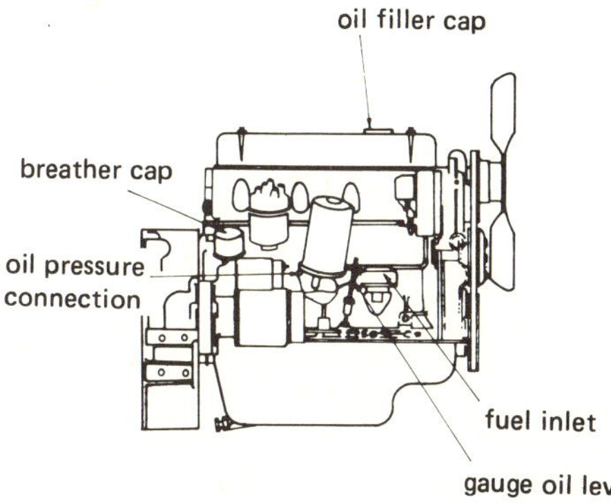

Model D-2000

RADICAL ENGINE GROUP
(Subsidiary of International Harvester)
Solar Division
2200 Pacific Highway
San Diego
California 92138
U.S.A.

Telex: 695045
Phone: (714) 238 5551

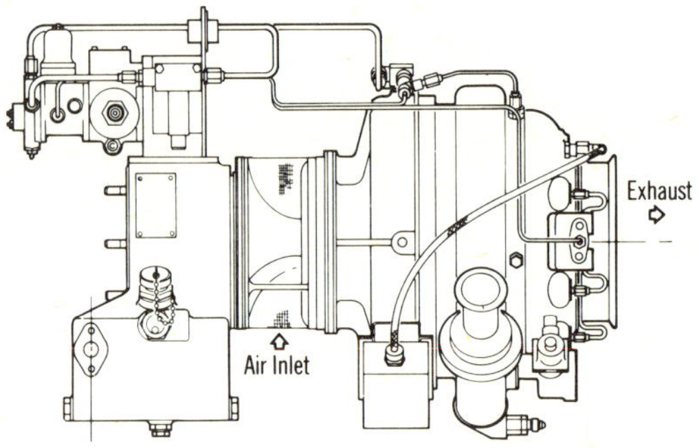

Solar Titan Gas Turbine Engine T-64T-40

A series of small gas turbines designed for aircraft auxiliary power unit and ground starter applications, but suitable for other generating applications, mechanical applications and for the production of pneumatic power simultaneously if required. All with single-stage centrifugal compressor and radial inflow turbine mounted back-to-back. All fuelled with aviation turbine fuels.

Model	*Max. Power Rating (hp)*	*Max. fuel consumption (lb/hr)*	*Output speeds (r.p.m.)*	*Environmental limits Temp. (°F)*	*Altitude (ft)*	*Weight (kg)*
T-62T-16B1	75	82	8000	– 65 to 125	15000	34
T-62T-11	75	92	(8000/ 7788)	– 65 to 130	6000	32
T-64T-40	90	140	(11760/ 12240)	– 65 to 130	30000	41
T-62T-16A1	95	108	8000	– 65 to 125	15000	32
T-62T-2A1	95	114	6000	– 65 to 125	15000	32
T-62T-12	105	115	(8216 & 8000)	– 65 to 130	20000	34
T-62T-27	150	140	(8216 & 8000)	– 65 to 130	20000	38

All of these operate for 2000 hrs between overhauls and have an expected service life of 10000 hrs. Typical start-up time to full power is 16 seconds.

6
External Combustion Engines

STEAM ENGINES
STIRLING ENGINES
THERMO-ELECTRIC CONVERTERS

In common with the internal combustion engines described in the previous section, external combustion (e.c.) engines convert heat into mechanical (or electrical) energy. The primary difference which characterises e.c. engines is that the heat is applied externally (and continuously) and transferred via suitable heat exchange surfaces to a working fluid within the engine.

There are three primary types of working fluid, a liquid/vapour combination (the steam engine), gas/hot air (Stirling engines) and electrons (thermoelectric generators). The first two systems formed the basis of the technological advances which led to the Industrial Revolution, while the latter is a relatively new development based on principles known for some time but not well understood until recently.

It is relevant to consider briefly the advantages and disadvantages of this category of heat engines, as they may in future come into increasing use again for small power applications.

The primary disadvantage of e.c. engines in general is that due to the fuel being burnt externally, considerable heat exchange area is required to transmit the heat into the working space(s) and to reject the waste heat at the end of the cycle. This normally implies numerous boiler tubes to transmit the heat from the flue gases to the water in a steam engine boiler, and a similar arrangement in the condenser which cools the exhaust steam. The same problem applies to the Stirling (or Erikson) engine which uses expansion and contraction of alternately heated and cooled air (or some other gas in some cases) to drive a piston. Thermoelectric generators, which utilise temperature differences to move free electrons across junctions between dissimilar metals (thermocouples) also require a large heat exchange area in order to admit sufficient heat flux from the burner or other sources of heat to the hot end and to keep the cold end cool.

The result is that e.c. engines are generally bulky and consequently expensive to construct compared with i.c. engines of the same power output. They also have considerable thermal inertia so that they cannot be switched on or off quite as quickly as an i.c. engine. As a result, i.c. engines rapidly became the primary autonomous power source early this century, largely replacing steam engines in medium power applications while Stirling engines which tended to be used in the nineteenth century for fractional horsepower applications, were rapidly replaced by electric motors once widespread electrification became the norm in industrialised countries. Steam engines succeeded in remaining dominant for railway and marine use until the middle of this century, as weight and bulk were less important in those applications. However, less labour intensive and more convenient installations became economically desirable and petroleum prices fell in real terms, so steam engines were superseded by the diesel in those roles too. The only area where steam engines are still favoured is for large-scale centralised electric power production and for propelling large ships requiring 10,000's of horsepower. Here steam turbines are used, as they can achieve better efficiencies than any other type of heat engine. In the lower power ranges, diesel engines have tended to reign supreme as a result of their efficiency and lower first cost resulting from mass-production.

However, the main disadvantage of diesel and other i.c. engines is that they are usually fed with petroleum-based fuels, which for the first time since the start of the industrial revolution are becoming more expensive in real terms. Since fuel costs are a primary expense with any heat engine used for anything other than very intermittent or stand-by applications, this development is likely to offer a better future to external combustion engines which have the characteristic of requiring heat rather than fuel – in other words, they can be run on anything that burns providing suitable combustion and heat exchange facilities are designed in. Coal and other solid fuels are becoming relatively cheaper than oil, and this trend is likely to continue due to depletion of oil reserves at a far faster rate than the world's far larger coal reserves. In the very long run, e.c. engines are the only heat engines that can readily make direct use of solar energy once all fossil fuels become scarce – Section 1 includes examples of solar heat engine systems of this kind developed for pumping water in desert areas.

Having listed the major shortcomings of steam and other e.c. engines, it is worth noting that they offer a number of advantages in addition to their multi-fuel capability. The main ones are that they tend to be longer-lasting than equivalent i.c. engines (which goes a long way to make up for their intrinsically higher capital costs); they are possibly better suited to small-scale manufacture – so that a suitable design could more likely be manufactured economically in small numbers within less developed countries; they are much quieter and more vibration-free than diesels in particular; and their exhaust emissions are much cleaner in terms of poisonous carbon monoxide and oxides of nitrogen than i.c. engine exhaust. Technical maintenance requirements and spare part inventories tend to be reduced with e.c. engines compared with i.c. machines, but (depending on the fuel), a considerable amount of routine ash removal and other low level maintenance may be required.

The general demise of e.c. engines throughout the first half of this century has meant that relatively little effort (or funding) has been invested in improving the technology. However, various new developments offer some prospect that much improved e.c. engines will emerge within the next few years – for example, heat pipes may be used to transfer heat much more cheaply and rapidly, small fluidised bed burners may improve the ease and efficiency with which fuels may be burnt, solid-state electronics offer the possibility of reliable, low-cost control, and so on. Work is in hand in many research centres on this kind of development, and in the meantime, a number of well-tried traditional steam engines have enjoyed a bit of a comeback, particularly for use in small boats, because of their quietness and reliability.

It remains to give a brief description of the principles and features of the three main categories of e.c. engines.

Steam engines

The steam engine's initial development in the 18th century, mainly for pumping water out of coal mines, led directly to the Industrial Revolution. In the engine water is boiled to produce steam under pressure, which is then allowed to expand either in a cylinder or through the rotor of a turbine, to do work. In traditional railway and traction engines the steam was then exhausted to the atmosphere doing nothing more useful on the way than to create a good draught for the fire by inducing the flue gases to flow more effectively out of the chimney. This is very wasteful, as a lot of heat is lost in the exhaust system; as a result the overall efficiency of steam engines of this kind tends to be less than ten per cent, which with present day fuel prices (even for coal) is generally unacceptable. The other disadvantage is that the lost steam has to be replaced with fresh water, which ideally needs to be softened and filtered to prevent boiler corrosion or scaling. Therefore it is preferable to recycle the exhaust system through a condenser (the latter of course adds to the weight, size and cost of the system). This allows a limited pure water supply to be used (which reduces boiler maintenance) and, more important, the efficiency is greatly improved as the condensate can be readmitted to the boiler at closer to boiling temperature. Engines of this kind can be from 15 to 30% efficient, comparing favourably with i.c. engines in this respect. However, fans or high smoke stacks are required to create a good draught.

Most of the units in this section are conventional reciprocating engines of traditional design, but there are also some steam turbines. The latter are particularly appropriate for direct electricity generation as they can couple directly to a standard alternator, whereas the reciprocating piston engines tend to be slow speed devices which would need a lot of gearing up to achieve synchronous speeds.

Small steam turbines are quite widely used to generate electricity from the waste products of such tropical industries as sugar refining and palm oil production, while reciprocating steam engines are best used for applications demanding high torque and slow speeds, such as for driving small boats, slow-speed machinery, etc. It is believed that a number of new products may be appearing in this field within the next year or two, and it is hoped to include details in future editions of this publication.

The principles of the steam engine can be applied using working fluids other than water. A number of engines have been developed which run on organic heavy vapours of the kind often used in refrigerators in order to operate efficiently at lower temperatures than is possible with a water boiling engine. Engines of this kind are of course always of the condensing type so that there is no loss of working fluid. Low temperature vapour engines using organic fluids or ammonia are of particular interest for solar powered applications as they can use the low-grade heat from flat-plate solar collectors or low concentration factor solar collectors. Although low temperature heat engines are inherently inefficient, low temperature solar collectors are the most efficient. As a result, the best system efficiency is often obtained with a low temperature vapour engine, rather than by trying to concentrate solar energy sufficiently to run a normal water-based steam engine. Renault/SOFRETES offer systems of this kind – see Section 1.

Stirling engines

This engine, originally developed by the Rev. Robert Stirling, in 1816, uses the pressure changes caused by alternatively heating and cooling air (or some other gas) to do work. A variety of different configurations have been developed. The primary advantage over the steam engine is that no boiler with its dangerous contents of high pressure steam is involved, but nevertheless, early Stirling engines tended to be bulky and heavy in terms of their power outputs and they were readily superseded by electric motors once electricity became more widely available.

In recent years a number of sophisticated new devices utilising the Stirling hot air (or more often another gas) cycle have been under development. Only one of these, the Aga Harwell Thermo Mechanical Generator, has appeared on the market, but as can be seen from the section on equipment under development, United Stirling in Sweden and Sunpower Inc. in the USA are in the process of developing Stirling engines which will eventually also be commercially available.

The main attractions of the Stirling engine are that being an e.c. engine it can operate on almost any heat source (the Sunpower version can be run on solar heat) and, if worked with a pressurised working gas, high efficiencies and a compact and reasonably light weight design can be achieved that appears competitive or even superior in some respects to diesel engines. To achieve this, however, the working gas needs to be at quite high pressures and until recently it was technically impossible to achieve effective sealing under the temperatures and conditions prevailing. Modern technological advances have made these problems soluble. The other major attractions are relatively quiet operation, less vibration and a cleaner exhaust (depending on fuel used) than comparable i.c. engines. It is difficult for Stirling engines to compete economically with i.c. engines initially, except in applications where their unique advantages, such as multi-fuel capability, give them an edge. This is because the engine has in many cases a more complex mechanism; the sealing problems already mentioned; and until now its development has been limited. Obviously, a coal or wood fuelled Stirling engine of comparable thermal efficiency to an i.c. engine would seem to be potentially very useful in areas where petroleum based fuels are particularly expensive and it is expected that machines of this kind will be developed within the next few years.

Thermo-electric generators

These are not external combustion engines in the strictest sense of the word, although they do convert heat energy directly into electricity, which is a more desirable and versatile energy form compared with a mechanical output. Obviously a mechanical output can be arranged simply by using an electric motor or servo.

The thermo-electric generator depends on a principle known since about the time when the steam and Stirling engines were invented, namely the thermocouple or Seebeck effect. That is, a voltage develops when the junctions of an electrical circuit consisting of dissimilar metals are maintained at different temperatures. The reasons for this phenomenon were only understood quite recently with the development of solid-state physics, and this knowledge has allowed the process to be optimised to some extent.

Two ranges of devices of this kind that are commercially available are listed in this section (from Teledyne Energy Systems in the USA and Global Thermoelectric Power Systems of Canada. These are designed for similar unattended remote low power applications as the previously mentioned Aga Harwell TMG and produce outputs in the range up to 90 watts while running on gaseous hydrocarbon fuels such as propane, butane or natural gas. The units can of course be combined to satisfy larger power requirements. As the power output is proportional to the temperature difference between the hot and cold end, these devices tend to give greater power outputs in cold ambient conditions.

The main advantages of the thermo-electric generator are that it has no moving parts at all (being a solid-state electronic device). So it is completely silent and vibrationless and the only maintenance needed is in connection with the burner or heat sources. Hence one example (from Teledyne Energy Systems) is claimed to have been operated without maintenance for 6 years and the claimed operational life is of the order of 75 years. Obviously like the other external combustion devices it potentially has a multi-fuel capability or could be solar heated, although currently available models generally run on propane or natural gas (methane).

Its disadvantages are that it has a high capital cost for its power output and the thermal efficiency is rather poor, so the fuel consumption tends to be high. Obviously, where long life, freedom from attendance or maintenance and great reliability are essential the high first and running costs become worthwhile. It is claimed that thermo-electric generators become economic in competition with other devices in remote applications where a continuous reliable power source is needed in the range from 1 to 300 watts.

There is some hope that if thermo-electric devices become more widely used so that they can be manufactured in larger quantities, the first cost may come down and further technical improvement may well lead to better thermal efficiencies. But even now there is little doubt that the existing commercially-available devices can be applied more economically than any other alternative in a number of specialised remote small-power applications.

Steam Engines

APE-ALLEN LTD
P.O. Box 43
Queens Engineering Works
Bedford MK40 4JB
U.K.

Telex: 824686
Cables: Apeng Bedford Telex
Phone: (0234) 67400

A typical 500kW ALLEN-KKK Turbine Generator

A major steam turbine manufacturer with a wide range of multi-megawatt steam plant, the following represent the very smallest end of the range and many units are in use generating power from the heat of waste products of various tropical industries, such as the palm oil industry, or from sugar cane bagasse.

Model	*Max. Output* (bhp)	*Max. Output* (kW)	*Temp. limit* (°C)	*Max. inlet steam pressure* (psi)	*Max. inlet steam pressure* (kg/cm^2)	*Max. exhaust steam pressure* (psi)	*Max. exhaust steam pressure* (kg/cm^2)	*Drive*
BF2.5	60	45	500	1800	125	70	5	direct
BF3.5	60	45	420	570	40	70	5	..
BF4	270	200	500	640	45	70	5	..
AF3.5D	410	300	500	1425	100	230	16	..
AF3.5G	410	300	500	1425	100	230	16	geared
AF4D	820	600	500	1425	100	140	10	direct

AF4G is a geared version of AF4D and AF4GV is a vertical shaft geared version.

All the above machines have overhung, single-stage rotors and all are fitted as standard with hydraulic overspeed and control governors. Some systems can be supplied as purpose built alternator sets.

PETER BROTHERHOOD LTD
Lincoln Road
Peterborough PE4 6AB
U.K.

Telex: 32154
Cables: Brotherhood Peterborough
Phone: (0733) 71321

Manufacture a range of small to medium steam turbines and turbo-alternators intended for waste process heat, marine waste heat and similar waste heat utilisation as well as for direct power production.

Smallest range of single-stage turbines are type SS MD. These are intended for any industrial mechanical drive application.

Power range: 200 to 3000 b.h.p. at speeds from 800 to 200 r.p.m.
Rotor: Impulse type, overhung.
Gearing: Built-in reduction gearing.

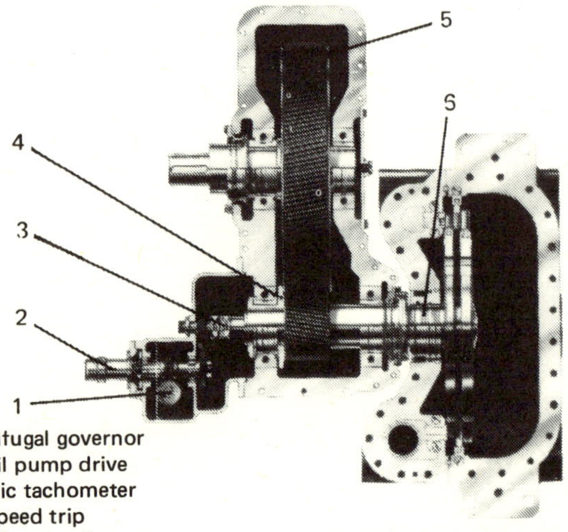

1. Centrifugal governor and oil pump drive
2. Electric tachometer
3. Overspeed trip
4. Gear thrust collar
5. Hardened and ground reduction gearing
6. Labyrinth gland seal

Overhung Turbine with Integral Geared Drive

M.B. ENGINE COMPANY
Watlington
King's Lynn
Norfolk
U.K.

Phone: (055) 381 282

Single-cylinder double-acting engine weighing 26kg (with flywheel). Uses steam at 85psi to produce 2 b.h.p. at 500 r.p.m. from a monotube boiler.

Suggested for combined domestic heating/lighting (to 1½kVA), small boats, pumping water, etc.

Available complete with monotube boiler as portable power unit weighing approximately 100kg.

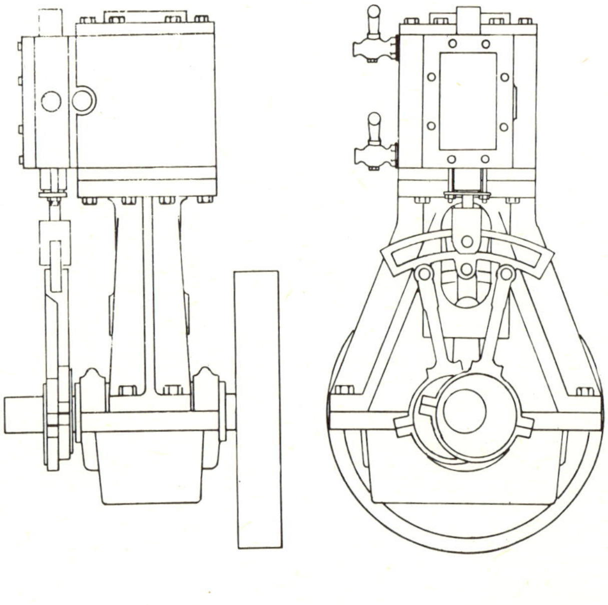

Single Cylinder Steam Engine

O'CONNOR ENGINEERING LABORATORIES INC.
100 Kalmus Drive
Costa Mesa
California 92627
U.S.A.

Telex: 685-641
Phone: 714 979 3993

Two and three cylinder steam engines. Power range from 5 to 25hp. Suitable for marine or industrial use. (See also for boilers).

Uniflow Steam Engine

PONTIS STEAM PLANT
The Gate
Cotterstock
Oundle
Peterborough PE8 5HD
U.K.

Phone: (083) 26 277

Manufacturers of small steam boilers and steam engines, most suitable for marine applications. Also adaptable, with suitable modifications, for stationary situations.

Vertical Coal Fired Boiler

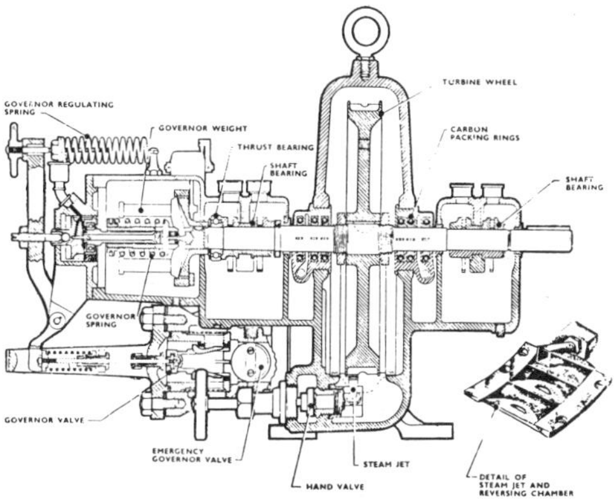

Sectional Arrangement: T100 Turbine

HAYWARD TYLER & CO. LTD
P.O. Box 2
Luton
Bedfordshire
U.K.

Telex: 82158
Phone: (0582) 31144

A range of simple, single-wheel steam turbines of well proven design. All units are fitted with a governor (various types available) and a hand synchroniser allowing 20% manual variation in the constant speed setting.

The makers claim that poor condition steam can be used for some models and a wide range of operating pressures and speeds are possible. As power depends on both steam pressure and speed of rotation, the use of a reduction gearbox to permit higher wheel speed often results in both a higher power output and improved efficiency.

Typical applications include pumps, mixers, alternators, compressors, blowers, etc. Immediate start up is possible.

Model	*Inlet psi.*	*Temp. Limit °C*	*Exhaust Press psi*	*Power Range h.p.*	*Max. Output h.p.*	*Max rpm*	*Net. Weight kg*
T100	200	274	50	5-150	250	4500	318
TS200	450	400	90	5-200	250	4500	354
T400	250	274	50	75-450	750	4500	795
TS600	450	400	90	75-650	750	4500	886
TC18	450	260	90	5-150	250	3600	375
TC18H	640	400	90	5-200	250	3600	450
TC24	450	260	90	75-450	750	3600	886
TC24H	640	400	90	75-650	750	3600	977

SEMPLE ENGINE COMPANY INC.
P.O. Box 6805
St. Louis
Missouri 63144
U.S.A.

Phone: (314) 961 6244

Range of small steam engines ranging from 5 to 20hp. Suitable for steam boats of 18 to 35ft. Units are either single or compound, operating at 600 r.p.m.

Units can also be used in stationary applications but the boiler may need an additional firebox if used with low grade fuels. The engine may also require a governor.

34-DW	Double-acting, single-cylinder	5hp at 600 r.p.m.	Weight 39kg
34-DL	Double-acting, single-cylinder	5hp at 600 r.p.m.	Weight 42kg
354-DL	Two-cylinder vee compound	10hp at 600 r.p.m.	Weight 84kg

FT-40 Vertical tubular boiler, operates at 150psi, steams in 14 minutes from cold; produces 245lbs/steam hr. weight 304 kg (suitable for above engines). Complete marine power units available comprising above engines/ boiler complete with pump, heater and condenser. Other boilers available to suit engines of 10, 20, 30, 60 and 120hp. The 354-DL engine can be twinned to produced a four-cylinder 20hp power unit.

354DL (10hp) Compound Unit

SPILLING CONSULT AG
Sonnenweg 4
CH-5610 Wohlen
Switzerland

Telex: 57 939
Cables: Spill Ch

Manufacture two different single-cylinder steam engine modules which can be assembled into multi-cylinder compound units, allowing a large number of power outputs from 10 through to 1200hp. These modules are designed to run at 750, 1000 and 1500 r.p.m. The units may be used for multi-fuel or waste-burning installations.

A consultancy service is available to ensure the most efficient use of basic resources.

Spilling 2 Cylinder Steam Motor

"STEAM POWER"
106a Derby Road
Loughborough LE11 OAG
U.K.

This organisation publishes a quarterly magazine on steam power, plus numerous plans of different types of steam engine and associated equipment. They also offer the "Panther" Steam Engine to the following specification:

"Panther" Steam engine:	90° Vee twin cylinder double-acting compound expansion engine.
Maximum power:	35 b.h.p. at 2000 r.p.m. with steam at 100 psig (approx. 70 bar) Minimum steam pressure 40 psig (approx. 3.5 bar) for continued operation.
Overall weight:	76.2kg (168lb)

It is understood that a comprehensive brochure is available for £1.70 or US $5.00.

Panther Steam Engine

Stirling Engines

Harwell TMG
AGA Navigation Aids
77 High Street
Brentford
Middlesex
U.K.

Telex: 935956
Cables: Agafaros Hounslow
Phone: 01-950 6465

The Aga Harwell Thermo Mechanical Generator is an unconventional device invented at the Harwell nuclear research centre. It works on the Stirling engine principle, but has no rotating parts. The output is via a linear, oscillating electric generator that produces a 25 watt, 100Hz AC electrical primary output which is then rectified to maintain 10 12V nickel cadmium cells capable of sustaining a 36 watt electrical load for a mean duty of 14 hours in each 24.

Its function is to produce a reliable electrical output from a propane (or other heat) energy input source, consumption being 200kg of propane per year. The device is intended to run for ten years without maintenance other than cleaning the propane burner annually when the fuel supply is replenished. Typical applications are remote unattended low power electrical devices such as marine buoys, microwave repeaters.

It is likely that further, higher powered models of this device will become available later.

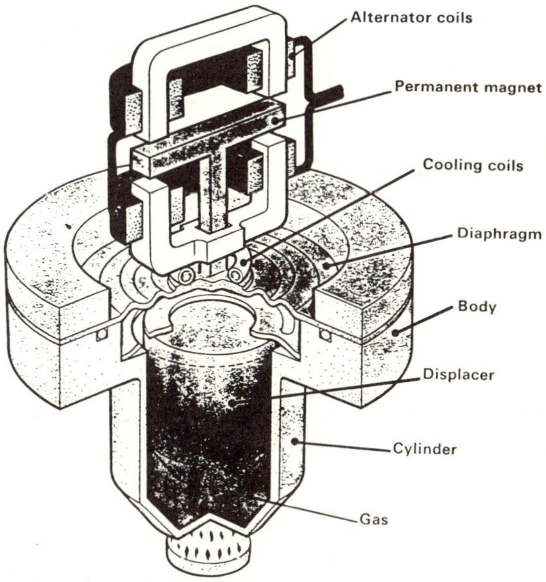

Aga Harwell Thermo Mechanical Generator

Thermo-Electric Converters

Global Thermoelectric Units

GLOBAL THERMOELECTRIC POWER SYSTEMS LTD
P.O. Box 459
Bassano
Alberta Canada T0J 0B0

Telex: 03 848141
Phone: 403 472 3512

Manufacturers of thermoelectric generators for use in remote and unattended locations — recommended preventive maintenance operations for these systems average ½ man-hour per year. With no moving parts, these systems are claimed to last for over 20 years of continuous operation.

The following models are available:

Model	*Voltage** (volt)	*Current** (amp)	*Power** (watt)	*Fuel consumption propane* (kg/day)	*natural gas* (m^3/hr)	*Weight of unit* (kg)
5020	1.8	11.1	20	0.90	0.05	19.5
5020-12	12	1.3	15			
5020-24	24	.65	15			
5030	1.8	16.7	30	1.50	0.087	19.5
5030-12	12	2.0	24			
5030-24	24	1.0	24			
5060	6.5	9.2	60	2.9	0.18	41
5060-12	12	4.2	50			
5060-24	24	2.1	50			
5060-48	48	1.0	50			
5120	7	17.2	120	5.8	.36	64
5120-12	12	9.0	108			
5120-24	24	4.5	108			
5120-48	48	2.25	108			

*Matched load @ 24°C.

Voltage adjustments 12-18, 24-30 or 48-60V DC; regulation ± 0.25% no load to full load.

Voltage ripple peak to peak under 250mV; ripple frequency 400Hz.

TELEDYNE ENERGY SYSTEMS
110 W. Timonium Road
Timonium
Maryland 21093
U.S.A.

Telex: 8 7780
Cables: Telises
Phone: 301 252 8220

Telan Thermoelectric Generator

Manufacturers of thermoelectric generators for use in remote and unattended locations. The manufacturers claim that installation is very easy, the system is ultra reliable and has a very long life of up to 75 years. As with all devices of this kind, performance improves under cold ambient conditions – the rated performance is given for ambient 70°F (21°C):

Model	*Voltage** (volt)	*Current** (amp)	*Power** (watt)	*Fuel consumption* propane (kg/day)	*Fuel consumption* natural gas (m^3/week)	*Weight of unit* (kg)
Telan 2T1	12	.67	8	5.1	6.52	15
	24	.33	8			
	48	.17	8			
Telan 2T2	12	1.58	19	10.2	13.00	24
	12	1.41	17			
	24	.71	17			
	48	.35	17			
Telan 2T3	12	2.50	30	15.3	19.55	33
	24	1.05	25			
	48	.52	25			
Telan 2T4	12	3.0	36	20.4	26.1	42
	24	1.5	36			
	48	.7	34			
Telan 2T5	12	4.16	50	25.4	32.6	51
	24	2.08	50			
	48	.87	42			
Telan 2T6	12	5.0	60	30.5	39.1	59
	24	2.5	60			
	48	1.06	51			
Telan 2T7	12	5.67	68	35.6	45.6	68
	24	2.83	68			
	48	1.23	59			
Telan 2T8	12	6.58	79	40.7	52.2	77
	24	3.08	74			
	48	1.52	73			
Telan 2T9	12	7.50	90	45.8	58.7	86
	24	3.71	89			
	48	1.85	89			

*Matched load @ 21°C.

7
Electricity Generation

ALTERNATORS & GENERATORS
GENERATING SETS
OTHER ELECTRICAL EQUIPMENT

This, of course, is not the only section in this Guide to deal with electricity generation, since many of the devices in the specialised sections on energy conversion produce electricity as an end product. Also, as the main prime-mover for small-scale electricity generation is the diesel engine, (and likely will remain so for the next decade or two at least), it is inevitable that a lot of diesel engine manufacturers offer their products with an alternator attached, while a lot of alternator manufacturers offer theirs with a diesel engine attached. As a result, quite a number of the entries in this section consist of alternators with diesel engines, while quite a number of the diesels in Section 5 are listed as being available with alternators. In fact there are so many different diesel generating sets available on the international market, from such a wide variety of manufacturers, representing an enormous number of permutations and combinations of different makes of engine and electrical equipment, that it would be quite impossible in a book of this scope to attempt to be comprehensive. It is hoped that the selection included provides a good cross-section of typical products and shows most of the more widely used machines; we will of course welcome suggestions from readers and manufacturers who would like to recommend further generating sets and associated equipment for inclusion in future editions.

Also, it is beyond the scope of this book to go into detail on internationally-available electrical transmission and control equipment. Agents and suppliers of this kind of equipment are well established and well known in most parts of the world and the best practice will generally be to use equipment that is locally available and which will, no doubt, conform to local electrical standards.

Electricity

This is not the place to discuss the technicalities of electricity generation — but a number of fundamental points are worth a brief mention as they affect the choice of equipment that might be used.

Electricity is primarily of course an energy transmission medium possessing the virtue of being readily convertible, with high efficiency and at low cost, into mechanical, light or heat energy. It can be easily transmitted some distance, but the greater the distance of transmission the higher are the voltages required to prevent undue losses in the transmission lines. High voltage lines need well-spaced conductors on large insulators carried by high towers — so long distance electricity lines are expensive and need to carry a heavy load to be economic. As a result it is not economic to send electricity to distant small communities — it is usually cheaper to send diesel fuel in drums and generate it locally — which is what is normally done in such situations.

Most electricity generation for anything other than battery charging involves the production of alternating current (a.c.), mainly because a.c. generators (alternators) are simpler, cheaper and more reliable than direct current (d.c.) generators (sometimes known as dynamos). In recent years, even cars have generally been fitted with alternators to save cost and improve reliability — the a.c. current then being rectified to d.c. to charge the battery.

Direct current (d.c.) consists of a steady flow of electric current at a fixed voltage, analogous to a steady stream of water flowing down a pipe, while a.c. fluctuates in voltage, changing polarity either 50 times a second (under most European standards) or 60 times a second (U.S. standards). The frequency of a.c. current is controlled by the speed of the generator — most generators produce 50 cycles/second (or 50 Herz) at a driven speed of 1500 rpm, while American machines are often driven at 1800 rpm to produce the US normal output of 60 Herz. Most European electrical appliances are designed for voltages in the range 210 to 240, (50hz) while American equipment tends to be 110V. A lot of equipment on the international market tends to be dual standard and can be adjusted before being put into operation to use either type of supply, or it can be ordered in the first place to suit either type.

Therefore the most basic decision in choosing a generating set is the voltage and frequency that will be required. Here it is probably best to use whatever is normal for the areas of your country with mains electricity for two reasons; firstly appliances and electrical equipment, cables, etc., will be readily available to suit and secondly, if mains power is ever extended to wherever the local generating set is located, then the local wiring and appliances will all be immediately usable.

With direct current, the current (electron flow) and voltage (motive force that moves the electrons) are constant and in the same direction, and the power being transmitted is simply calculated by the product of the voltage and the current; i.e. volts x amps = watts, (for example, a current of 5 amps (A) at a voltage of 240 volts (V) = 5 x 240 = 1200 watts (W) or 1.2 kW). With a.c., this is only true if the voltage and current fluctuate simultaneously, or in phase with each other — in practice, for reasons that need not be detailed here, any electrical appliances with electro-magnetic components (i.e. electric motors, transformers, relays) will have the effect of making the current tend to lag behind the voltage, whereas any components with capacitance (i.e. electrical circuitry that can store an electrical charge) will have the reverse effect and tend to make the current variations lead the voltage ones. Long transmission lines are capacitive and a lot of electric motors in the load are inductive — the two could cancel each other out and make the voltage and current vary in phase, but in practice they are usually a little out of phase. The effect of this is to reduce the transmitted power by a certain amount which is measured by the Power Factor. For example, if the phase difference is such that the transmitted power is 90% of the product of amps times volts, we say that the power factor is 0.9 and the actual power available will be the volts times the amps times the power factor. As a result, it is normal to specify the output of a generator in kilo-volt-amps written as kVA and the actual power in kW delivered will be the kVA times the power factor of the circuit; often perhaps 0.8 or 0.9 times the kVA.

Obviously it is desirable to get the power factor as near to unity as possible. The economics of the equipment necessary usually dictate that this is only done with reasonably large, continuously running generating sets — it is hardly worth bothering with a small intermittently used portable set or stand-by unit. The manufacturers of larger, more sophisticated generating equipment (i.e. generators of 5 to 10kVA and upwards, rated for continuous operation) will usually specify in some detail what kind of circuitry should be used in conjunction with their equipment for different purposes and will often supply any special items as a package deal.

Another feature of a.c. electrical power generation is that it is common practice to generate three electrical pulses of voltage and current on the same machine, but equally-spaced, out-of-phase with each other — these voltages and currents are sent down three separate live conductors, but the return (neutral) lines are combined into one conductor. This is known as three-phase transmission (and it is almost universally used for all electrical transmission systems of any significant power and size). The reason is economic — three-phase generators and motors are smaller and cheaper than single-phase ones of the same output. On the other hand, it is far simpler to connect ordinary small domestic appliances, fractional-horsepower motors and lights to a single-phase, two-wire circuit — so most three-phase generating systems supplying a large number of small loads tend to have as evenly balanced a collection of small single-phase loads as possible on separate circuits connected to each of the three phases. Large motor loads, on the other hand, are usually three-phase as well and have four terminals to allow connection direct to four-conductor, three-phase systems.

The connecting up of three-phase equipment can be quite complicated, as there are several different methods for linking the terminals on a three-phase generator or motor. It should be noted that the voltages between the three lines are considerably higher than between each line and the neutral return. It is of course essential that installations of this kind are completed, or at least supervised, by a qualified electrician or electrical engineer, so no attempt will be made here to instruct in the details of connecting up electrical circuits.

It must of course be stressed that electricity is extremely dangerous if mishandled, and the smallest low-powered generating sets with tiny 49cc two-stroke power units can administer a fatal electric shock if of 240V output, while even a 110V output can be dangerous for someone with a weak heart if touched with wet hands or in some other careless way that ensured a good electrical connection. Therefore, always switch off the generator or at least isolate all circuits before undertaking any maintenance, repairs or modifications.

Alternators and generators

As already explained, almost all electricity generation is in a.c. using alternators. There are two primary types of alternator; so-called rotating armature slip-ring machines and fixed armature, rotating field, brushless machines. Slip ring machines have a lower first cost but they need more maintenance as they have carbon brushes running on brass slip rings which eventually wear out and need replacement. Brushless machines are really two alternators in one (which explains their greater cost) as they have a small rotating armature which activates a rotating d.c. field coil (via built in rectifiers) on the same shaft, which in turn generates the output within the stator armature. As a result no electrical connections are needed between rotating and static components and brushes can be eliminated from the design. This makes brushless machines almost maintenance free, the bearings being the only wearing components.

Dynamos for d.c. generation are relatively unusual except for low-power battery charging applications such as in motor vehicle electrical systems, and a few of the smaller windmills intended for low voltage, low power outputs. These have brushes and brass commutators (which resemble segmented slip rings) which tend to wear faster than slip rings and generally need slightly more maintenance.

Most modern electrical machines are enclosed sufficiently to be "drip proof" — (early machines tended to be exposed and vulnerable to damage from any water splash or drips). Control equipment is sometimes built in, but can usually be supplied when it is not, for regulating the voltage and frequency within close enough limits to permit the use of standard electrical equipment that might be sensitive to voltage or frequency "excursions". This applies particularly to electronic devices, but excessive voltage can burn out almost all electrical appliances, light bulbs in particular.

Generating sets

An enormous variety of different types of generating set is available. They range from very small portable units with a pull-start and outputs of only a few hundred watts, for use with portable electrical equipment that is used intermittently, such as small power tools or temporary lighting, to large permanent installations that are in effect miniature power stations, complete with automatic control equipment and designed for continuous operation.

Many of the larger diesel generating sets are available either as standby units to provide power during periods of mains supply failure, or as continuous power units. Standby machines have automatic start-up facilities which activate them if the mains voltage falls below some preset limit.

A lot of the remarks in the introductory passages to Section 5 on internal combustion engines apply to the choice of power units for generating sets. It is particularly important to make sure that the chosen unit is sufficiently derated for the operation envisaged. It is invariably cheaper in the long run to use a machine which is slightly too powerful for the job than to use one that has a struggle to deliver sufficient power to satisfy the demand. A vital factor, of course, is the availability of spare parts and service, particularly for the engine, so before looking too closely at technical features, it is often as well to investigate whether a reliable agency is close at hand.

When choosing a power unit it is particularly important to consider the likely load demands and their nature. A generator used to drive electric motors capable of absorbing a significant proportion of the generator output must be capable of sustaining a quite heavy over-load for a few seconds when starting an electric motor, as motors draw a much increased current while running up to speed. Even refrigerators and other domestic appliances with electric motors can cause an appreciable extra load when their motors start up. Obviously it would be very wasteful to use electricity from a generating set for heating purposes (either for space heating, water heating or cooking) as the same or similar fuel that is used to run the engine could be used much more efficienly and cheaply by burning it in suitable appliances directly, and renewable resources such as solar heating or wood fuels (where available) will often be still more economical. If heat is required for space or water heating and a water cooled generating set is in use, it is possible to use engine waste heat, via a heat exchanger, to heat water or to space-heat using equipment similar to a car's heater. Indeed, an interesting challenge for all who use regularly operated generating sets is to find use for the waste heat.

A number of generators intended for continuous operation can be supplied with extra silencing and soundproofing. This can be worthwhile specifying in many cases as in addition to the environmental improvement, it may permit the generator to be located nearer to its load and save on the costs of transmission lines and save transmission losses.

Other electrical equipment

This section includes a number of items connected indirectly with electricity generation, such as batteries and inverters.

Batteries are at present only generally available as lead-acid or nickel-cadmium for most practical and economic purposes. Other types exist, and some quite promising new types are under development, but none of these are yet economically competitive. Lead-acid batteries, as used for most motor vehicle electrical systems, tend to cost around £25 to £30/kWh, which makes them much cheaper than their only real rival, nickel-cadmium at around £100/kWh. Nickel cadmium batteries are worth the extra cost for some applications, as they can withstand deep discharge and overcharging much better than lead-acid batteries, and they can be short-circuited or deliver very heavy currents for a brief time with less likelihood of damage occurring.

Batteries in their more common form require regular replenishment of the electrolyte with distilled water. However, nickel cadmium batteries have been available for some time in a maintenance-free sealed version and new lead-acid batteries of a similar kind, with a jellied electrolyte that needs no topping up, are coming onto the market. It is quite important to ensure that the capacity of a lead-acid battery or batteries allows them to avoid excessively deep discharge between recharging intervals, which may necessitate a greater storage capacity having to be provided in certain circumstances when lead-acid are chosen than if nickel cadmium had been used, which could partially offset some of the cost differential.

For heavier power applications, lead-acid batteries would normally be used for cost reasons, while for specialised small power applications (such as in conjunction with solar cells) the lower maintenance requirement of sealed lead acid or of nickel cadmium batteries can justify their extra expense.

For power applications involving energy storage in excess of a few kWh, it generally becomes uneconomic to use batteries, and a standby diesel or petrol engine with a fuel supply is likely to be cheaper. The running costs for the standby engine in such situations will be higher, but its lower capital cost could more than outweigh this.

In any case, where energy storage by battery is being contemplated, it would be well to write to the manufacturers' technical advisory services to seek advice on correct sizing to suit the need. This is particularly important with lead-acid batteries to avoid excessive discharging which could greatly shorten their life.

Where batteries are required for storage, but a conventional a.c. output is needed to run mains voltage equipment such as domestic electrical appliances, the modern method for performing the conversion from d.c. to a.c. is via a solid-state electronic inverter, some examples being included in this section. A number of suppliers of electricity-generating windmills also supply inverters, since batteries are commonly used with small and medium-sized windmills as an energy store for use in windless periods.

Once an a.c. output is obtained, different voltages can be produced using a suitable transformer, although most inverters are designed to convert to normal domestic a.c. voltages. To go from a.c. to d.c. requires a rectifier; modern practice being to use solid-state diodes to perform this function. It is beyond the scope of this book to detail suppliers of transformers and rectifiers; they are fairly readily available from most wholesalers of general electrical equipment.

Note on code letters in the following section:

B	=	brushless
S	=	slipring
4	=	4-stroke engine
2	=	2-stroke engine
D	=	diesel engine
P	=	petrol engine (spark ignition)
G	=	gas engine
W	=	Water (liquid) cooled engine
A	=	air cooled engine
T	=	turbo-charged engine
P.T.O.	=	power take off drive

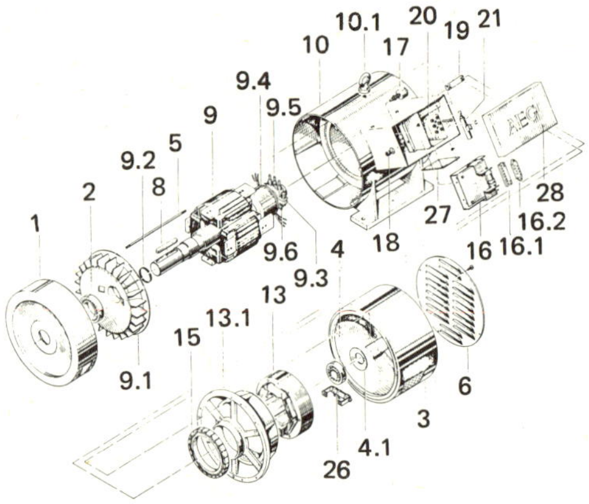

Type 4221-4230/04

1. bearing plate (driving end – AS)
1.1 louvre plate (driving end – AS)
1.2 air baffle
2. bearing (driving-AS)
3. bearing plate (non-driving end – NAS)
4. bearing (non-driving end – NAS)
5. stud bolts
6. cover plate
8. feather
9. rotor, complete
9.1 fan
9.2 spacer ring
9.3 diode assembly
9.4 exciter rotor
9.5 rotating diode (+)
9.6 rotating diode (–)
9.7 varistor, complete
10. stator complete
10.1 eyebolt
13. exciter stator
13.1 exciter support
15. pilot exciter stator
16. regulator, complete
16.1 terminal strip L2
16.2 terminal strip L3
17. reference voltage adjuster
18. locking screw
19. terminal strip L1
20. main terminal board
21. condenser
26. series resistor
27. cover plate
28. terminal box cover

AEG-TELEFUNKEN
(U.K.) LTD
202 Kensington Church Street
London W8 4DP
U.K.

Telex: 22795
Phone: 01 229 9244

Mounting: Flanged feet
Range: 64 models in 2 ranges from 30 to 2520kVA.
Connections: Keyways are standard, conical and optional.
Control Options: Automatic voltage control, with overload and over-speed protection. Parallel running facility is optional.
Voltage options: 400V 50Hz and 450V 60Hz three-phase are standard, others on request.
Power factor: 0.8
Protection: Drip proof and screen protected.
Agencies: Subsidiaries world wide.

Type	*kVA Single-phase 50Hz*	*kVA Single-phase 60Hz*	*kVA 3-phase 50Hz*	*kVA 3-phase 60Hz*	*V.*	*Gen. Spec.*	*Weight (kg)*
DKBH4221/04	Not		30	36	400/	B	290
.. 4222/04	applicable		35	42	450	B	300
.. 4223/04			40	48	..	B	320
.. 4224/04			45	54	..	B	330
.. 4225/04			51	61.2	..	B	350
.. 4226/04			60	72	..	B	380
.. 4227/04			70	84	..	B	420
.. 4228/04			80	96	..	B	450
.. 4229/04			90	108	..	B	460
.. 4230/04			97	115	..	B	510
.. 4254/04			120	144	..	B	575
.. 4255/04			140	168	..	B	655
.. 4256/04			150	180	..	B	735

AC5-24 Alternator

CAV LTD
P.O. Box 36
Warple Way
London W3 7SS

Telex: 27881
Cables: Vanteria London
Phone: 01 743 3111

Mounting: Cradle mount or wing mount (as per vehicle generator)
Range: 7 models in 3 ranges from 0.78kVA to 1.2kVA
Connections: Keyed output shaft
Alternator make: CAV
Control options: Automatic full wave regulation to give d.c. output. Solid state regulator for voltage options
Voltage options: 12.24 or 32V d.c.
Power factor: 1
Protection: Corrosion resistant
Special features: Small range of electrical control and protection devices. Different models in the same range have varying cut-in speeds.

Type	*Engine power absorbed (kW)*	*Single-phase*	*3-phase (kVA)*	*V d.c.*	*Gen. Spec.*	*Weight (kg)*
AC512	1.8	65-100 amps	0.78-1.2	12	S	5.9
AC524	1.8	33-50 ..	0.79-1.2	24	S	5.9
AC532	1.8	23 ..	0.74	32	S	5.9
	@ 3500 r.p.m.					

Brush SCA/SDA Generator

BRUSH ELECTRICAL MACHINES LTD
Light Machines Division
P.O. Box 18
Loughborough
Leicestershire LE11 1HJ
U.K.

Telex: 341091
Cables: Brush Loughborough
Phone: Loughborough 63131

Export address:
Hawker Siddeley Electrical Export Ltd.
P.O. Box 18
Loughborough
Leicestershire LE11 1HJ
U.K.

Mounting: Flanged feet
Range: 6 models in 2 ranges from 1.2kVA to 14.4kVA
Connections: Keyed shaft output suitable for V belt drive
Control options: Self regulating, top mounted terminal box. Loose leads optional
Voltage options: Specify from 115/230/240 single-phase, 210/380/420/440, 3-phase. 2 to 4 outputs available from each frame.
Power factor: 0.8
Protection: Drip proof and screen protected
Special features: Regulation and radio suppression may be omitted.
Agencies: Through Hawker Siddeley and Lister agents

Type	*kVA*				*V*	*Gen. Spec.*	*Weight (kg)*
	Single-phase		*3-phase*				
	50HZ	*60Hz*	*50Hz*	*60Hz*			
SCA 117	1.2–3.2	1.2–3.6	2.4–4.0	2.4–4.8	115 up	S	59
SCA 225	3.2–6	4–7.2	4.8–8	5.6–9.6	to 440	S	105
SCA 320	6.4–8.8	8–10.4	8.8–12	11.2–14.4	in steps	S	181
SEA 1A	1.2–2.4	1.4–2.9	2–3.2	2.4–3.8		S	55
SEA 1B	3.2–4.8	3.6–6	4–6.4	4.8–7.7		S	59
SEA2	6.8–8	8–9.6	8–10.4	10.–12.4		S	105

Portable Gensets Petrol and Diesel Engine

No detailed specifications available.

ELMOT ENGINEERING CO. PRIVATE LTD.
2 Udyog Nagar
Swami Vivekanand Road
Goregaon (West)
Bombay 62 NB
India

Telex: 011-3389
Cables: Alternator Bombay 400 023
Phone: 696351,2

Mounting: Smaller units are frame mounted, larger ones are skid mounted. Trolley and trailer options are available
Range: From 0.5kVA to 250kVA
Engine make: Made in India, diesel and petrol types available
Engine features: Hand or electric start available.
Connections: Keyed output shaft on generators supplied separately
Control options: Machine mounted as standard. Mains failure and parallel running facilities are available.
Voltage options: 230V single-phase, 415V three-phase, or to order, 50Hz
Protection: Drip proof Class A insulation. Canopies are optional
Special features: All spare parts manufactured in India.
Agencies: Branches through India

BRF 280 Drip-proof Machine

ELECTRIC CONSTRUCTION (W'TON) LTD
(Hawker Siddeley Group)
Bushbury Engineering Works
Wolverhampton W10 9LE
U.K.

Telex: 339618
Cables: Electric Wolverhampton
Phone: 0902 21455, 27831

Mounting: Flange mounted
Range: 12 models from 15kVA to 1200kVA
Connections: As specified or with SAE or Ford flanged end bracket
Alternator make: Electric Construction
Control options: Built in voltage regulator, machine mounted terminal box.
Voltage options: 415/240 option is standard. Specify phase and other voltages.
Power factor: 0.8
Protection: Tropical insulation. Drip proof or screen protected as specified.
Special features: Machine or wall mounted control panel optional
Agencies: Concessionaires in 36 countries.

Type	*Single-phase*		*kVA 3-phase*		*V*	*Gen. Spec.*	*Weight (kg)*
	50Hz	*60Hz*	*50Hz*	*60Hz*			
200	15	18	25	31	415/	B	225
200L	21	25	37	44	240	B	310
225	24	29	40	50	..	B	310
225L	30	35	50	60	..	B	360
250	42.5	51	72	86	..	B	450
250L	58	68	100	120	..	B	535
280	70	84	120	150	..	B	600
280L	100	120	170	204	..	B	715
355	120	144	210	250	..	B	1090
up to							
500L			1000	1200	..	B	3390

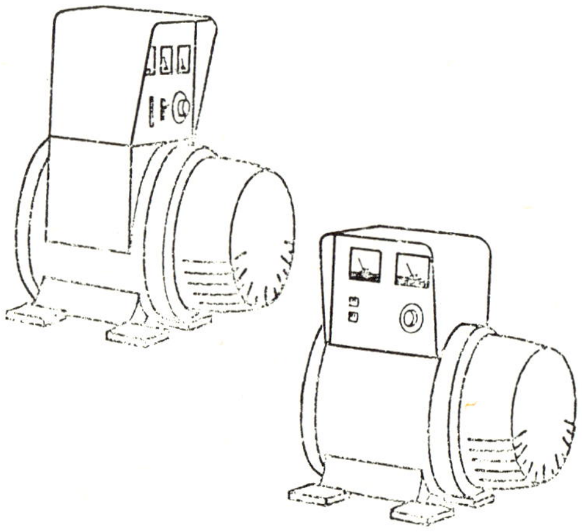

Model H Loadster Brushless Generator

FIDELITY ELECTRIC COMPANY
332 N. Arch Street
Lancaster
Pennsylvania 17603
U.S.A.

Phone: 717 397-8231

Mounting: Flange feet
Range: 7 models from 5kVA to 50kVA
Connections: SAE 2, 3, 4, and 5 are standard, others available.
Control options: Inherent voltage regulation. Machine mounted terminal box as standard. Variety of control options available.
Voltage option: 120 to 480 in steps by connecting terminal box
Power factor: 0.8
Protection: Drip proof. Tropical insulation and sustained short circuit protection available.
Special features: Limited range of control and instrumentation options available. The generator is lightweight and (it is claimed) has good electric motor starting characteristics.
Agencies: U.S.A.

Frame size	*kVA Single-phase*		*3-phase*		*V*	*Gen. Spec.*	*Weight (kg)*
	50Hz	*60Hz*	*50Hz*	*60Hz*			
284	5	7.5	6.25	9.4	120	B	not
284	7.5	10	9.4	12.5	up	B	avail-
284	10	12	12.5	18.75	to	B	able
284T	12	15	18.75	25	480	B	
286T	15	20	25	31.25	..	B	
286T	20	25	31.25	37.5	..	B	
286T	25	30	37.5	50	..	B	

"D" Range D8 Alternator

NEWAGE ENGINEERS LIMITED
P.O. Box 17
Barnack Road
Stamford
Lincolnshire PE9 2NB
U.K.

Telex: 32268
Cables: Leopower
Phone: 0780-2552

Mounting: Flange mounted
Range: 20 models in 2 ranges from 2kVA up to 630kVA
Connections: Standard range of flange adaptors available
Alternator make: Newage
Control options: Automatic voltage regulation, output is loose wires or terminal box
Voltage options: Most standard voltages available by connection
Power factor: 0.8
Protection: Drip proof on larger range. Optional on small range. Tropical insulation
Special features: Short circuit protection and parallel running units are available
Agencies: Comprehensive world wide agency list

Type	*Engine Input h.p.*	*kVA single-phase 50Hz*	*kVA single-phase 60Hz*	*kVA 3-phase 50Hz*	*kVA 3-phase 60Hz*	*V*	*Gen. Spec.*	*Weight (kg)*
D8A	5	3.5	4	5	6	220	S	59
D8B	6.5	4	5	6	8	up	S	81
D8C	12-17	7.5	9	11	13.5	to	S	102
D11A	15-27	10	12	15	18	440	S	145
D11B	20-37	13.5	16	20	24	..	S	159
C20A		12.5	16	20	25	100	B	216
C20B		20	25	31	37.5	up	B	254
C30A		30	44	50	62.5	to	B	381
C30B		50	62.5	80	94	600	B	457
C40A		65	81	100	125	..	B	622
C40B		85	106	137.5	162.5	346	B	686
C434A				150	180	up	B	775
up to						to		
C634B				630	750	600	B	2123

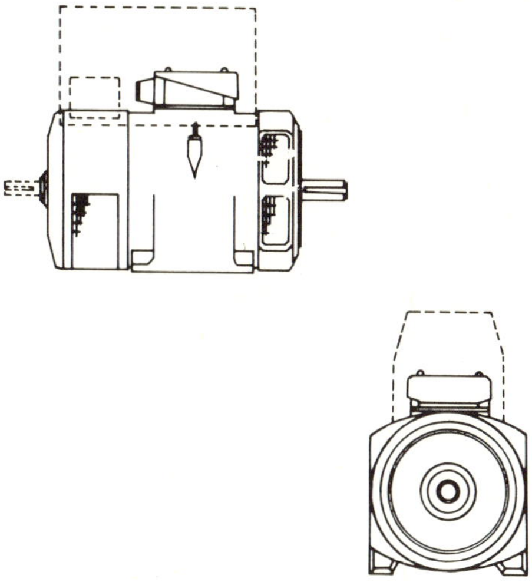

GEC Industrial A.C. Generators

GEC MACHINES LTD
Diesel Generator and Machine Sales
Mill Road
Rugby
Warwickshire CU21 1BD
U.K.

Telex: 31671
Cables: Assocelect Rugby
Phone: (0788) 2121

Mounting: Flanged feet
Range: 22 models in 6 frames from 6.25kVA to 162.5kVA
Connections: Variety of splined output shafts to order
Alternator make: GEC (as used on Dorman Diesels)
Control options: Self-regulating, machine or wall mounted switchboards can be supplied. Parallel running, short circuit protection and voltage build up accessories are available.
Voltage options: Any standard voltage up to 480V by reconnection
Power factor: 0.8
Protection: Tropically insulated
Special features: A variety of control units are available specially suited to a more industrial use
Agencies: Through Dorman agents

Type	*Engine*	*kVA single-phase 50Hz*	*kVA single-phase 60Hz*	*kVA 3-phase 50Hz*	*kVA 3-phase 60Hz*	*V*	*Gen. Spec.*	*Weight (kg)*
C20A	16	7.5	6.25	12.5	12.5	up	B	216
C20B	35	17.5	22	25	31	to	B	254
C30A	50	27.5	37.5	45	56	480	B	381
C30B	80	45	50	70	81	..	B	457
C40A	100	60	81	100	125	..	B	622
C40B	175		106	137.5	162.5	..	B	686

B Range Series 2 Alternator

MARKON ENGINEERING CO. LIMITED
Long Row
Oakham
Leicestershire LE15 6LW
U.K.

Telex: 34472
Cables: Markon Oakham
Phone: (0572) 3811/5

Mounting: Flanged base
Range: Various ranges 0.5kVA to 1125kVA. 1 range listed
Connections: ¾" or ⅞" standard SAE taperbore shafts. Turning @ 3000 r.p.m.
Alternator make: Markon
Control options: In-built automatic voltage control
Voltage options: Specify 220-240 or 110-220V
Power Factor: 0.8
Protection: Drip proof
Agencies: World wide manufacturers, subsidiaries and agencies

Type	*Engine Input h.p.*	*kVA single-phase 50Hz*	*60Hz*	*3-phase 50Hz*	*60Hz*	*V*	*Gen. Spec.*	*Weight (kg)*
SC21a	4-5	2	2.5	Not applicable		110	S	20.5
SC21b	4.5-7	3	3.75			up	S	23
SC21c	6-8.5	4	5			to	S	27
SC21d	7-10	5	6			240	S	30

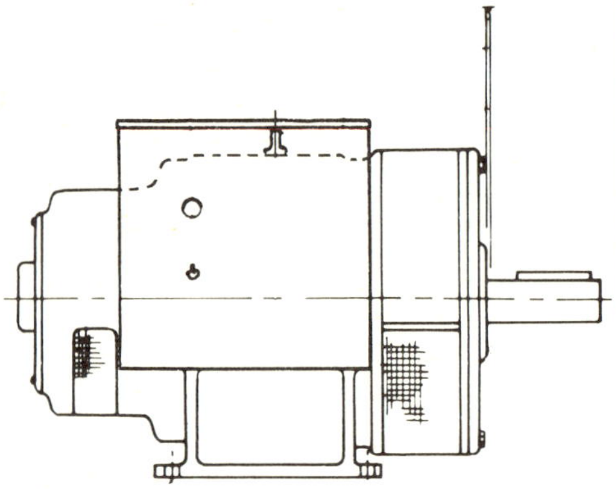

Brushless A.C. Generator

NEWTON DERBY LIMITED
Alfreton Road
Derby DE2 4AG
U.K.

Telex: 37580
Cables: Dynamo Derby
Phone: 0332 47676

Mounting: Skid mounted (generators have flanged feet).
Range: 24 models in 2 ranges from 7.5kVA to 400kVA. 3 ranges (generator only) 1.5kVA to 330kVA. Full range not shown.
Engine make: Smaller range uses Lister, larger uses Cummins
Engine features: Crank start for smallest engines, remainder are self starting.
Connections: Smallest generators have a female taper output shaft. Larger ones are spiggotted to accept a range of SAE adaptors.
Alternator make: Newton Derby
Control options: Machine mounted control box for complete sets. Generators have automatic voltage control.
Voltage options: 380/440 Volts for complete sets. Generators offer 115/230, 220/440 or 380/440 or others as specified.
Power factor: 0.8
Protection: Drip proof and tropically insulated
Special features: Limited accessory range of basic items (e.g. fuel tank)
Agencies: Agencies in the Middle East, North Africa and Far East Asia.

Type	*Engine*	*kVA Single-phase 50Hz*	*60Hz*	*3-phase 50Hz*	*60Hz*	*V*	*Gen. Spec.*	*Weight (kg)*
RC3LA6	4DA	5	6	7.5	9	380/	S	Not
RC3LA7.5	4DA	7.5	9	10	12	440	S	avail-
RC3LA11.5	4DA	10	12	15	18	..	S	ble at
RC3LA15	4DA	12.5	15	18.75	22.5	..	S	time
RC3LA20	4DA	17.5	21	25	30	..	S	of
RC3LA28.8	4DA			36		..	S	print-
RC3LA32	4DA			40	48	..	S	ing.
RC3LA40	4DA			50	60	..	S	
RC3LA48	4DA			60	72	..	S	
SER3CW56	4DW	36	40	70	80	..	B	
SER3CW64	4DW	40	47	80	90	..	B	
SER3CW80	4DW	50	60	100	110	..	B	
SER3CW104.5	4DW	70	80	135	156	..	B	
SER3CW124.8	4DW	80	90	156	175	..	B	
SER3CW140	4DW	90	100	175	200			
up to								
SER3CW320	4DW			400				

Generating Sets

Type PBF Generating Set

ALLAM GENERATORS
Arterial Road
Eastwood
Leigh-on-Sea
Essex
U.K.

Telex: 995127
Cables: Epaltrical Southend
Phone: 0702 526551

Mounting: Petrol engines are frame mounted, diesel engine is skid mounted. A Barrow mount is available for the diesel.
Range: 5 models in 2 ranges, from 1.5kVA to 7.5kVA
Engine make: Petrol engines are Briggs and Stratton, diesel engine is Petter.
Engine features: All engines are hand started and run at higher r.p.m.
Control options: Self regulating with machine mounted panels. Short circuit protection is standard.
Voltage options: Each unit offers 1 or 2 voltages from a range of standard voltages.
Power factor: 0.8
Protection: Drip proof
Special features: Long run fuel tank available for diesel set
Agencies: Middle East, Africa and Indian sub-continent.

Type	*Engine*	*kVA Single-phase*		*kVA 3-phase*		*V*	*Gen. Spec.*	*Weight (kg)*
		50Hz	*60Hz*	*50Hz*	*60Hz*			
PBD1.5	4DA	1.5	1.5			115/230	B	35
PBE3	4DA	3		3		.. .. / 415	B	70
PBF5	4DA	5		5		.. .. ..	B	124
PBF7	4DA	7		7		.. .. ..	B	128
DPF7.5	4DA	7.5		7.5		.. .. ..	B	176

3304 Generator Set

CATERPILLAR OVERSEAS S.A.
118 Rue du Rhone
P.O. Box 408
1211 Geneva
Switzerland

Telex: 22706, 22833
Cables: Catoversea Geneva

Mounting: Skid mounted
Range: 20 types, 63kVA to 1125kVA
Engine make: Caterpillar
Engine features: Variety of cooling options, self-start, all sets are test run.
Control options: Machine mounted, variety of control board configurations, automatic start and remote monitoring available. Parallel running facility
Voltage options: Short steps from 115 to 600
Power factor: 0.8
Protection: Drip proof
Special features: Small accessory range (includes parts of starting equipment)
Agencies: World wide, plus printed guides to aid generator selection.

Type	*Engine*	*kVA Single-phase*		*kVA 3-phase*		*V*	*Gen. Spec.*	*Weight (kg)*
		50Hz	*60Hz*	*50Hz*	*60Hz*			
3304	4DW			63	63	115	B	1200
3304	4DWT			75	75	to	B	1204
3304	4DWT			100	106	600	B	1259
3306	4DW			125	131	..	B	1905
3306	4DWT			163	163	..	B	2000
up to								
0399	4DWT			875	1063	..	B	10297

44T28 – 28kVA Diesel Generator

BERMOTOR LIMITED
21 London Road
Tunbridge Wells
Kent
U.K.

Telex: 95446
Cables: Bermotor Tunbridge Wells
Phone: 0892 37588

Mounting: Frame and skid mounted, larger units on trailer or trolley as option.
Range: 30 models in 6 ranges from 0.5kVA d.c. to 28kVA a.c. Specifications are given for a part of the range only.
Engine make: Barnard (a Renault Company)
Engine features: Recoil starting on smaller engines, larger ones are self starting. All engines turn at 3000 r.p.m.
Control options: Smaller engines have integral control panels, larger ones use wall mounting. Remote start and stop. Automatic start and mains failure facilities are available.
Voltage options: Voltage to be specified on larger sets from 400/230, 380/220 or 220/127.
Power factor: Given as 1 for smaller alternators; 0.8 for larger
Protection: Most models are drip proof
Special features: Propane may be used as fuel on some models
Agencies: Through Renault agents.

Type	*Engine*	*kVA*				*V*	*Gen. Spec.*	*Weight (kg)*
		Single-phase		*3-phase*				
		50Hz	*60Hz*	*50Hz*	*60Hz*			
18ACB66	4DA	0.3 d.c.				6/12	C	16
117CBD68	4DA	0.75 d.c.				12/24	C	37
21CB305	4DA	1.5 d.c.				12/24	C	97
318AMAP502	4DA	0.75 d.c.				110	B	32
318AMAP603	4DA			1		up to	B	32
117-6M16	4DA	1.6				220	S	48
217M2	4DA			2.4		220	S	85
217M24	4DA	2				220	S	50
810M4	4DA	4				220	S	100
71M5	4DA	5				220	S	165
112TM553	4DA	5				220	B	160
71M491	4DA	6				up to	B	165
						110		
239AM606	4DA	1.8				220	S	70
610AM606	4DA	2.5				220	S	51
810M606T	4DA	3.3				220	S	53
610AT	4DA			3.75		400/		120
112tT	4DA			6.25		230		205
71T495	4DA			7.5		380/		230
						220		
42T497	4DA			12.5		220/		290
44T28	4DA			28		127		430

D-K Generating Plant

DAWSON-KEITH ELECTRIC LTD
Deekay House
North Street
Havant
Hants PO9 1QH
U.K.

Telex: 86491
Cables: Autogen Havant
Phone: (07012) 74122

Manufacturers of a wide range of generating plant as follows:

	Base-load plant	*Transportable Plant*	*Auto-standby plant*
Power range:	12.5 to 1125kVA	20 to 1125kVA	12.5 to 1125kVA
No. of models:	37	34	37
Engine types:	Lister, Perkins, Ford, Dorman, Rolls, Cummins, and GM	Perkins, Ford, Dorman, Rolls, Cummins and GM	Lister, Perkins, Ford, Dorman, Rolls, Cummins and GM

Alternators: Brushless self-exciting, self-regulating – all to BS2613
Special features: A wide range of extras available for control, silencing mounting, enclosing, etc. A 24-hour round the clock advice service is offered to customers.

Model ADV100

AUTO DIESELS BRABY LTD
Cowley Mill Road
Uxbridge
Middlesex UB8 2QG
U.K.

Telex: 263835
Cables: Audodiesels Uxbridge
Phone: 0895 38262

Mounting: Skid mounted. Trailer mounting optional
Range: 40 models in 3 ranges from 20 to 700kVA. Full range not shown.
Engine make: Volvo, Cummins, and Dorman, in ascending size order.
Engine features: Self starting industrial engines.
Control options: Machine or wall mounted controls. Self regulating. Comprehensive range of starting, control and protection controls are available.
Voltage options: Any standard voltage in the limits specified can be obtained.
Power Factor: 0.8
Protection: Drip proof, screen protected
Special features: Standard specification includes many 'options'. Comprehensive list of control and mechanical extras
Agencies: Distributors in nearly 50 countries.

Type	*Engine*	*kVA*				*V*	*Gen.*	*Weight*
		Single-phase		*3-phase*			*Spec.*	*(kg)*
		50Hz	*60Hz*	*50Hz*	*60Hz*			
ADP20	4DW	Not available		20	24	380	B	726
ADF30	4DW			30	40	up	B	907
ADP37	4DW			37	45	to	B	930
ADF50	4DW			50	60	440	B	1021
ADF69	4DW			69	85	..	B	1265
ADP80	4DW			80		..	B	1261
AS/F85	4DW				85	..	B	1265
ADV100	4DW			100		..	B	2000
AS/V120	4DW				120	..	B	2000
ADL107	4DW			107		..	B	1987
AS/L130	4DW				130	..	B	1987
ADS160	4DW			160	167	..	B	2404
ADC130	4DW			130		220	B	2550
ADC165	4DWT			165		up	B	2800
ADC700	4DWT			700		to	B	7706
ADC280	4DWT			280		650	B	4536
ADC700	4DWT			700		..	B	9304

Kawasaki KG700B Generator

KAWASAKI HEAVY INDUSTRIES LIMITED
World Trade Centre Building
Hamamatsu-cho
Minatoku
Tokyo
Japan

Telex: J-22672J26888
Cables: Kawasakiheavy Tokyo

Mounting: Smaller units on feet, larger units in tube frames
Range: 4 types from 0.5 to 2.3kVA
Engine make: Kawasaki
Engine features: Recoil start, engine runs at high r.p.m.
Alternator make: Kawasaki
Control options: None, comes as complete unit
Voltage options: Either 110 or 220 volts available
Power factor: 1.0
Protection: Unit totally enclosed, no external moving parts. Small d.c. output for battery charging.
Special features: Smaller units may be carried by one man
Agencies: As for motorcycles.

Type	*Engine*	*kVA*				*V*	*Gen.*	*Weight*
		Single-phase		*3-phase*			*Spec.*	*(kg)*
		50Hz	*60Hz*	*50Hz*	*60Hz*			
KG600	4PA	0.5	0.6			110	B	25
KG900	4PA	0.8	0.9			to	B	30
KG1300	4PA	1.1	1.3			220	B	42
KG2300	4PA	2	2.3			..	B	53

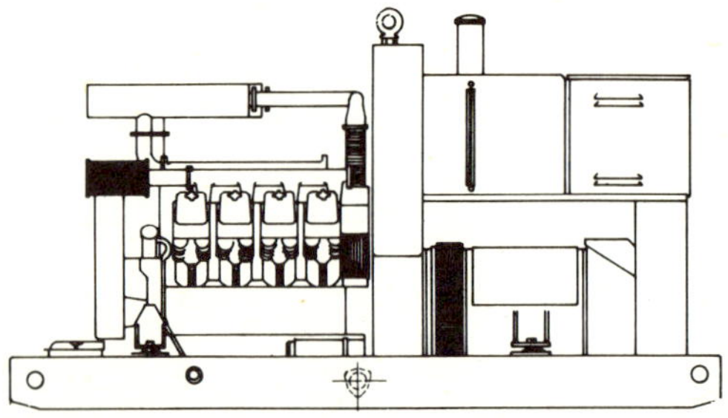

Model 22J

CGE INTERNATIONALE
U.K. LIMITED
Castle Works
Station Road
Hampton
Middlesex TW12 2BY
U.K.

Telex: 928735
Cables: Electexpel Hampton
Phone: 01 941 2525

Mounting: Skid base standard. Road and handling trailers optional
Range: 10 models from 14kVA to 210kVA
Engine make: Alsthom-Tarbes (division of CGE)
Engine features: Self-starting
Alternator make: Unelec (Alsthom subsidiary)
Control options: Machine mounted with built in safety cut-out. Automatic voltage control
Voltage options: 380 or 220V
Power factor: 0.7
Protection: Drip proof. Tropical insulation. All weather cover optional
Special features: Optional starter motor and cold start equipment. Silent running cover available
Agencies: World wide

Type	*Engine*	*kVA*				*V*	*Gen. Spec.*	*Weight (kg)*
		Single-phase		*3-phase*				
		50Hz	*60Hz*	*50Hz*	*60Hz*			
22J	4DA			14			B	1065
..	4DA			23		220/	B	1125
..	4DA			32		380	B	1155
..	4DA			50		..	B	1350
..	4DA			70		..	B	1730
..	4DA			85		..	B	2250
..	4DA			110		..	B	2850
..	4DA			140		..	B	3250
..	4DA			170		..	B	3700
..	4DA			210		..	B	3900

DIPL. -ING. HITZINGER & COMPANY
Helmholtzstrasse 56
4020 Linz
Austria

Telex: 021769
Cables: Tebehi Linzdonau
Phone: 07222/81681

Mounting: Skid mounted, covered and trailer versions available.
Range: Large range up to 500kVA
Engine make: Haty, UW Lister, MWM Jenbacher, MAN and Deutz engines all used
Engine features: Electric start
Alternator make: Hitzinger
Control options: Machine or wall mounted, manual or automatic start synchronising equipment for parallel operation available.
Power factor: 0.8
Protection: Drip proof – complete cover optional
Special features: Range of accessories and equipment for emergency power plant erection. Continuous supply models available.
Agencies: Africa, N. Africa, Mid-East Asia, Far East, Pacific, Central America and Caribbean.

No specifications available at time of printing.

Ventilated and Sound-proofed Hitzinger Installation

1. grille
2. outgoing air-sound absorber
3. tank
4. flap for outgoing air
5. bellows
6. generating set
7. switch-panel
8. silencing door
9. flap for ingoing air
10. ingoing air-sound absorber
11. exhaust-silencer
12. rain protection cap
13. exhaust-pit

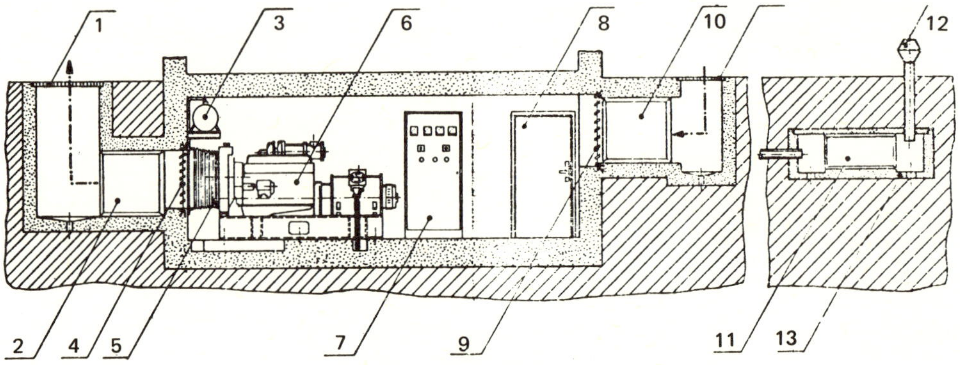

DA Series Generating Set

DORMAN DIESELS LTD
Tixall Road
Stafford ST16 3UB
U.K.

Telex: 36156
Cables: Dorman Stafford
Phone: 0875 3141

Mounting: Skid mounted. Trailer mount available on some models
Range: 27 types in 4 ranges from 18.5 to 573kVA. Full range not shown
Engine make: Dorman
Engine features: Seif start, sets may be run slow
Control options: Control panel mounted on or off machine. Machine mounted single/parallel running, auto start/stop, alarm shutdown available.
Voltage options: Any single standard voltage up to 480V; 4-wire
Power factor: 0.8
Protection: Drip proof, overall cover available
Special features: Refer to manufacturer for special requirements
Agencies: World wide

Type	*Engine*	*kVA* Single-phase 50Hz	*kVA* Single-phase 60Hz	*kVA* 3-phase 50Hz	*kVA* 3-phase 60Hz	*V*	*Gen. Spec.*	*Weight (kg)*
4DA	4DA			37	44	120	B	934
4DA	4DA			55	66	up to	B	1161
6DAT	4DA			66	77	480	B	1184
8DA	4DA			70	82	..	B	1447
4LD	4DL			61	71	..	B	1691
5LD	4DL			76	89	..	B	2011
6LD	4DL			89	100	..	B	2385
6LE	4DL			110	124	..	B	2385
4LDT	4DLT			88	100	..	B	1863
6LDT	4DLT			137	156	..	B	2476
up to								
6LETCA	4DLT			188	211	..	B	2958
6QK	4DL			184	204	..	B	4355
8Q	4DL			237	264	..	B	6096
up to								
12QBTCW	4DLT			658	515	..	B	8709

Alternators are available on their own. (See xxxx)

Model ZA9

PETBOW LTD
Ramsgate Road
Sandwich
Kent
U.K.

Telex: 96329
Phone: (030 46) 3311

Mounting: Skid mounted as standard. Site or main road trailers are optional.
Range: 4 models from 25 to 93.75kVA
Engine make: Deutz
Engine features: Self-starting
Control options: Automatic voltage regulation with set mounted controls are standard, mains failure and other start options available. Also various safety stopping options. Parallel running control panel available.
Voltage options: From 110V to 550V in steps, as specified.
Power factor: 0.8
Protection: Tropical insulation. Weatherproof canopies are optional. Drip proof.
Special features: Small accessory range. Heavy motor starting option.
Agencies: In Middle East, Africa, West, Central and East Asia.

Type	*Engine*	*kVA* Single-phase 50Hz	*kVA* Single-phase 60Hz	*kVA* 3-phase 50Hz	*kVA* 3-phase 60Hz	*V*	*Gen. Spec.*	*Weight (kg)*
ZA5	4DA			25	31	120	B	704
ZA6	4DA			35	42.5	up	B	795
ZA8	4DA			54	65	to	B	950
ZA9	4DAT			79	94	550	B	1090

Ruston/Dunlite Generating Set

DUNLITE
(Div. of PYE Industries Sales PTY Ltd)
28 Orsmond Street
P.O. Box 100
Hindmarsh
South Australia 5007

Telex: 82893 Aust
Cables: Daylite
Phone: 46 3832

Mounting: Small range in tube frame, larger ranges on skids with canopy and trailer options
Range: 3 ranges, 12 models, 2.5kVA to 185kVA (with engines) plus range of alternator only.
Engine make: Briggs & Stratton, Ford and Volvo (range each)
Engine features: Small range: recoil start and high r.p.m.: larger ranges: self-start
Alternator make: Dunlite
Control options: Machine mounted, local/remote control; mains failure auto start (on larger ranges)
Voltage options: Either 240 or 415 volts
Power factor: 0.8
Protection: Drip proof, weatherproof canopy available
Special features: Various voltage options on request, short circuit protection, tropicalised insulation
Agencies: Covering Australia

Type	*Engine*	*kVA* *Single-phase* *50Hz*	*60Hz*	*3-phase* *50Hz*	*60Hz*	*V*	*Gen. Spec.*	*Weight (kg)*
31311	4PA	2.5				240	C	65
31335		5				240	C	88
020B3FDE	4DW			20			B	824
030B3FDE	4DW			30		415	B	830
035B3FDE	4DW			35		up to	B	335
055B3FDE	4DW			55		240	B	1207
070B3FDE	4DW			70		..	B	1243
075B3UP–E	4DW			75		..	B	1794
090B3UP–E	4DW			90		..	B	1931
130B3UP–E	4DW			130		..	B	1950
150B3UP–E	4DW			150		..	B	2083
185B3UP–E	4DW			185		..	B	2767
BLF10		7.5				110	B	
BLF13		10				up to	B	
B3F13				20		440	B	
B3F11				30		in	B	
B1F11		40				small	B	
B3F12				90		steps	B	
B3F14				200			B	

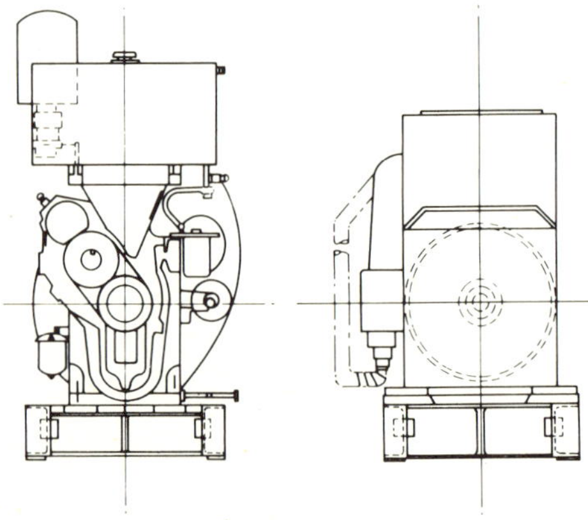

Ruston YWAK Alternator Set

RUSTON & HORNSBY (INDIA) LIMITED
Chinchwad
Poona-411 019
India

Telex: 303 Greaves (PN)
Cables: Rustonind

Mounting: Skid mounted
Range: 3 machines in 1 range 5 to 16 kVA
Engine make: Ruston & Hornsby (India) Ltd.
Engine features: Crank starting, self starting optional.
Control options: Cubicle mounted separately, mains failure control and parallel running are optional
Voltage options: 240 volts single-phase, 415 volts 3-phase.
Power factor: 0.8
Protection: Drip proof
Special features: Basic accessories available
Agencies: Dealers across India.

Type	*Engine*	*kVA* *Single-phase* *50Hz*	*60Hz*	*3-phase* *50Hz*	*60Hz*	*V*	*Gen. Spec.*	*Weight (kg)*
1YWAK		5.2		5.3		240		570
2YWAK		10.6		11.3		or		710
3YWAK		16		17.2		415		900

Buffalo 'E' Range

DALE ELECTRIC OF GREAT BRITAIN LTD
Filey
Yorkshire YO14 9PJ
U.K.

Telex: 52163
Cables: Dalectric
Phone: 0723 514141

Mounting: Smaller units are frame mounted. Larger units are skid mounted with trailer options.

Range: Over 100 models in 15 ranges, from 0.5kVA to 2585kVA. Neither all models nor all ranges shown.

Engine make: Engines include, in ascending order of power, Villiers, Briggs and Stratton, Petter, Dorman, Ford, Volvo and Deutz

Engine features: Smaller engines are recoil or rope start and run fast. Larger ones are self starting

Connections: 1 range is specifically designed for tractor p.t.o. operation

Control options: Automatic voltage regulation is standard. Controls are machine mounted on smaller units, and wall mounted on larger ones. Larger models have safety shut down switches as standard mains failure and other control options are available.

Voltage options: A variety of standard voltages is available, specify phase and voltage when ordering.

Power factor: 0.8

Protection: Drip proof, all weather protection is optional

Special features: Small standard accessory range, refer to factory for comprehensive options range.

Agencies: Distributors world wide.

Type	*Engine*	*kVA* Single-phase 50Hz	*kVA* Single-phase 60Hz	*kVA* 3-phase 50Hz	*kVA* 3-phase 60Hz	*V*	*Gen. Spec.*	*Weight (kg)*
T15/7/SR	For	7				200 up	–	–
T16/8.5/SR	tractor		8.5			to 480	–	–
T35/12.5/SR	p.t.o.			12.5		220 up	–	–
T36/15/SR	use				15	to 440	–	–
T16/12/SR	..		12			200 up	–	–
T15/20/SR	..	20				to 480	–	–
T35/30/SR	..			30		220 up	–	–
T36/40/SR	..				40	to 480	–	–
DM/0.5/PUB	4DA	0.5				12/24	–	–
						d.c.	C	41
DM/0.8/PUH	4DA	1	1			230/ 115	S	41
DM/1.5/PUF	4DA	1.9	1.9			.. ..	S	43
DM/3.5/PUF	4DA	3.5	3.5	3.5	3.5	115 up	S	28
DM/S/PBSF	4DA	5	5	5	5	to 415	S	120
DM/2.25/DPT	4DA	2.25	2.25			115/ 230	S	115
DM/3/DPF	4DA	3	3			.. ..	S	117
DM/4/DPT	4DA	4	4	4	4	115 up to 415	S	116
EM/DDZX	4DA			14	17.5	220 up	B	594
EM/25.5/DDZX	4DA			25.5	30.5	to 440	B	699
EM/44.5/DDZX	4DA			44.5	53	.. ..	B	965
EM/80/DDZX	4DA			80	93	.. ..	B	1420
EM/157.5/DDZX	..			157.5	191	.. ..	B	2071
EM/35/DFE	4DW			35		.. ..	B	827
EM/50/DFE	4DW			50		.. ..	B	1018
EM/68/DFE	4DW			68		.. ..	B	1127
EM/37/DDX	4DA			37.5	44.5	.. ..	B	981
EM/65/DDX	4DA			65	77.5	.. ..	B	1190
EM/72/DDX	4DA			72	85	.. ..	B	1325
EM/62/DD	4DW			62	70	.. ..	B	1542

BM Type Portable Diesel Generating Set

GEORGE COHEN MACHINERY LTD
600 Wood Lane
London W12 7RL
U.K.

Telex: 21288
Cables: Omniplant London
Phone: 01 743 2070

Mounting: Smaller sets mounted in a frame, larger sets are trolley mounted.
Range: 21 models in 3 ranges from 1kVA to 15kVA
Engine make: Petrol engines are Norton Villiers and smaller diesels are Petter. Larger diesel are Lister or Ruston.
Engine features: Small engines are rip starting and run at high r.p.m. Larger engines are rip start or hand cranked. Self starting is optional.
Alternator make: Newton Derby, Brush or Stamford
Control options: Self regulating with machine mounted control panel standard. Safety shutdown and mains failure start are optional.
Voltage options: 110V or 240V or both
Power factor: 0.8
Protection: Drip proof, tropically insulated. Weatherproof canopies are optional
Special features: Small accessory range
Agencies: Agents in Middle East and North Africa

Type	*Engine*	*kVA*				*V*	*Gen. Spec.*	*Weight (kg)*
		Single-phase		*3-phase*				
		50Hz	*60Hz*	*50Hz*	*60Hz*			
AMF1503	4DA	1				110/	S	Not
AMF1506	4DA	1.5				240	S	avail-
AMC3009	4DA	2				..	S	able
AMC3012	4DA	2.5				..	S	
AMC3015	4DA	3				..	S	
BM/AA1/02	4DA	1.5				..	S	
BM/AB1/06	4DA	2				..	S	
BM/AB1/09	4DA	2.5				..	S	
BM/AC/12	4DA	3				..	S	
CM/ST1	4DA	4.38				..	S	
CM/ST2	4DA	8.75				120/	S	
CM/ST3	4DA	13.13				240	S	
CM/SR1	4DA			3		..	S	
CM/SR3	4DA			12.5		..	S	
CM/HR2	4DA			16.25		..	S	
CM/HR3	4DA			25		..	S	
CM/HR4	4DA			34.4		..	S	
CM/HR6	4DA			52.2				
CM/HRS6	4DA			65.5				
CM/IA6	4DA			90		..	S	
CM/IAS6	4DA			117		..	S	

Model 6S110

PRAGOINVEST
Foreign Trade Corporation
Ceskomoravská 23
180 56 Praha 9
Czechoslovakia

Telex: 122 379
Cables: Pragoinvest Praha
Phone: 822 741-6

Mounting: Skid mounted
Range: 1 model shown, refer to manufacturers for details of others
Engine features: Self starting from continuously running flywheel
Control options: The machine is built to give automatically an uninterrupted power supply in the case of a mains failure.
Voltage options: 231 or 400
Power factor: 0.8
Special features: Generator driven as a motor by the mains under normal conditions
Agencies: Refer to manufacturer.

Type	*Engine*	*kVA*				*V*	*Gen. Spec.*	*Weight (kg)*
		Single-phase		*3-phase*				
		50Hz	*60Hz*	*50Hz*	*60Hz*			
6S110	4DW			50		231/ 400	S	

Flyweight Dieselite AC1 Series Generator

G & M POWER PLANT LTD
Magnet Works
Whitehouse Road
Ipswich
Suffolk
U.K.

Telex: 98216
Cables: Green Acre Ipswich
Phone: 0473 41795

Mounting: Smaller one frame mounted, larger one skid mounted. Trailer options available.

Range: Over 50 models in 4 ranges from 0.3kVA to 231kVA. Full range not shown. Also similar choice in military generators and marine generators — not shown here.

Engine make: Briggs & Stratton, Onon, Petter, Lister, Perkins, British Leyland, Volvo.

Engine features: Smaller engines are recoil started and high revving. Larger ones are self starting.

Control options: Smaller machines machine mounted, larger models machine mounted as standard, wall mounting, remote control, demand starting and mains failure options are available.

Voltage options: 120/240 on small machines. 120/440 in steps on larger ones.

Power factor: 0.9

Protection: Small models drip proof, larger are screen protected only, weatherproof canopies are optional. Tropical insulation.

Special features: Limited range of protection equipment available.

Agencies: World wide (except S. America), especially in Europe, Middle East, and Australia.

Type	*Engine*	*kVA Single-phase 50Hz*	*kVA Single-phase 60Hz*	*kVA 3-phase 50Hz*	*kVA 3-phase 60Hz*	*V*	*Gen. Spec.*	*Weight (kg)*
03B80–2QE	4DA	0.3d.c.				12/15	C	27
06B80–224E	4DA	0.6				24/30	C	34
105B10–224E	4DA	1.5				24/30	C	55
015AA1–232E	4DA	1.5				32/40	C	74
2B17–53P	4DA	2				120/240	S	59
3B17–53P	4DA	3				120/240	S	65
305B17–63P	4DA			3.5		120/240	S	106
2AB1–53P	4DA	2				120/240	S	90
3AC1–53P	4DA	3				120/240	S	106
305AC1–63P	4DA			3.5		120/240	S	106
ST1	4DA	3.5	4	4	4.8	220/415	S	244
ST2	4DA	7	8	8	9.5		S	290
ST3	4DA	10.5	12.5	12.5	15	220/415	S	405
MR2	4DA	15	17.5	16	19	220/415	S	549
MR3	4DA			25	29.5	220/415	S	648
MR4	4DA			34.5	40	220/415	S	843
MR6	4DA			52.5	61	220/415	S	1103
JA6	4DA			90	103.5	220/415	B	1615
JAS6	4DA			117.5	125		B	1710
10RDP	4DW	10				120/440	B	483
12RDP	4DW	12	12	12		120/440	B	488
15RDP	4DW		15	15	15	120/440	B	493
24/42RDP4	4DW	24/35	29/42	25/38	31/42	110/440	B	766
45/62RDP6	4DW			45/54	52/62	110/440	B	920
70/82RDP6T	4DW			70	82	110/440	B	1125
RDL4	4DW	33.5	39	35	40.5	120/480	B	1000
RDL	4DWT			72	86	120/480	B	1474
RDV100B	4DW			100	115	127/440	B	2117
RDVT100A	4DWT			150	166	127/440	B	2390
RDVT120A	4DWT			200	231	127/440	B	3043

Trailer-mounted Tractor Drive Unit

KATOLIGHT CORPORATION
3201 Third Avenue
P.O. Box 3229
Mankato
Minnesota 56001
U.S.A.

Telex: 29-0787
Cables: Interkato – NY
Phone: 507-625-7973

Mounting: Basic machines are skid mounted, small units are frame mounted and P.T.O. drive generators have flanged feet with a trailer option.
Range: Over 30 models in 5 ranges from 2kVA to 1250kVA, full range not shown.
Engine make: Wisconsin, Hercules, International Moline, Cummins Wausessa, General Motors, Allis Chalmers, Briggs & Stratton, Lister
Engine features: Small units are rip start and faster running. Larger units are self starting. Diesel, petrol and liquid petroleum gas are optional fuels.
Connections: 1 range is designed for operation by a tractor P.T.O.
Control options: Machine mounted as standard with safety cut outs. Wall mounting, mains failure remote and automatic start are optional
Voltage options: Order from 120/208, 277/480 or 115/230, or connect output to suit.
Power factor: 0.8
Protection: Screen protected, weatherproofed canopies are optional.
Special features: Fuel conversion and improved silencing are available.
Agencies: Middle East, N. Africa, and Far East.

Type	*Engine*	*kVA* Single-phase 50Hz	Single-phase 60Hz	3-phase 50Hz	3-phase 60Hz	*V*	*Gen. Spec.*	*Weight (kg)*
10FW4CEP	4GA		12.5			115/230	S	590
15FW4CEP	4GA		18.8			..	S	658
25FW4CEP	4GA		31.5			..	S	794
35FH4CEP	4GW				43.7	120/240	S	1067
55FH4CEP	4GW				68.6	..	S	1112
140FM4CEP	4GW				175	..	S	2860
10FW4	4PA		12.5		12.5	120 up	S	431
15FW4	4PA		18.75		18.75	to 480	S	522
23FH4	4PW		29		29	..	S	522
55FH4	4PW		68.6		68.6	..	S	1089
8SF14	4PW		106		106	..	S	1633
N140FM4	4PW		175		175	..	S	2359
D15FL4	4DA		18.8			..	S	522
D35FH4	4DW		43.7		43.7	..	S	817
D60FH4	4DW		75		75	..	S	1270
D110FH4	4DW		137		137	..	S	1860
D1000FX4	4DW		1250		1250	..	S	9271

Model 100N3 Generating Set

GRAHAME PUTTICK LIMITED
Sandwich
Kent
U.K.

Telex: 96366 Puma UK G
Cables: Puma Sandwich
Phone: 03046 2901

Mounting: Skid mounted as standard. Trailer mounting is optional on all except the smallest range.
Range: Over 100 models in 10 ranges. From 3.5kVA to 1430kVA. Full range not shown.
Engine make: Rolls Royce, Cummins, Perkins, Kirloskar, Ford, Deutz, Bedford, Dorman, G.M. (Detroit)
Engine features: Smallest range by crank. All others are self-starting.
Alternator make: Grahame Puttick
Control Options: Set mounted as standard. Wall mounting optional. Also available are parallel running, auto starting, short break mains standby and mains failure control options.
Voltage options: Choice from 346/440V or 190/240V.
Power factor: 0.8

Protection: Drip proof. Tropical insulation. Weatherproof canopy optional
Special features: Comprehensive accessory range
Agencies: World wide.

Type	*Engine*	*kVA*				*V*	*Gen.*	*Weight*
		Single-phase		*3-phase*			*Spec.*	*(kg)*
		50Hz	*60Hz*	*50Hz*	*60Hz*			
M3.551	4DA	3.5				220/	S	230
M7.552	4DA	7.5				240	B	370
20U1	4DA			20		208/	B	780
25U1/J	4DA			25	25	240	B	780
30U1J	4DA				30		B	960
30U2	4DA			30			B	960
40U2J	4DA				40	346/	B	1010
50U3	4DA			50		440	B	1160
55U3J	4DA			55	55		B	1160
65U4	4DA			65			B	1385
80U4J	4DA				80		B	1385
80U8	4DA			80			B	1516
200U11	4DA			200			B	3680
225U10G/J	4DA			225	225		B	2815
20F1	4DW			20			B	870
25F1/J	4DW			25	25		B	870
40F1	4DW			40			B	1040
45F1J	4DW				45		B	1040
60F3	4DW			60			B	1320
83F4J	4DW				83	208/	B	1325
80F3G/J	4DW			80	83	240	B	1360
30G1	4DW			30			B	736
55G2J	4DW				55		B	1059
70GS	4DW			70			B	1100
100G5G/J	4DW			100	100	346/	B	1281
100N3	4DW			100		440	B	2170
137N3J	4DW				137		B	2300
145N4	4DW			145			B	2130
160N4J	4DW				160		B	2130
up to								
1550P22	4DW			1550			B	

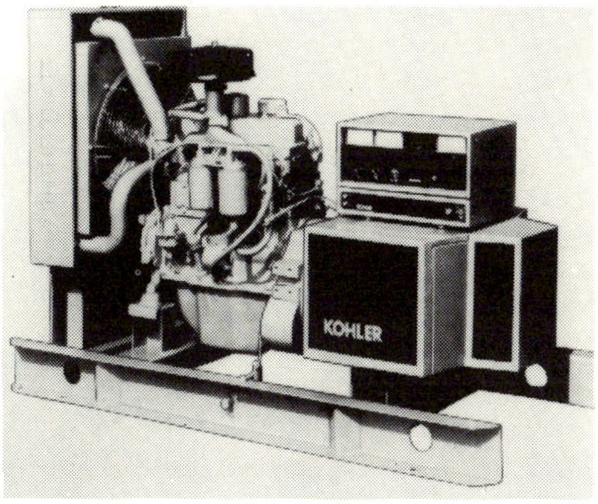

Kohler Fast Response Generator

KOHLER INTERNATIONAL LTD
High Street
Kohler
Wisconsin 53044
U.S.A.

Telex: 2 6888
Cables: Kohlerint
Phone: 414 457 4441

Mounting: Skid mounted as standard. Trailer options available.
Range: Over 60 models in 8 ranges, from 1.5kVA to 1000kVA. Full range not shown.
Engine features: Hand starting on smallest models, which are higher revving. Most models are self starting.
Control options: Machine mounted controls as standard. Variety of remote controls available.
Voltage options: 120/240 on small machines. 120 up to 500 on larger ones.
Power factor: 0.8
Protection: Weatherproof and soundproof covers are available
Special features: Larger models feature rapid response to load change
Agencies: Dealers through N. America.

Type	*Engine*	*kVA*				*V*	*Gen.*	*Weight*
		Single-phase		*3-phase*			*Spec.*	*(kg)*
		50Hz	*60Hz*	*50Hz*	*60Hz*			
1.5MM25	4DA		1.5			120	B	44
3MMV55	4DA		3			120/	B	72
4CM021	4DA		4			240	B	136
6.5CM021	4DA		6.5			..	B	197
10CM021	4DA		10			..	B	316

Type	Engine	kVA Single-phase 50Hz	kVA Single-phase 60Hz	kVA 3-phase 50Hz	kVA 3-phase 60Hz	V	Gen. Spec.	Weight (kg)
15CM021	4DA		15			..	B	457
8.5CM021	4DA		8.5			..	B	364
12CM021	4DA		12			..	B	451
–	4DW		7.5			120	B	245
–	4DW		12.5			up	B	341
–	4DW		20			to	B	457
–	4DW		30			600	B	709
–	4DW		45			..	B	926
–	4DW		55			..	B	926
–	4DW		70			..	B	907
–	4DW		85			..	B	1205
–	4DW		115			..	B	2290
–	4DW		170			..	B	3050
–	4DW		12.5			..	B	421
–	4DW		20			..	B	515
–	4DW		30			..	B	841
–	4DW		45			..	B	897
–	4DW		60			..	B	1025
–	4DW		100			..	B	1357
–	4DW		125			..	B	1936
–	4DW		150			..	B	2341
–	4DW		170			..	B	2611
			up to 1000			..	B	10832

Single Sentry in Vibration Isolated and Noise Suppressed Cabinet

KING ENGINEERING LTD
Greenland Mills
Bradford-on-Avon
Wiltshire BA15 1BZ
U.K.

Telex: King Bradonavon
Phone: (02216) 2709

Mounting: Complete units are skid mounted with a trailer option, tractor driven units have flanged feet with a tractor mounting frame optional.
Range: 30 models in 2 ranges from 5kVA to 70kVA. Full range not shown.
Engine make: 1 range is powered by British Industrial Diesels
Engine features: Self starting
Connections: 1 range is driven by a tractor P.T.O. through special shaft
Control options: Controls are machine mounted as standard, complete units feature automatic voltage regulation. Speed drift warning safety shut down, mains failure, auto start, remote start are optional.
Voltage options: 110/220 on P.T.O. units. 240/480 single phase, 240/415 three phase on complete units. 60Hz on request.
Power factor: 0.8
Protection: Drip proof, anti-condensation and tropical insulation as standard. Various stages of protection up to noise-proof, weather-proof, vibration-proof and dust-proof.
Special features: Control and environment options available also P.T.O. output. Any particular requirement considered by the factory.
Agencies: Order from factory.

Type	Engine	kVA Single-phase 50Hz	kVA Single-phase 60Hz	kVA 3-phase 50Hz	kVA 3-phase 60Hz	V	Gen. Spec.	Weight
6S	P.T.O.	5.5	5.5			110/220	B	–
10S	..	10				240	S	–
15S	..	15				..	S	–
19S	..			19		240/415	S	–
25S	..			25		..	S	–
25SBP	..	25				240/480	S	–
40SBP	..	40				..	S	–
38SBT	..			37.5		240/415	S	–
65SBT	..			65		..	S	–
N6BS/T/P	4DA	5.5		5.5		..	S	–
10SS/T	4DA	10		10		..	S	–
15SS/T	4DA	15		15		..	S	–

Type	Engine	kVA Single-phase 50Hz	kVA Single-phase 60Hz	kVA 3-phase 50Hz	kVA 3-phase 60Hz	V	Gen. Spec.	Weight (kg)
19ST	4DW			19		..	S	–
25S/BT	4DW	25		25		..	B or	–
							S	–
33BS	4DW	33				..	B	–
38BT	4DW			38		..	B	–
50BT	4DW			50		..	B	–
70BT	4DW			70		..	B	–
D25ST	4DW			25		240/480	S	–
D30BS	4DW	30		30		..	S or	–
							B	–
D38ST	4DW			38		..	S	–
D40BT	4DW			40		..	B	–
D50ST	4DW			50		..	S or	–
							B	–

'P' Type Generating Set

PETTERS LTD
Hamble
Southampton S03 5NJ
U.K.

Telex: 47626
Cables: Petter Hamble
Phone: 042 122 2061

Mounting: Smaller units have flanged bases as standard with skid bases and frames as optional. Larger units have skid mounting as standard.
Range: 32 models in 8 ranges from 1.5kVA up to 860kVA. Full range not shown.
Engine make: In size order. Petters, Perkins, Dorman and Cummins
Engine features: Smallest engines are higher revving and rope started as standard. Larger units are self starting. Most air cooled units have a water cooling option.
Control options: Smaller units have machine mounted control boxes. Larger ones are wall mounted. All alternators are self regulating. Automatic start, mains failure and remote start/stop are optional. Larger units have safety shutdown devices.
Voltage options: 110 or 220 single phase, 127/220 or 230/400, 3-phase, specify frequency and phase where ordering.
Power factor: 0;8
Protection: Screen protected, tropical insulation
Special features: Refer to factory for optional extras. High level of parts standardisation in the range.
Agencies: Through associated agents.

Type	Engine	kVA Single-phase 50Hz	kVA Single-phase 60Hz	kVA 3-phase 50Hz	kVA 3-phase 60Hz	V	Gen. Spec.	Weight (kg)
AA1	4DA	1.5	1.8			230/110	S	78
AB1	4DA	2.5	2.75			.. ..	S	78
AC1	4DA	3.2	3.5			.. ..	S	78
PH1	4DA/W	3.5	4	4.5	5.5	110 up	S	356
PH2	4DA/W	7.2	8.7	9	11	to 400	S	483
PJ1	4DA/W	4.9	5.8	6.25	7.25	.. ..	S	368
PJ2	4DA/W	10	11.6	12.5	15	.. ..	S	630
PJ3	4DA			20	23.75	127 up	S	703
PJ4	4DA			27	32	to 400	S	976
P4.236	4DW			35	42.5	127/220	S	821
P6.354	4DW			52.5	60	or	S	1043
PT6.354	4DWT			72.5	85	230/400	S	1139
6LD	4DW			90	102.5	127/200	S	2563
6LE	4DW			110	122.5	or	S	2586
6LDT	4DWT			147.5	160	230/400	S	2812
8JT	4DWT			187.5	210	127/220 or	S	3107
8JTCA	4DWT			250	287.5	230/400	S	3155
N855–G	4DW			127.5	152.5	240/415	B	2550
NT855–G1	4DWT			166	187.5	.. ..	B	2700
up to						or 220/127		
KTA2300–G	4DWT			720	860	240/415	B	6100
						or 220/127		

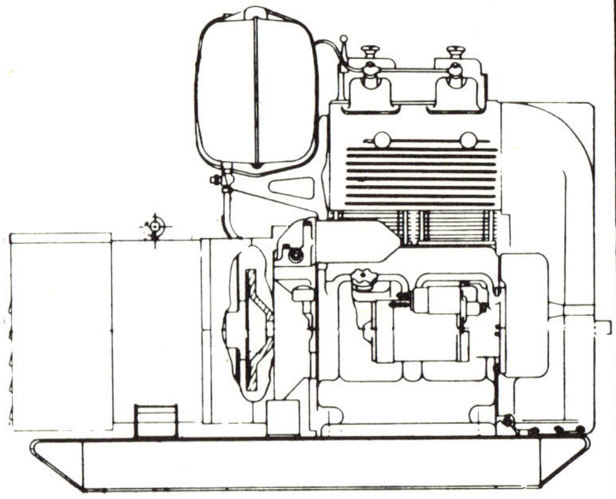

Model DE12P4 Generating Set

WINPOWER CORPORATION
P.O. Box 99
Newton
Iowa 50208
U.S.A.

Telex: TWX 910 520 1557
Cables: Winpower
Phone: 515 792 1301

Mounting: Skid mounted, carrying frames and carts optional.
Range: 28 models in 3 ranges from 1.5. to 25kVA. Full range not shown.
Engine make: Includes Petter, Briggs & Stratton, and Wisconsin
Engine features: Smaller units are higher revving. Most units are hand started, others self start.
Connections: 1 range is designed to be driven by a tractor P.T.O.
Control options: Machine mounted, remote start is optional.
Voltage options: 120 or 120/240
Power factor: 0.8
Protection: Drip proof, high quality insulation.
Special features: Starting battery and equipment is optional. Automatic idling facility available.
Agencies: North, Central and South America.

Type	*Engine*	*kVA*				*V*	*Gen.*	*Weight*
		Single-phase		*3-phase*			*Spec.*	*(kg)*
		50Hz	*60Hz*	*50Hz*	*60Hz*			
GM105B2	4PA		1.9			120	S	39
GM2B2	4PA		2.5			120/ 240	S	43
GM205B2	4PA		3.1			..	S	49
GM4W2	4PA		4.7		4.7	..	S	91
GM5B2	4PA		6.25			..	S	101
GM6W2	4PA		6.9			..	S	111
GM205B4	4PA		3.2			120	S	97
GM308W4	4PA		4.7			120/ 240	S	134
GE6W4	4PA		7.5			..	S	261
GR10W4	4PA		12.5			..	S	285
GR15W4	4PA		18.75			..	S	427
GR20W4	4PA		25			..	S	536
DE12P4	4DA		15		15	..	S	481

12hp Air-cooled Diesel Engine

R.A. LISTER POWER PLANT LIMITED
Thrupp
Stroud
Gloucestershire GL5 2BW
U.K.

Telex: 43559
Cables: Machinery Dursley
Phone: 045-388 5166

Mounting: Skid mounted, with trolley option on smaller models.
Range: 32 models in 3 ranges from 2.5kVA to 119kVA. Full range not shown.
Engine make: Lister (Hawker Siddeley subsidiary)
Engine features: Crank start on smaller models. Self start is optional on smaller, and standard on larger models.
Alternator make: Brush (Hawker Siddeley subsidiary)
Control options: Machine or wall mounted. Comprehensive range of remote automatic and mains failure control available.
Voltage options: Voltage choice to be specified on larger sets from 220/380, or 400/230, or 415/240, or 220/127, or 230/133
Power factor: 0.9
Protection: Drip proof, tropical insulation. Canopy available on smaller models.
Special features: Silencers, mountings, fuel pumps and control accessories. Large selection. Advice available in booklet.
Agencies: World wide.

Type	*Engine*	*kVA*				*V*	*Gen.*	*Weight*
		Single-phase		*3-phase*			*Spec.*	*(kg)*
		50Hz	*60Hz*	*50Hz*	*60Hz*			
ST1	4DA	3.5	4	4	5	133	S	244
ST2	4DA	7	8	8	10	up	S	290

Type	Engine	kVA Single-phase 50Hz	kVA Single-phase 60Hz	kVA 3-phase 50Hz	kVA 3-phase 60Hz	V	Gen. Spec.	Weight (kg)
ST3	4DA	10.5	12.5	12.5	15	to	S	405
HR2	4DA	15	17.5	16	19	415	S	549
HR3	4DA			25	29	..	S	648
HR4	4DA			34	40	..	S	843
HR6	4DA			52.5	62	..	S	1103
HRS6	4DA			66	78	..	S	1196
IA6	4DA			90	104	..	S	1615
IAS6	4DA			117	125	..	S	1710
HRW2	4DW	14	17	15	18	..	S	450
HRW3	4DW			24	28	..	S	620
HRW4	4DW			33	38	..	S	784
HRW6	4DW			50	58	..	S	1082
HRWS6	4DW			62.5	75		S	1167
IW6	4DW			86	99		S	1684
IWS6	4DW			111	119		S	1734

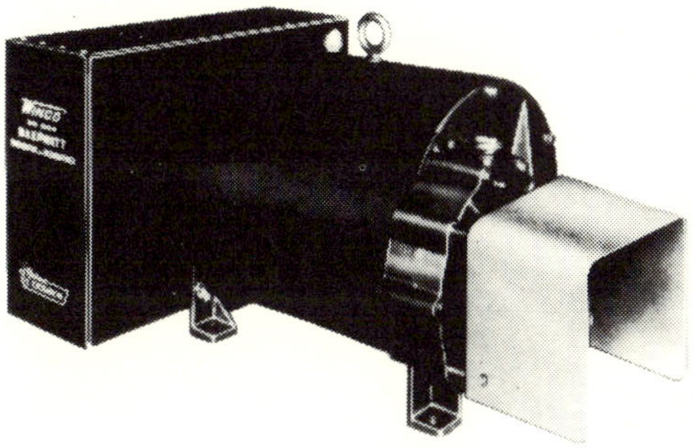

Series PTOC Generator

WINCO
Division of Dyna Technology Incorporated
East 7th and Division Streets
P.O. Box 3263
Sioux City
Iowa 51102
U.S.A.

Telex: 487110
Phone: (712) 252-1821

Mounting: Frame and skid mounted as standard. Trailer mounting is optional.

Range: Over 70 models in 20 ranges from 1.9kVA to 100kVA. Full range not shown.

Engine make: In size order for complete units, Briggs and Stratton, Wisconsin, Lister, Petter.

Engine features: Smaller engines are recoil or rope start with self starting options and run at higher speeds. Larger ones are self starting and run at normal speeds.

Connections: 1 range of generators specifically designed for operation from a tractor P.T.O. 1 range has a keyed shaft connection.

Control options: Machine mounted as standard. All units have built in overload protection. Mains failure is available on some larger models.

Voltage options: 115/230

Power factor: 0.8

Protection: Drip proof

Special features: Automatic idling when off load, optional use of 2 voltages simultaneously and exceptional ability for heavy motor starting load. 180 cycle generators and fire fighter generators are available.

Agencies: Agents in Africa, Middle East, Central and South America.

Type	Engine	kVA single-phase 50Hz	kVA single-phase 60Hz	kVA 3-phase 50Hz	kVA 3-phase 60Hz	V	Gen. Spec.	Weight (kg)
15PTOF	from			15		120/	B	124
20PTOF	tractor			20	20	240	B	138
30PTOC	P.T.O.			30		..	B	224
35PTOC	..				35	..	B	234
50PTOC	..			50		..	B	307
65PTOC	..				65	..	B	311
P1TH–1M	4DA			1		115	B	27
P205BH–1M	4DA			2.5		..	B	39
C109BH–1M	4DA			1.9		..	B	41
C202WRH–1M	4DA			2.25		..	B	48
305BH–FE	4DA			3.5		115/	B	64
P307BH–3E	4DA			3.7		230	B	55
403WH–FM	4DA			4.3		..	B	122
P503BH–3M	4DA			5.3		..	B	88
P703WH–3E	4DA			7.3		..	B	133
10WH–3E	4DA			10		..	B	221
6WS–3E	4DA			6		..	B	188

Type	Engine	kVA Single-phase 50Hz	kVA Single-phase 60Hz	kVA 3-phase 50Hz	kVA 3-phase 60Hz	V	Gen. Spec.	Weight (kg)
1205WS–4R	4DA			12.5		120/ 208	B	290
305LDS–1E	4DA			3.5		115	B	195
7LDS–3E	4DA			7		115/ 230	B	316
1005LDS–4E	4DA				10.5	120/ 208	B	365
6PDS–3E	4DA			6		115/ 230	B	324
1205PDS–17E	4DA				12.5	..	B	427

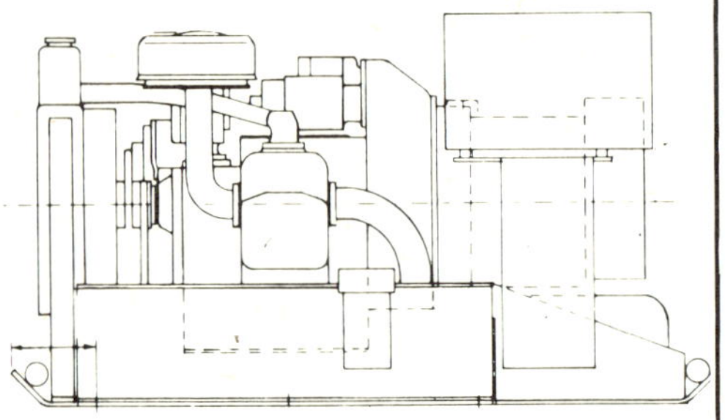

Model 120/RDP

WITTE ENGINE CORPORATION
P.O. Box 386
555 East 56th Highway
Olathe
Kansas 66061
U.S.A.

Telex: 042-6262
Phone: 913 764 2512

Mounting: Skid mounted. Trailer mounting is optional.
Range: 6 models in 2 ranges
Engine make: Witte (Hawker Siddeley subsidiary)
Engine features: Self starting. Engines available without generator. 260 model available with propane gas fuel conversion (2nd range belt driven).
Alternator make: Hawker Siddeley subsidiary
Control options: External voltage regulator in machine mounted control box. Automatic and mains failure starting optional.
Voltage options: 120/240V single-phase or 120/208V 3-phase
Power factor: 0.8
Protection: Drip proof
Special features: Close voltage regulating requirements can be met. Tropical cooling and cold weather starting equipment optional.
Agencies: A Hawker Siddeley Company, U.S.A. subsidiary.

Type	Engine	kVA Single-phase 50Hz	kVA Single-phase 60Hz	kVA 3-phase 50Hz	kVA 3-phase 60Hz	V	Gen. Spec.	Weight (kg)
100	4DW				12.5	120/	B	–
120	..				16.5	208	B	–
G260	..				15	..	B	–
100	..		10		10	120	B	590
120	..		15		15	up to	B	590
G260	..		15		15	240	B	590

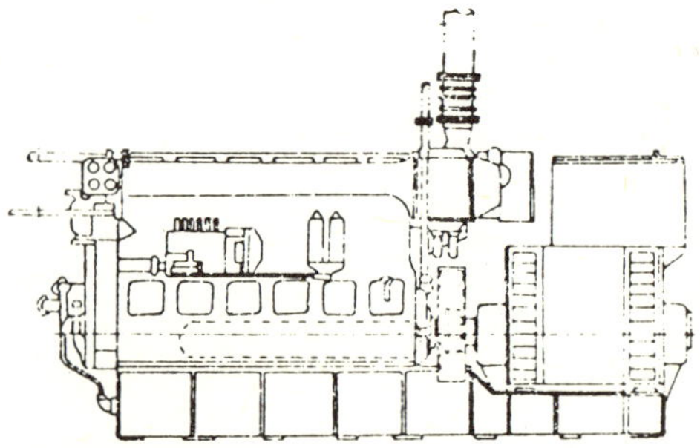

Model 6RK Series

YANMAR DIESEL ENGINE CO. LIMITED
1-11-1 Marunouchi
Chiyoda-ku
Tokyo
Japan

Telex: 0222-2310
Cables: Yanmar Tokyo

Mounting: Skid mounted
Range: Over 100 models in 9 ranges from 12.5kVA to 1600kVA. Full range not shown.
Engine make:
Engine features: Self-starting. Engines originally designed as marine diesels. Larger models are slower running.
Control options: Machine or wall mounted control. Parallel running, auto start, load shedding, remote control and protection devices available.
Voltage options: Specify on ordering
Power factor: 0.8
Protection: Drip proof
Special features: Cooling and fuel supply options available
Agencies: Service depots in 130 countries.

Type	Engine	kVA Single-phase 50Hz	kVA Single-phase 60Hz	kVA 3-phase 50Hz	kVA 3-phase 60Hz	V*	Gen. Spec.	Weight (kg)
2TLE	4DW			12.5	15		not	–
3TLE	4DW			20	20		yet	–
3ESDL	4DW			30	40		known	–
4ESDL	4DW			50	56		at	–
3KDL	4DW			65	56		time	940
4KDL	4DW			90	75		of	1150
5KDL	4DW			100	90		print-	1345
6KDL	4DW			140	120		ing	1500
6KDL–T	4DW			187.5	150			1610
6KFL	4DW			140	120			1780
6KFL–HT	4DW			225	185			1930
6RL	4DW			140	130			3360
6RL–T	4DW				160			3340
6RAL	4DW			200	187.5			3510
up to								
12GLET	4DW			1600	1600			18650

**As specified by purchaser*

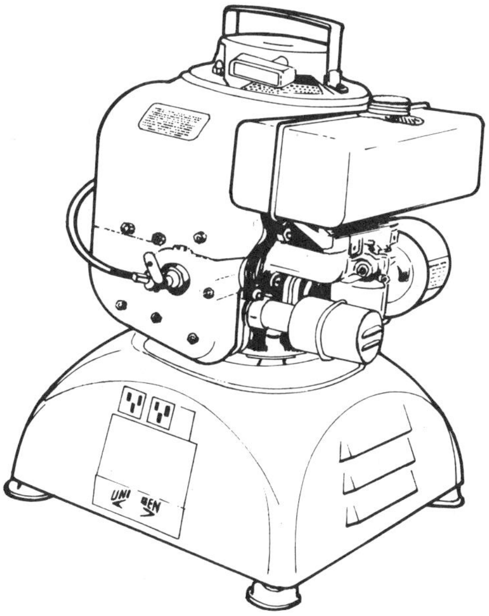

Unigen Model 420

UNIGEN INCORPORATED
194 West Stone Street
Almont
Michigan 48003
U.S.A.

Telex: 810-235-1100
Phone: (313) 798 3150

Mounting: Smaller units are free standing, larger units are skid mounted. All feature carrying handles.
Range: 15 models in 3 ranges from 0.8kVA to 4kVA
Engine features: Rip start with electric or remote options on larger models. All models are higher revving. Small units only have a vertical axis of rotation with engine over alternator.
Control options: Machine mounted controls
Voltage options: 12 or 24 d.c. on battery chargers. 115 or 230 on a.c. models. 60Hz frequency shown, 50Hz available on request
Power factor: 1
Protection: Drip proof tropicalised insulation
Special features: Instrumentation, battery charging facility and automatic voltage regulating are optional
Agencies: Write to factory.

Type	Engine	kVA Single-phase 50Hz	kVA Single-phase 60Hz	kVA 3-phase 50Hz	kVA 3-phase 60Hz	V	Gen. Spec.	Weight (kg)
712	4DA	0.85 d.c.				12	B	19
720	4DA	0.85 d.c.				12/24	B	34
1412	4DA	1.7 d.c.				12	B	32
724	4DA	1.7 d.c.				24	B	32
1224	4DA	1.7 d.c.				12/24	B	34
410	4DA	1 a.c. 400Hz or d.c.				110	B	19
420	4DA	2 a.c. 400Hz or d.c.				110	B	34
LA12			1.2			115/230	B	–
LA20			2			115/230	B	–
LA30			3			115/230	B	–
LA40			4			115/230	B	–

Other Electrical Equipment

C & D BATTERIES DIVISION
3043 Walton Road
Plymouth Meeting
Pennsylvania 19462
U.S.A.

Phone: (215) 828 9000

This company claims to be the largest manufacturer of lead-acid batteries in the U.S.A. and has a very comprehensive product range covering almost every type of lead-acid battery, plus an assortment of industrial battery chargers. Applications include batteries for industrial trucks, switchgear, communications equipment, mine locomotives, railroad loco starters, all kinds of engine starters, electric vehicles, photovoltaic and wind energy storage systems, etc. Various lead-calcium and lead antimony batteries are included in the range.

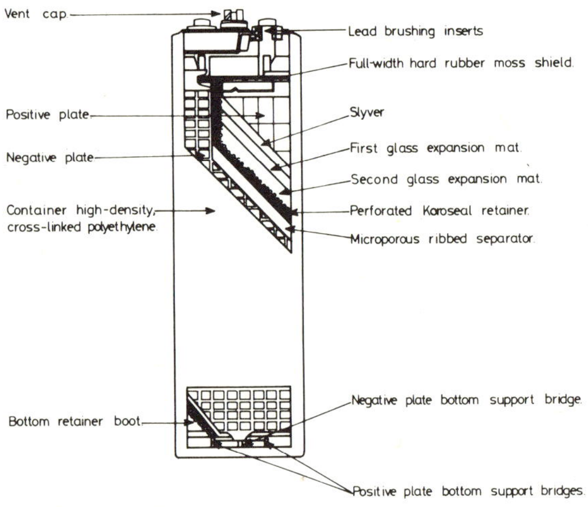

C & D Battery

CHLORIDE ZAMBIA LTD
Corner Dr Agrey Ave/North End Road
P.O. Box 1892
Kitwe
Zambia

Telex: Clorid ZA52170
Cables: Chloride Kitwe
Phone: 2200/2209/3631

Manufacturers of a wide range of lead acid batteries and ancillary equipment for automotive, traction and other purposes. Also suppliers of nickel cadmium battery cells, battery chargers, rectifiers, and emergency lighting sets.

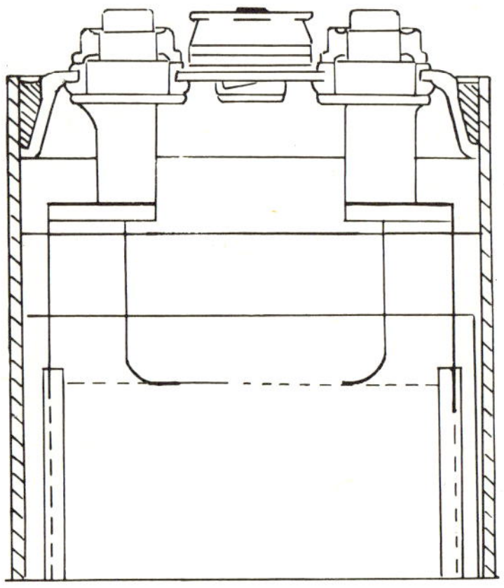

Kathanode Battery

CHLORIDE ALCAD
P.O. Box 4
Union Street
Redditch
Worcs. B98 7BW
U.K.

Telex: 33816
Cables: Alcad Redditch
Phone: Redditch (0527) 62351

Manufacturers of a wide range of nickel cadmium batteries (companies in the group produce other types of batteries and there are world wide distribution channels and associated companies).

RV and RVP series for deep cycle applications in 16 – 1040Ah sizes.

EP series for low rate discharge and float charging in 5 – 1040Ah sizes.

DLS and DLP for high rate discharge in 11 900Ah sizes.

'Unibloc' types for low rate discharge requiring virtually maintenance free operation in sizes from 7.5Ah up to 27Ah. PS series of sealed cell batteries require no maintenance in 7-27Ah sizes.

A choice of steel or plastic containers is available for many of the above batteries.

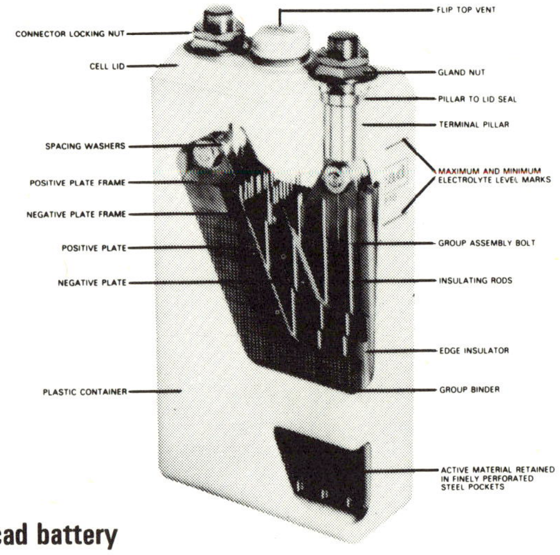

Alcad battery

QUIRK'S VICTORY LIGHT COMPANY PTY LTD
33 Fairweather Street
Bellevue Hill
New South Wales 2023
Australia

Cables: Quirklite Sydney

Sets of 2V or 6V cells, either rubberoid or polystyrene case made up to a size to meet requirements from 130 amps to 330 amps. Heavy duty house lighting batteries with an effective working life of from 10-15 years. 13 to 29 – Plate.

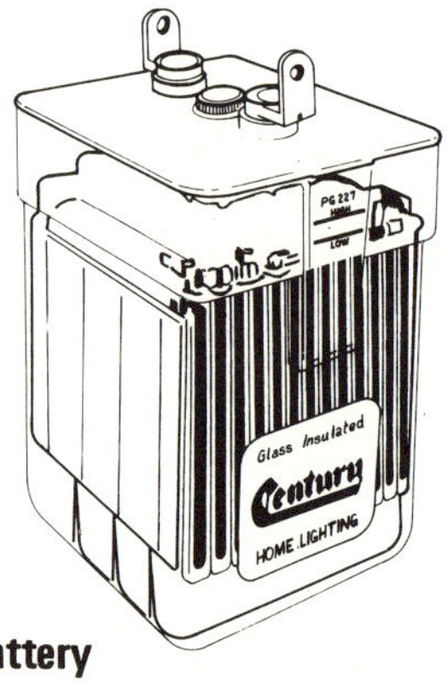

Glass Insulated Battery

DELATRON SYSTEMS CORPORATION
553 Lively Boulevard
Elk Grove Village
60007
U.S.A.

Phone: (312) 593 2270
(312) 438 9235

Manufacturers of lead acid deep discharge, 6V at either 245Ah or 375Ah rating, batteries and d.c. to a.c. inverters. The inverters are specifically designed for application in wind, hydro and solar electric generating systems and are up to 6kVA with 90% efficiency. The batteries are sold with a 60 month warranty on a pro-rata basis after 12 months.

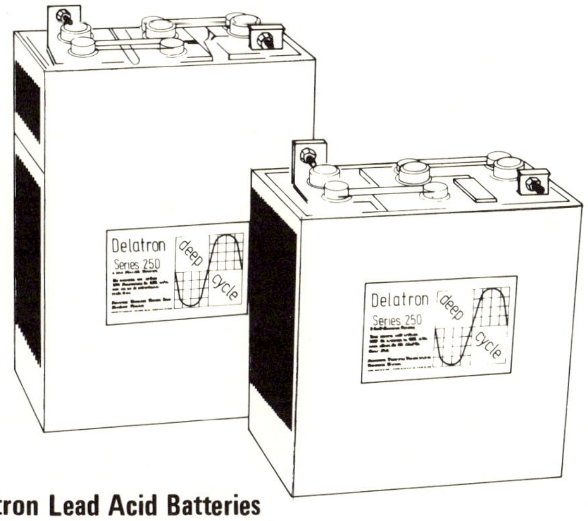

Delatron Lead Acid Batteries

ITT MERIDIAN
West Road
Harlow
Essex CM20 2BP
U.K.

Telex: 817202
Phone: (0279) 35351

Supplier of 'Gelyte' sealed and spillproof lead/lead dioxide batteries with gelled acid electrolyte as follows:

Model	*20hr rating Amp. hr @ 20°C*	*Nominal voltage*	*Weight lb*
PB626	2.6	6	1.4
PB1226	2.6	12	2.8
PB645L	4.5	6	1.9
PB660L	6.0	6	2.8
PB660	6.0	6	2.3
PB1260	6.0	12	4.6
PB690	9.0	6	3.7
PB1290	9.0	12	7.4
PB6180	18.0	6	7.4
PB12180	18.0	12	14.8
PB6270	27.0	6	11.1
PB12270	27.0	12	22.2
PB2300	30.0	2	4.2

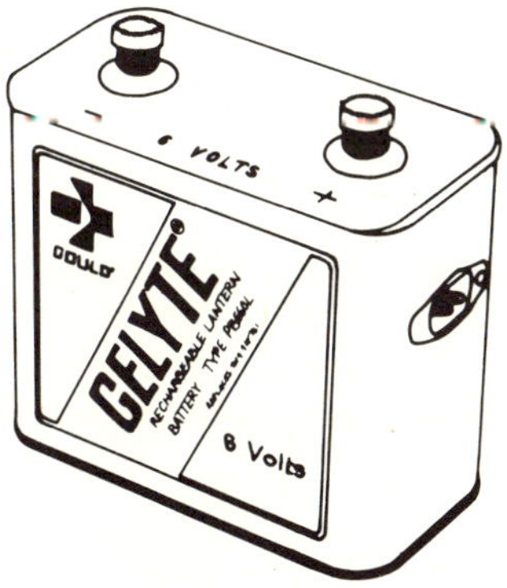

Model PB660L Battery

SAFT-SOCIETE DES ACCUMULATEURS FIXES ET DE TRACTION
Storage Battery Division
156 Avenue de Metz
93239 Romainville
France

Telex: 220100
Cables: Saftalcalin-Paris
Phone: (1) 843 93 61

Manufacturers of a large range of steel or plastic cased nickel-cadmium storage batteries and associated crates and stands for high power applications.

For example:

KPM steel cased cells range from 30 to 520Ah at 5 hour discharge

KPMP plastic cased cells from 10 to 320Ah at 5 hour

KPH series steel cased cells range from 30 to 310Ah at 5 hour

KPHP plastic cased cells from 14 to 250Ah at 5 hour

Crated battery assemblies in a variety of sizes are available.

Completely sealed miniature and small nickel cadmium batteries requiring no maintenance are also available, some as replacements for non-rechargeable dry-cells. These are as follows:

Sealed cylindrical cells 0.1 to 10Ah (20 different sizes)

Sealed button cells 40 to 600Ah (5 sizes)

Sealed rectangular cells 3.8 to 26Ah (8 sizes)

All the above ratings being taken over a 5 hour discharge cycle.

SAFT Crated Battery Assembly

This assembly is particularly suitable for mobile applications (railway, marine . . .). The crates are made from impregnated timber.

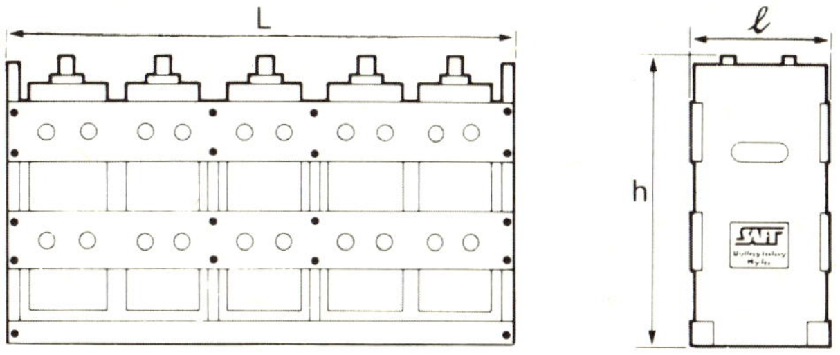

ACCUMULATORENFABRIK SONNENSCHEIN GmbH
D 6470 Buedingen/Hessen
Federal Republic of Germany

Telex: Sonnenschein Export No. 4184 637
Cable: Accusonne

U.K. Distributor:
F.W.O. Bauch Ltd
49 Theobald Street
Boreham Wood
Herts WD6 4RZ
U.K.

Phone: (0 60 42) 91
Telex: 27502
Phone: 01 953 0091

Manufacturer of Dryfit range of maintenance-free portable lead-acid storage batteries and Ulimatic power supply units and battery chargers.

A wide variety of Dryfit batteries are available with nominal voltages of 2, 4, 6, 8 and 12V and capacities of 1.0 up to 36Ah (20hr discharge rate). Many of these have high power/weight ratios compared with standard cells.

Two versions are available, the A200 and A300 series. The A200 is designed for high cyclic applications, the A300 for use in stand-by and float service applications.

The Ulimatic chargers come in four sizes, 0.8, 2.5, 5 and 10A nominal for use with 110 or 240V, 50 or 60Hz mains supplies and are designed for use under ambient conditions from −20°C up to 45°C.

Dryfit batteries would be particularly useful in conjunction with unattended solar photovoltaic systems due to their maintenance free specification.

Dryfit A300

CHLORIDE TRANSIPACK LTD
Stanley Road
Bromley
Kent BR2 9JE
U.K.

Telex: 896071 code: Transipack London
Cables: Transipack, Bromley
Phone: 01 460 9861

This company's main product line is a range of static inverters and battery chargers many of which are intended to provide continuity to essential electrical equipment during mains failure or to provide a good quality voltage and frequency with fluctuating supplies. Some are also designed for military applications to run communications equipment in the field and are therefore particularly robust. Some typical products of interest are:

Transipack 200D: Low cost domestic battery charger/inverter which charges two 12V car batteries while mains power is available and provides 240V at 50Hz sinusoidal up to 200VA when mains is off.

Transipack 350: Similar to above but 350VA capacity and in carrying case.

Transipack 606: 350VA unit in military Creeth case or for Post Office rack.

A wide selection of other inverters and stabilised power supplies up to as much as 14kW or more are produced.

Also available is the Alcos range of general purpose nickel-cadmium batteries with capacities from 2 to 500Ah per unit. Member of same group as Chloride Zambia and Chloride Alcad.

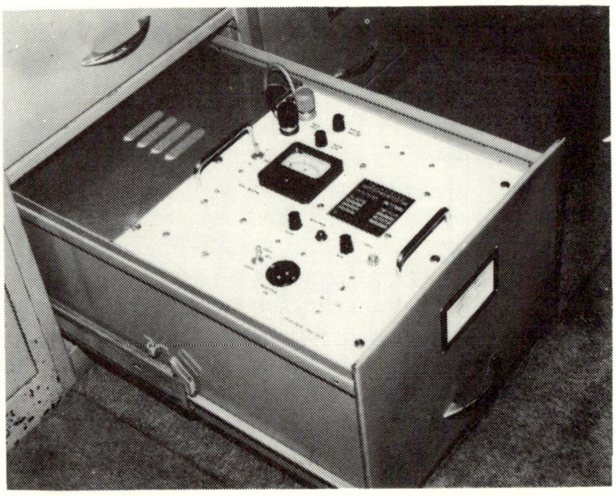

Transipack 200D

VARLEY DRY ACCUMULATORS LTD
Alfred's Way
Barking
Essex 1G11 0TB
U.K.

Telex: 8951668
Phone: 01 594 3346

Manufacturers of a range of lead-acid batteries with the electrolyte absorbed in a porous separating medium. This allows the batteries to be used in any position and renders them spill-proof. Other advantages are that bridging of the cells cannot occur, which allows much thinner plates to be used, so size-for-size they have a greater capacity and short-circuited current capability than free-acid counterparts.

Batteries in this range are available at voltages of 2, 6 and 12V with capacities ranging from 5 to 28Ah over a 20 hour cycle. An encased "Shoulder Pack" battery and a "Mini Power Pack" of 12V and 12Ah capacity (20hr) is also offered.

Typical Varley System Battery construction

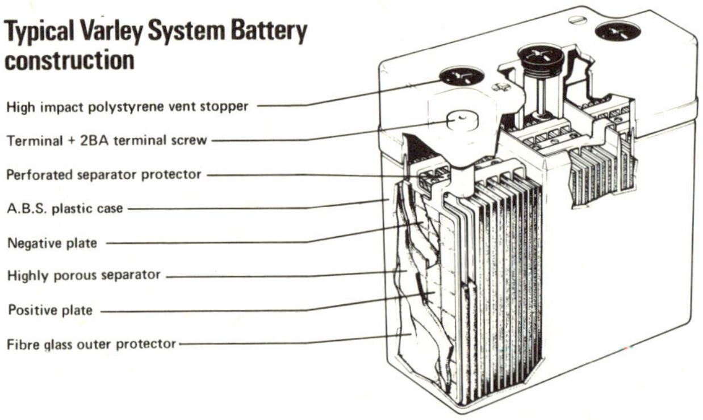

NORTH WIND POWER COMPANY INC.
P.O. Box 315
Warren
Vermont 05674
U.S.A.

Phone: 802 496 2955

This company distributes the Gemini Synchronous Inverter System developed by the Windworks research group.

The Gemini inverter is designed to interface between a variable voltage d.c. power source such as a windmill generator and an a.c. grid system. All the d.c. is converted to a.c. and any excess available after supplying the local load is fed into the grid, while if insufficient is available for the load from the wind-generator or variable source, the necessary balance is taken from the grid.

Synchronous inverters are less expensive than most conventional inverters and they allow a wind system to be fully utilised to yield the maximum possible output. In some American states the electricity utilities are prepared to credit the system with any power fed in to the grid and the system could equally be applied in conjuction with a small grid powered by say a diesel generating set as a fuel saving device, since any surplus from a windmill appears on the grid as a load reduction.

The basic Gemini unit is rated at 8kW, with d.c. inputs of 0-200V and 0-40A, a.c. connection being 120 or 240V, maximum efficiency 95%. Units of up to 20kW single-phase or up to 1500kW three-phase are possible.

EARLE ENGINEERING
P.O. Box 850
Alpine
California 92001
U.S.A.

The Watchman regulator combines an electronic voltage regulator with switching and charge sensing circuits to provide automatic battery charging for wind driven generators fitted with automotive alternators. The group also produce a.c. servo regulators for use with waterwheels.

The Watchman regulator incorporates a special delay to allow a windmill to free-wheel until the alternator reaches sufficient speed to sustain a charge to the battery and prevents continuous "hunting". The system can carry a maximum of 70A at 14V (980W). It is available at a reduced price in kit form for self-assembly and weighs 2lb (1kg).

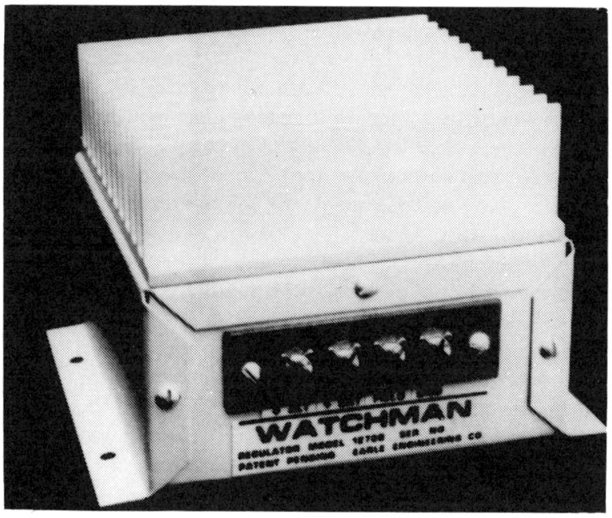

Watchman 1270B Regulator

RATELCO INC — Telex: 32 1200 RATELCO SEA
610 Pontius Avenue N — Phone: 206 624 7770
P.O. Box C-19080
Seattle
Washington 98109
U.S.A.

Manufacturers of an extensive range of d.c. to a.c. inverters with standard input voltages of 24, 32, 48 and 120V d.c. (others available to special order). Output voltages are regulated 120V ± 5% at 60Hz to ± 0.5% and 240V ± 5% is also available. The manufacturers claim less than 15% harmonic distortion on all loads and less than 7% for loads from 75 to 100%. Models available with outputs from 500VA up to 5000VA.

Also offer an extensive range of battery chargers for single-phase 120 or 240V a.c. operation and for three-phase 208, 240 or 480V a.c. Normal outputs available as follows:- 12V 10 models from 3 to 150A; 24V 12 models from 3 to 200A; 32V 9 models from 3 to 100A; 48V 11 models from 3 to 300A; 120V 11 models from 3 to 250A and 240V 10 models from 3 to 150A.

Other products include ranges of uninterruptible power supplies, a.c. to d.c. converters, various communication power equipment systems and components, transistorised voltage regulators, battery testers, battery balancers, etc.

Ratelco SCR Inverter

INDEPENDENT POWER DEVELOPERS — Phone: (406) 847 2315
Box 1467
Noxon
Montana 59853
U.S.A.

Distributors of Creative Electronics inverters intended for use in conjunction with wind-electric and small hydro-electric systems. These inverters have a very low "idle current" and are claimed to be over 90% efficient from a load of 275W up to maximum, reaching a mid-load efficiency of as high as 98%. They can handle 500% overload to cope with starting electric motor loads.

These inverters are available in seven different models from 1kW up to 12kW.

This company also markets their own model of small turbine, Dunlite wind generators and Delatron deep cycle storage batteries.

8 Instrumentation, Monitoring and Control Equipment

There is an enormous variety of monitoring and control equipment available to suit mass-produced conventional power plants. As already mentioned in the introductions to the engine sections and generator section, it is obviously advisable to install sufficient control equipment to protect a plant from damage by loss of coolant, overheating, or other possible catastrophic malfunctions. However this section will not deal with this kind of equipment since information on what is recommended and available is readily available from the suppliers of engines and it is generally best to adhere to their advice.

Renewable energy conversion devices, such as windmills, solar heaters or hydro-turbines generally require a lower level of monitoring since these devices are usually relatively less complex than engines and less susceptible to the kind of damage that might have been prevented by some form of automatic monitoring device. However, a major unknown with many of these devices is the potency or otherwise of the energy resource being tapped at a particular locality. How much solar, wind or water energy is actually available? This is a very important question to resolve, since the choice of equipment and its economic viability will often hinge on the energy availability.

Therefore, we have included mainly a selection of meteorological equipment that would be applicable in particular for the assessment of wind and solar energy availability. At present, much of the equipment on the market has been developed purely for conventional meteorological applications and it tends to be incidentally useful for energy assessment, but it is expected that increasing usage and research into the application of solar-based energy will result in the development of more specialised equipment requiring a lower level of interpretation of their data to arrive at a reliable conclusion.

Since this section at present is also a place to present the odd items that would be of interest but which do not fit into other categories, it also includes controllers for wind and water turbines.

BELFORT INTERNATIONAL CORP
P.O. Box 5268
Baltimore
Maryland 21224
U.S.A.

Cables: BELF
Phone: 301 342 2626

A full range of meteorological measuring and recording equipment including anemometers, chart drives, liquid level recorders, barometers, rain gauges, including many specialised items of equipment such as dew balances, evaporation recorders, pyranographs (for measuring and recording variations in solar radiation), etc.

Belfort Raingauge

C.F. CASELLA & CO LTD
Regent House
Britannia Walk
London N1 7ND
U.K.

Telex: 26 16 41
Phone: 01 253 8581

Manufacturers of a variety of meteorological instruments including several alternative types of high accuracy anemometer that are widely used by weather stations throughout the world.

Casella Cup Counter Anemometer

LULL MARINE LTD
11 Suckling Green Lane
Codsall
Wolverhampton WV8 2BL
U.K.

Phone: (09074) 2700

This company offers a range of instruments as follows:

Sims hand-held anemometers (ranges available from 0-15 knots up to 0-60 knots in units of knots, m.p.h. m/s, k.p.h. or Beaufort force).

Sims remote reading anemometer (ranges 0-30kt or 0-80kt).

Sims depthometer — a depth sounder which measures water depths from 0-10 or 0-100ft (could be used for investigating flow rates of a river). Requires 12V battery but draws only 40mA.

Sims Hand-held Anemometer

SMITHS INDUSTRIES LTD
50 Oxgate Lane
Cricklewood
London NW2 7JB
U.K.

Telex: 22671
Cables: Speedofac London
Phone: 01 452 3333

Manufacturers of a range of marine and engine monitoring instruments. Products include oil pressure gauges, tachometers, temperature gauges, voltmeters, ammeters, engine vacuum gauges, speedometers (vehicle and marine), engine hours counters, fuel level gauges, windspeed indicator, stop watches, etc.

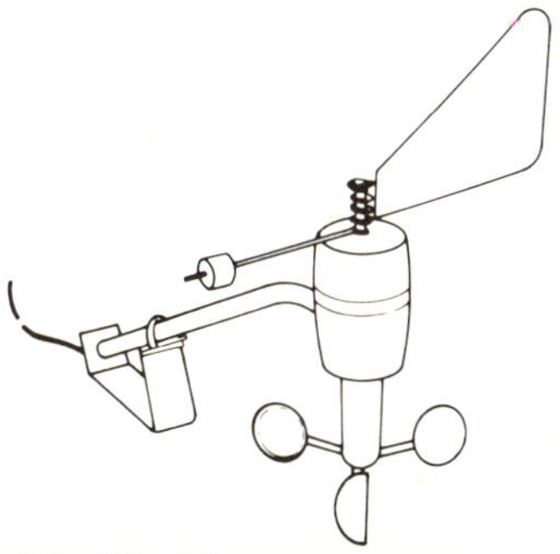

Smiths Wind Speed Indicator

SCIENCE ASSOCIATES INC
Box 230
230 Nassau Street
Princeton
New Jersey 08540
U.S.A.

Phone: 609 924 4470

Suppliers of a wide selection of instrumentation for the measurement of wind and solar energy. The products available from this company range from cup anemometers with dial indicators or automatic chart recorders, to hot wire type windspeed recorders. Numerous forms of solar radiation measuring equipment are also available together with sophisticated chart recorders and integrators. The range offered includes equipment intended for marine, airport and meteorological departments as well as experimenters and educational establishments.

Weather-proof Pyranometer

SOLAR POWER CORP
5 Executive Park Drive
N Billerica
Massachusetts 01862
U.S.A.

Telex: 710 347 6792
Phone: 617 667 8376

and:

SOLAR POWER LTD
110-111 Strand
London WC2E 0AA
U.K.

Telex: 975247
Phone: 01 386 8918

A new device on the market for measuring solar radiation. The idea is that a large number of sun stations could be put out in an area where there is little or no insolation data and, using only one portable readout unit, all the sun station records could be monitored.

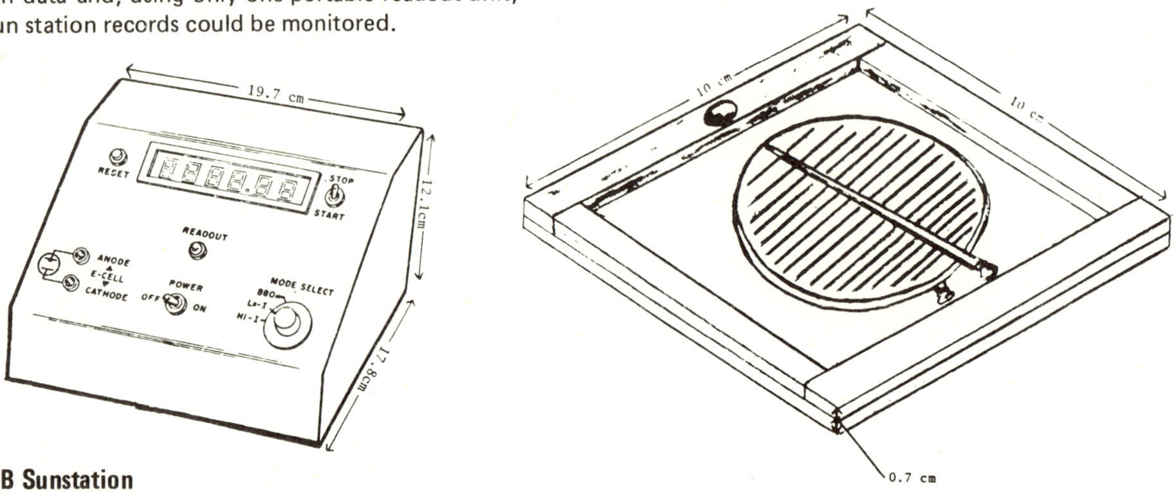

Model 3B Sunstation

TECNICO ELECTRONICS
53 Carrington Road
P.O. Box 12
Marrickville
New South Wales 2204
Australia

Telex: AA21490
Phone: 55 0411

This battery powered device is designed to monitor solar radiation, outputs from solar panels, wind generators, etc. It is intended for long unattended operation away from mains electricity supplies.

It has two separate integrator channels each having two input ranges of 10mV and 100mV and the readout is displayed on two six digit counters. A 12V battery supply such as two 509 lantern batteries will last in excess of 12 months operation.

Model R 2TL/101A/WP/B dual recorder integrator has a similar function to Model 2TL but has an additional chart recorder built in to give a continuous record of variations in weather conditions, solar cell output, etc. Hence overcast periods or windless periods, in the case of solar or wind monitoring respectively, can be clearly identified.

The integrator battery requirements are as for the other model. The chart recorder requires four No.509 cells per month of operation and the paper runs 31 days at 1in/hr with 1000 counts per hour.

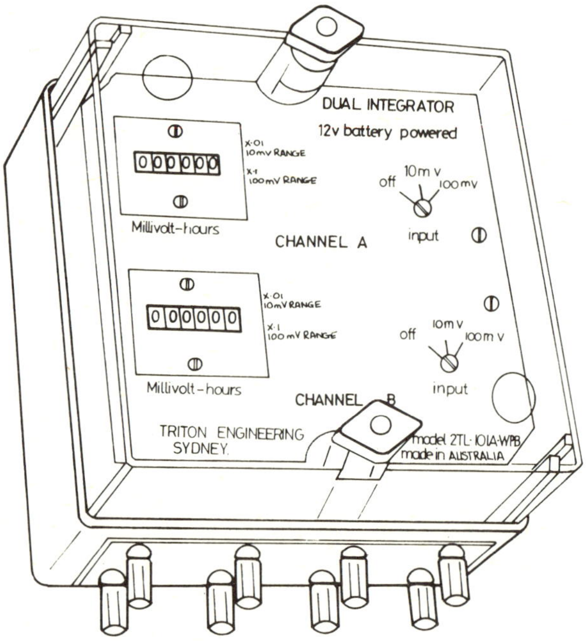

Model 2TL/101A/WP/B
Dual Integrator

CONTROL TECHNOLOGY LTD
Bolney Avenue
Peacehaven
Sussex BN9 8HQ
U.K.

Telex: 877623
Phone: (079 14) 5841

This company, which is associated with the Wind Energy Supply Co., (WESCO), offers various items of electrical control equipment and instrumentation. It has made notable developments including induction generators, for use with wind powered systems. Its products are also designed for usage with other intermittent and variable power supplies.

In particular its products include "Self Powered DC voltage transformers"; AC and DC current transformer; AC and DC voltage and current relays; automatic voltage regulators, battery chargers, main injection units; inverters etc.

It can also supply, by quotation, windmill instrumentation for controlling and monitoring performance, together with a statistical wind analyser for use in site location. Also available is a new "Load Priority Controller" which enables loads to be added in individual priority as increased power becomes available from aerogenerators.

In addition, the company offers, by enquiry or quotations, a comprehensive advisory service on all electrical controls and applications associated with wind powered systems.

Enquiries from Europe & N. America to:

EVANS ENGINEERING
Land & Leisure Services Ltd
Priory Lane
St. Thomas
Launceston
Cornwall PL15 8DQ

Phone: (0566) 3982

Enquiries from rest of the world to:

Intermediate Technology
Industrial Services
3rd Floor Myson House
Railway Terrace
Rugby CV21 3HT
U.K.

Phone: 0788 70126
Cables: ITIS Rugby

Electronic load controller to provide a continuous stabilised power supply from small hydro-electric plants, and other generating systems by switching unrequired power into a ballast circuit to keep the alternator load correct for the required voltage and frequency. This is a much less expensive and more reliable control system than traditional mechanical or hydraulic governing systems for a variety of applications.

Available to handle up to 100kW or more, single or three-phase with a variety of safety devices to suit customer's needs.

Evans Electronic Load Controller

9 ITDG Power Project

THE INTERMEDIATE TECHNOLOGY DEVELOPMENT GROUP'S POWER PROGRAMME

ITDG exists to try and transfer technology appropriate to the needs of people in the underdeveloped parts of the world so that it can be instrumental in reducing poverty in those regions. In effect the Group seeks to redistribute technology as a means towards the redistribution of wealth.

Market forces dictate that equipment is developed for people who can afford to pay for it, so its scale and cost is geared to suit the needs of the customer with money. Sometimes, equipment developed by the purchasing pressures from the affluent market will, by chance, be suitable and helpful if applied in the poorer regions using aid or development money to subsidise its movement there. So one function of ITDG is to communicate knowledge of available technology that might be useful in development — this book being part of that function.

However, another function of our Group is to develop new equipment to fill specialised technological gaps in the under-developed world. Since the Group is a non-profit Registered Charity we are not motivated by any need to produce profit-making products. Indeed, our aim, with most of our own technology, is to see it transferred and taken over, perhaps in modified form in some cases, by the people whom it might benefit. As lack of job opportunities is one of the major sources of poverty in underdeveloped areas, job creation as well as end-use inevitably affects our choice of technology, and most of the equipment we develop is intended for local manufacture at or near the place of usage, using local skills and resources. In many cases appropriate local skills and resources are initially absent, in which case the aid element involves the setting up of the necessary manufacturing unit and the training of local people to run it, as soon as possible, independently and self-sufficiently. An important criterion of whether or not aid is effective is the question of how it affects the self-sufficiency of the receivers — arguably, technology that undermines their self-sufficiency by making them more dependent on outside help is generally counter productive in all except the short-term.

To this end the Group is actively promoting a number of R&D programmes, usually with the backing of various aid donor organisations, and often in co-operation with other organisations both in the U.K. and overseas. The Group's R&D effort is implemented mainly by its Project Officers, who at the time of writing include staff engineers with experience in agricultural engineering, water supplies, rural transport, building, building materials and of course energy and power and they are supported by numerous panels of specialists operating voluntarily as advisors and who occasionally serve as consultants for short-term projects. In addition, a newly formed wing of the Group called Intermediate Technology Industrial Services exists to develop and transfer the technology for small-scale industrial processes by advising on and setting up small industries in under-developed parts of the world.

This section contains a brief outline of the Power Project's current research and development work.

Windmill system for pumping water

Water lifting is clearly a highly appropriate application for wind power. Many arid regions where the need is greatest have adequate or even good wind regimes, and usually water lifting is not time-dependent, so that the intermittent nature of the wind poses no great problem. Where it is a problem, water storage is relatively cheap compared with other forms of energy storage. As already explained in the section on wind power, windpumps have been vital to the development of the livestock industries of the arid USA and Australia.

Many developing countries have contemplated licence

manufacturing of American, Australian or other industrial countries' windpumps, but most of those products require sophisticated and expensive manufacturing facilities which would be prohibitively expensive in terms of the small initial market for the product. As a result, ITDG has developed a windpump specifically intended to be suitable for small-scale low-volume production using no more than basic machine shop facilities of the kind that are commonly available throughout the world. It is however designed to technical specifications competitive with equipment available in the industrial countries and bearing in mind that the main running cost of a windmill is the interest on the investment, it is intended to have a long life to permit a long amortisation period.

Two distinct applications for windpumps have become apparent to the Group. Low-lift, high volume pumping for irrigation purposes and high-lift small volume pumping (from tube wells) for village and live-stock water supplies. These two applications require different rotor and pump combinations to obtain optimum outputs, so the ITDG windmill system consists of a common tower, transmission, tail and governing system, with two alternative rotors and pumps. The low lift version has a relatively high speed rotor, with a special pump designed to obtain better than usual windmill to pump matching for the conditions. The high speed rotor for the prototype can exploit the advantages (structural and aerodynamic) of using aerofoil section blades (aluminium filled with polyurethane foam), as it only needs six blades on account of its low solidity, but the high lift version carries 24 "paddle" blades of sheet metal (or fibreglass) to obtain the necessary high starting torque and low speed of operation for a deep well pump. Both versions have 6m (20ft) diameter rotors.

ITDG prototype fast-rotor windpump.

ITDG developed the UK prototype for this system with the support of the Engineering Department of Reading University. The U.K. charity Christian Aid has funded this work and nine institutions (some commercial and some R&D orientated) in eight countries having appropriate applications for the system are participating in the overseas development phase as a joint venture with ITDG. The plan is that a variety of production prototypes adapted from the UK prototype are to be produced in

ITDG prototype multi-bladed windmill.

Botswana, Kenya, Egypt, Oman, Pakistan, Sri Lanka, Zambia and Antigua in order to evaluate and improve the design both for ease and economy of local manufacture and for operational reliability and performance. After this phase it is planned that it will go into production in those and any other countries wanting it, probably around 1980.

Lowering the vertical axis river turbine into the water for a test run.

Run-of-stream river turbine

There is a substantial energy flux present in moving water. A 3 knot river current represents an energy flux of $1.8 kW/m^2$ ($0.23 hp/ft^2$) available 24 hours per day. Conventional engineering practice for tapping river energy is to dam the entire river in order to create sufficient head to allow the use of traditional encased turbines. With any sizeable, well-developed river this is an enormously expensive proposition only justified economically if there is a large market for electricity such as for an aluminium smelter (e.g. Volta) or a copper industry (Kariba), since "all or nothing" of the river flow energy must be used.

Traditionally, large undershot waterwheels were used in parts of the Middle East and the Sudan to tap small quantities of river current energy, but these are massive and material-intensive in relation to their power output, and therefore uneconomic. With this in mind, the ITDG Power Project has developed a "low solidity" device which runs completely submerged in a river current and is therefore potentially quite efficient yet does not need much construction material. Initial work indicates that it can convert up to 40% of the energy flux, amounting to 720W/m^2 (0.1hp/ft^2) from a 3kt current.

This device is the same in principle as a Darrieus vertical-axis windmill or a Voigth-Schneider ship propeller. The initial prototype tested over the front of a motor boat on the river Thames is only 1m in diameter by 0.5m deep. It is planned to develop a larger version, to be suspended beneath a pontoon, for pumping irrigation water from rivers passing through arid regions (such as the Nile, Niger, Euphrates, Indus, etc.). At the time of writing the Power Project is working on a design for a unit with a cross-sectional area of 3m^2 which could lift approximately 61m^3 (13500 UK gall) per hour against a 5m (16ft) head if submerged in a 3kt (1.5m/s) current. Large areas of fertile alluvial but arid soil that cannot at present be exploited economically for agriculture (due to the high cost of imported energy sources such as diesel or electric pumps) will become accessible to a device of this kind. We are hoping to test this prototype in Juba, Southern Sudan, in late 1979 or early 1980.

Later work will, it is hoped, result in a variant for electricity generation to permit village electrification along well-developed rivers with adequate currents.

The Humphrey pump.

Liquid piston Humphrey Pump

This is a four-stroke spark ignition engine which has no piston or mechanical output; instead, the cylinder is open to water and a column of water behaves as a piston so that the ignition and expansion of the charge drives water directly from one level to another. ITDG developed an initial prototype at Reading University and the University Engineering Department then took the work further under a Ministry of Overseas Development grant. ITDG has now commissioned the manufacture of two field prototypes at Reading which have been sent for testing to Egypt and Nepal.

The present level of development only allows the use of gaseous fuels (such as methane, propane, bio-gas or wood-gas), but work is proposed to develop methods of utilising liquid fuels. The prototype machine functions reliably with a conventional coil ignition system, but work is also planned to find alternatives not requiring a battery for irrigation; considerable promise is offered by a piezo-electric ignition system which has been under test with some success, although it is not yet as reliable as coil ignition.

This machine is suitable for small-scale manufacture, needs virtually no maintenance and has an almost indefinite life since there are no loaded moving parts. The Reading prototype has achieved an overall system efficiency of 10% maximum (which is not bad for a small system (i.e. engine plus pump) of this power – around 500w), and lifts about 25m^3/hr (5500 gall/hr) through 6m (20ft).

One attractive application is as an irrigation pump powered by bio-gas generated from agricultural wastes, sewage, animal dung and/or water weeds. The effluent from the bio-gas digester provides fertiliser in addition to the irrigation water. Efforts are being made in Nepal to test this concept in practice and it is hoped that a system working on water hyacinth may be evaluated in Egypt, where this water weed is currently a pest which could for this purpose form a combined fuel and fertiliser.

Micro-hydro-electric systems

Most of the power projects have been preoccupied with lifting water, one of the most basic prerequisites for improved agricultural production in most parts of the world. However, rural electrification is another important need.

To this end, ITDG has supported work by an innovative hydro-power engineer in the West of England, Mr Rupert

The Evans 6kW propellor turbine pilot installation (the belt guards are not yet fitted).

Armstrong Evans, to develop further some promising small-scale hydro-electric equipment he has evolved.

Some of this equipment is on the UK market and in use (see page 151), but the Group and Mr Evans have identified a need for low head turbines to succeed the Francis turbine, used traditionally for such applications in the 1930s and before. Mr Evans has also developed an electronic control system which eliminates many of the most expensive components in traditional turbines (e.g. no gates, spear valve, governor or mechanical linkage are needed) and which is more reliable and maintenance free with a quicker response than conventional hydro-mechanical control systems. For high and medium heads he has developed a family of Pelton Wheels which can also be electronically controlled.

As a result of identifying this need, ITDG has at its disposal a newly developed propeller turbine which can readily be machine-shop fabricated, economically in small quantities. Initial prototype testing gave an overall efficiency in the 65-70% range at a head of 4m (12ft), with a machine having an output of 6kW. It is planned to test-manufacture and field-test prototypes overseas in the near future and several electronic controllers have already been sent by the Group to places such as Fiji, Nepal and Pakistan. At the time of writing about 30 are in regular use in U.K. installations.

The Evans electronic turbine load controller (foreground) shown with a traditional hydro-mechanical turbine flow controller.

Solar energy

The Power Project is beginning work, in consultation with the Chairman of the Group's Solar Sub-panel, Mr Costis Stambolis, on several direct applications of solar energy.

These will include a simple basin solar still to be installed in conjunction with the windmill project in Oman. It is hoped that this work will lead later to the development of more sophisticated types of still utilising solar and/or wind energy in order to desalinate brackish and sea water.

It is proposed that the Group should develop specialised expertise in advising on the setting up of small manufacturing units in less developed countries to produce solar water heaters, a primary objective is the job creating aspect as well as that of energy saving. To this end we have made a financial contribution to a small project to develop suitable hardware at the University of Fourah Bay in Sierra Leone.

Improved stoves

The Power Project has recently commissioned an Australian engineer, Mr Stephen Joseph, to investigate the current situation with respect to "subsistence fuel" cooking stoves and to initiate a research and development programme aimed at the identification of improved efficiency low-cost cooking stoves. In terms of human needs, this is probably by far the most important project in the power field being undertaken by our group (as explained earlier in the section on bio-mass.

At the time of writing, Mr Joseph has received an encouraging response to a questionnaire from numerous institutions and individuals in developing countries. He has also developed a number of prototype stoves and is particularly interested in developing a type of light clay stove which can be turned on a potter's wheel and fired for strength prior to use. Also under investigation are small metal stoves with refractory linings that could be made and marketed in bazaars and small workshops in competition with mass-produced kerosene stoves.

This is such an important field for development that we hope and expect this programme to continue for some time. It should result in publications on stoves as well as collaborative field projects with other institutions in developing countries.

Steam engines

We have initiated investigative work in collaboration with staff from Reading University Engineering Department with a view to developing a 10kW steam powered electricity generating set. This system is to be developed to run on such fuels as wood, low grade coal, lignite, agricultural and other wastes, etc. At the time of writing a project proposal is being circulated in order to find funding to allow prototype construction and testing to proceed. It is expected that this work will get under way during the course of 1979.

Future projects and the new Energy Centre at Reading

The Engineering Department at Reading University has recently set up an Alternative Energy Research Centre off the main university campus, to conduct R&D into a number of new energy conversion systems, many of which are applicable for small-scale decentralised power production of the kind that is appropriate in under-developed areas. The department also proposes initiating post-graduate research and training courses in the fields of "alternative energy" based on the centre, particularly aimed at teaching technology relevant to the needs of the under developed regions. Obviously the department is keen to attract engineering and science graduates who are nationals of countries that could benefit from this kind of expertise and to co-operate with universities and similarly-minded institutes in such countries.

ITDG has an office and a staff member of the Power Project based at the new Centre and will of course be closely involved with the work there, which is directed by Professor Peter Dunn. Some of the work going on at the centre in addition to that involving ITDG includes development of heat pipes, fluidised bed combustors, stirling engine development, Dr Musgrove's vertical axis windmill (see page 54 for production model), small steam engines that could use solar energy, testing of solar collectors and work on bio-gas. Test facilities for reciprocating pumps, windmills and solar devices are also available. A permanent appropriate technology exhibition is being established at the Energy Centre by Intermediate Technology Industrial Services.

10

Information Sources

RESEARCH AND DEVELOPMENT INSTITUTIONS, CONSULTANTS, & GENERAL SUPPLIERS OF EQUIPMENT

The technology of small-scale decentralised energy conversion is in the transition period to a new epoch, in the sense that for several decades the internal combustion engine running on petroleum-based fuel was the only serious contender for the purpose. The i.c. engine superseded the previous generation of steam engines, small turbines and windmills that started back in the nineteenth century, and is now, for the first time, under serious challenge due to the threat of future (rather than present) petroleum shortages and price rises.

As a result, the new technology of small-scale energy conversion is in a relatively embrionic state, particularly so far as the economic exploitation of solar, wind and biomass energy is concerned. A healthy bit of evidence for this assertion is the large number of research institutions throughout the world that are actively looking for alternative solutions, (including of course ourselves at the Intermediate Technology Development Group). The first part of this section includes a far from exhaustive list of various institutions and consultants who may be able to offer technical advice on aspects relevant to the subjects covered by this Guide.

At the risk of seemingly stating the obvious to some readers, it should be emphasised that many of the organisations listed are small and their resources are stretched to the limit, so it is recommended that readers should send International Reply Coupons to cover expenses when writing to them and be patient if in some cases a reply is a few weeks coming. Obviously some organisations are better geared to coping with enquiries than others and this will not apply to all. From our own experience it is a great help if enquirers are as specific as possible as to what they ask — obviously it is not possible to do more than answer very specific technical questions in the space of a letter, or to recommend a reference that might provide a detailed answer.

It is believed that the consultants listed are prepared to undertake on-the-spot consultative commissions in their various specialisations.

Information Sources and General Suppliers of Equipment

Addresses, and where available, brief descriptions of organisations that can supply information, consultancy services or hardware, plus a selection of institutions active in research and development in fields related to the scope of this Guide. We should be grateful for suggestions and corrections for inclusion in the next edition of this Guide.

a

ALTERNATIVE ENERGY GROUP – READING UNIVERSITY
Department of Engineering
(Shinfield Centre)
Whiteknights
Reading RG6 2AY
U.K.
Phone: 0734 81269

Research and development work is in hand in such fields as windpower, stirling engines, biomass conversion, direct use of solar energy, heat pipes, fluidised bed combustion, etc. Much of ITDG's own power programme is being carried on at this centre (see Section 9 giving details).

THE AMERICAN WIND ENERGY ASSOCIATION
54468 CR 31
Bristol
Indiana 46507
U.S.A.
Phone: 219 848 4360

The primary purpose of the American Wind Energy Association is to promote the use of wind as a renewable energy source. It provides regular publications in the form of a newsletter and a technical journal and also organises conferences.

J. HILBERT ANDERSON, INC.
2422 South Queen Street
York
Pennsylvania 17402
U.S.A.
& SEA SOLAR POWER INC.
Phone: 717 741 0884

Specialists in vapour turbine development (Rankine Cycle) for applications such as geothermal energy utilisation and ocean thermal energy conversion. A variety of publications is available on various aspects of sea solar power and its applications, mainly describing the results of technical investigations. A publications list is available on request.

APPROPRIATE TECHNOLOGY DEVELOPMENT ORGANISATION
P.O. Box 1306
I-B, F-7/1 St. No. 57th
Islamabad
Pakistan
Phone: 42327

A government sponsored AT organisation with experience of constructing experimental methane digesters (Chinese types), wind-pumps and mini-hydroelectric plants, plus numerous activities in other non-energy fields.

APPROPRIATE TECHNOLOGY DEVELOPMENT UNIT
P.O. Box 311
Gandhi Bhawan
Lucknow
U P 226001
India
Cables: Aptech
Phones: 43451

An independent AT organisation with particular interests in small industry development and small scale processes such as small cement and sugar production units.

APPROPRIATE TECHNOLOGY DEVELOPMENT UNIT
University of Technology
P.O. Box 793
Lae
Papua New Guinea
Cables: Utech
Phone: 42 4999 Ext. 130

Working on various items of AT hardware and particularly active in conjunction with the University Department of Electrical Engineering on developing small-scale hydro-electric systems. Publishers of the admirable *Lik Lik Buk* on appropriate technology in rural Papua New Guinea.

APPROPRIATE TECHNOLOGY GROUP SRI LANKA
c/o P.O. Box 352
Colombo 1
Sri Lanka

A group with general interests in promoting small-scale industries and appropriate technology for development purposes.

ASIAN INSTITUTE OF TECHNOLOGY
Division of Community and Regional Development
P.O. Box 2754
Bangkok
Thailand
Cables: AIT Bangkok
Phone: 796415 or 796418

Studying solar radiation in Thailand, solar distillation systems, solar powered water pumps and solar refrigeration.

ASSOCIATION OF BRITISH GENERATING SET MANUFACTURERS
21 John Adam Street
London WC2N 6JH
U.K.
Phone: 01 839 6171

This is an association of some of the major British generating set manufacturers and can offer specialised technical advice, particularly with respect to generator set specifications and claims a comprehensive, impartial generating set consultancy service.

ASSOCIATION OF VOLUNTARY AGENCIES FOR RURAL DEVELOPMENT Phone: 678642
C-6 Safdarjang Development Area
Community Centre
New Delhi 110016
India

This group provide a consultancy service and information on the construction and operation of bio-gas plants in India.

ASTRA Phone: 30481
Cell for the Application of Science & Technology to Rural Areas
Dept. of Inorganic and Physical Chemistry
Indian Institute of Science
Bangalore 560012
India

Research and development work at a very practical level on simple low cost energy converters, including windpumps, methane generators, etc.

b

BELYEA COMPANY INC Phone: 201 653 3334, 212 732 7150
31-45 Howell Street
Jersey City
NJ 07306
U.S.A.

This company rebuilds electrical power equipment and machinery and carries a wide selection of reconditioned stock, much of which is in the mega-watt size range but there is usually a reasonable selection of motor-generating sets in the 25 to 100kW range.

BRACE RESEARCH INSTITUTE Phone: (514) 457 6580
Macdonald College of McGill University
Ste. Anne de Bellevue
Quebec HOA 1CO
Canada

Publishers of articles in English and French with some translations into Arabic, on the full range of solar heated appliances. They have a publications list available.

Some selected publications are:

How to Make a Solar Still	US$1.25
How to Make a Solar Steam Cooker	US$1.25
How to Build a Solar Water Heater	US$1.25
How to Construct a Cheap Wind Machine for Pumping Water	US$1.25
How to Make a Solar Cabinet Dryer for Agricultural Produce	US$1.25
Instructions for Making a Simple 8ft Square Solar Still	US$1.25
Plans for a Glass and Concrete Solar Still	US$4.50
Production Drawing for a Solar Cabinet Dryer	US$2.50
Notes on the Development of the Brace Airscrew Windmill as a Prime Mover	US$0.50

BUNDESVERBAND SOLARENERGIE e.V. (BSE) Telex: 0857 851 Phone: 0201 1853392
Kruppstrasse 5
D-4300 Essen 1
German Federal Republic

Numerous firms in the Federal Republic of Germany have developed techniques – especially in the field of solar energy for utilisation of regenerative sources of energy in different fields.

With the aim of promoting the investigation, testing and development of different possibilities for utilizing solar energy, these firms have founded the Bundesverband Solarenergie e.V. These aims also include working on technical problems of common interest, promoting co-operation with foreign countries and informing the public of the possibilites of using solar energy.

c

CECOCO Cables: Cecoco Ibaraki Phone: 0726 22 2441
(Chuo Boeki Goshi Kaisha)
P.O. Box 8
Ibaraki City
Osaka
Japan

This organisation develops and markets a wide selection of equipment for small-scale agriculture, energy conversion, cottage industries, etc. Their hydrams and wood gasification plants are listed in the relevant sections of this Guide; other products include a wide selection of pumps, small powered cultivators, plus small-scale manufacturing equipment that is too numerous to list here. Detailed lists are available from them (in English) and they also publish a booklet entitled *Guide Book for Small, Medium and Rural Cottage Industries, Paddy Rice Cultivation.*

CENTRO DE TECNOLOGIA PROMON (CTP) Phone: 021 205 0112
Praia do Flamengo 154
20000 Rio de Janeiro
RJ Brasil

This major research centre is studying solar energy, fuel alcohols, fluidized combustion, coal conversion, fuels from wastes, etc. Various publications are available on these subjects, mainly in Spanish, although a few are in English.

COPPER DEVELOPMENT ASSOCIATION Telex: 27711 Phone: Potters Bar 50711 Cables: CUDA Potters Bar
Orchard House
Mutton Lane
Potters Bar
Hertfordshire EN6 3AP
U.K.

Advice and publications available to those who wish to use copper in any form, for any purpose for solar collectors and for other power applications.

d

DESERT SUNSHINE EXPOSURE TESTS INC.
P.O. Box 185
Black Canyon Stage
Phoenix
Arizona 85020
U.S.A.

Telex: TWX910 950 4681
Phone: (602) 465 7525

Well instrumented and experienced outdoor exposure testing station, for the performance testing of solar devices (including photo-voltaics).

Also test any products particularly for weathering characteristics such as adhesives, building materials, paints, plastics, textiles, mirrors, etc.

DEVELOPMENT TECHNOLOGY CENTRE
Documentation Centre
Institute of Technology
P.O. Box 276
Jalan Ganesha 10
Bandung
Indonesia

Phone: 82051 or 82055

A series of simple designs for the construction of a simple solar dryer, a solar water heater and a simple water treatment system are available.

e

THE EAST-WEST CENTER
177 East-West Road
Honolulu
Hawaii 96822
U.S.A.

Cables: Easwescen

This centre is a private, non-profit, educational institution intended to promote better relations and understanding between the nations and peoples of the Pacific area, Asia and the U.S.A. It is engaged in co-operative research, training and education in the field of development technology. It is not engaged in the production of equipment or provision of consultancy services in technology.

A quarterly newsletter called *Infotech* is produced and various seminars, conferences and training courses on specialised subjects are organised from time to time.

ECOTOPE GROUP
747 16th East
Seattle
Washington 98112
U.S.A.

Phone: (206) 322 3753

The group offers a wide range of consulting, demonstration and research services in the field of "appropriate technologies" including solar, wind bio-conversion and energy conservation.

Also produce various publications and present workshops to teach people how to build solar collectors.

Publications include:

Methane Feasibility Study; Solar Retrofit Feasibility Study; Solar and Wind Power as Energy or Water Conservation Measures in the Columbia River Basin.

THE ELECTRICAL RESEARCH ASSOCIATION (ERA) LTD
Cleeve Road
Leatherhead
Surrey KT22 7SA
U.K.

Telex: 264045
Phone: 03723 74151 (from London 537 4151)

This association carries out research and development programmes on behalf of industry or government in a wide variety of electrical specialisations. They have long been associated with research into large scale wind-power usage and they have published numerous technical papers on aspects of wind-energy conversion. They also publish *Electrical Power Plant International,* which is a very comprehensive 700 page reference source on generating equipment, static inverters, generators and prime movers in the range from 1kVA to 1000mVA and would be particularly useful for readers of this Guide seeking higher powered equipment than covered here and for those seeking more detailed technical advice specifically on electrical generating plant and equipment. The ERA guide is US$60.00 or £35.00 surface mail included at the time of going to press of our Guide.

ENERTECH CORPORATION
Norwich
Vermont 05072
U.S.A.

Cables: Enertech Norwich
Phone: 802 649 1145

This company specialises in wind energy systems and offers systems design services in connection with wind or solar energy applications. They have agencies for Elektro, Dunlite, and Sparco windmills and also market Dwyer and Stewart wind measuring instrumentation.

They produce their own publication *Planning a Wind Powered Generating System* (US$2.00) and can offer various other useful publications on wind energy.

ENVIRONMENTAL ENERGIES INCORPORATED
Box 73
Copernish
Michigan 49625
U.S.A.

Phone: 616 378 2000

A group who conduct feasibility studies for solar heating and wind electric power systems on an international basis.

f

FICHES ECOLOGIQUES
Vingrau Village
66600 Rivesaltes
France

Phone: (68) 59 40 59

Publishers of over 400 leaflets many on health care, but some technical and general on aspects of self sufficiency. Leaflets are written in either French or English. Energy related topics include: *Autogas Conversion, Methane Fermentation, Solar Cabinet Dryer, Solar Water Heaters (200 litre and 50 litre versions), Wind Machines,* etc.

g

GOBAR GAS RESEARCH STATION
Ajitmal
Etawah (U.P.)
India

Cables: Gas Research

This organisation developed the Gobar Gas system of methane production for village and small farmers' use. They publish various information booklets and plans, notably:

Bio Gas Plant Generating Methane from Organic Wastes US$5.00
Bio Gas Plant Designs with Specifications US$7.00

The above are in English, but there are also various Hindi publications.

GOODMAN ASSOCIATES
4 Berryville Pike
Summit Point
West Virginia
U.S.A.

Phone: 703 667 7175

Publishers of information sheets and plans on *Vertical Windmill, Electric Car, Home Power Plant, Electric Bicycle, Electric Mini Bike, Fireplace, Log Splitter, Solar Heat, Water Wheel, Burn Waste Oil in Your Home Burner.*

These plans cost US$3.00 each or any four for US$10.00 or all for US$15.00.

GROUP SOCIALLY APPROPRIATE TECHNOLOGY
Afd. der Bedrijfskunde
Technische Hogeschool Eindhoven
Postbus 513
Eindhoven
Netherlands

Phone: 972443 (040)

Research is carried out at the University on various projects related to energy conversion in underdeveloped areas, e.g. water power, wind power, solar energy, methane gas, water pumping equipment, water supplies, small industries, food processing and building and construction. The group corresponds with other similar institutes and with field workers in many parts of the world and contributes to other Dutch groups working in this field such as TOOL and SWD.

h

HELIO ASSOCIATES INC
Interstate 10 & Vail Road
P.O. Box 17960
Tucson
Arizona 85731
U.S.A.

Telex: 666490
Phone: 502 792 2800

This company offers energy system analysis and management studies as well as consulting and design services for solar heating systems, particularly systems for use in buildings.

i

INSTITUTE OF SOLAR ENERGY
P.O. Box 6094
Khartoum
Sudan

Research centre currently active with research and development work.

INTERMEDIATE TECHNOLOGY DEVELOPMENT GROUP
9 King Street
London WC2E 8HN
U.K.

Cables: IT/DEV London WC2
Phone: 01 836 9434

The parent organisation of Intermediate Technology Publications, publishers of this Guide. Specialises in all fields of development technology; operates a technical enquiry service; publishes; carries out research and development and organises field projects and consultancies. An independent, non-profit-making U.K. registered charity. See Section 9 describing our energy related activities in detail.

INTERMEDIATE TECHNOLOGY INDUSTRIAL SERVICES
3rd Floor Myson House
Railway Terrace
Rugby CV21 3HT
U.K.

Cables: ITIS Rugby
Phone: 0788 70126

Newly formed wing of Intermediate Technology Development Group (above) specialising in the provision of technical, managerial and financial support for small industries in under-developed parts of the world. This organisation is run by ITDG staff and funded by the U.K. Ministry of Overseas Development.

INTERNATIONAL INSTITUTE FOR ENVIRONMENT & DEVELOPMENT (IIED)
10 Percy Street
London W1
U.K.
also:

Phone: 01 580 7656

1302 Eighteenth Street NW
Washington D.C. 20036
U.S.A.

Phone: (202) 462 0900

Research institute specialising in aspects affecting policy decisions in energy both in industrial and in underdeveloped regions. Areas studied include economic aspects of renewable energy technologies, tradeoffs between centralised and decentralised energy sources, energy costs of various agricultural systems, etc. Publishes papers and reports.

THE INTERNATIONAL RICE RESEARCH INSTITUTE (IRRI)
P.O. Box 933
Manila
Philippines

Telex: 7425365 via ITT
7222456 via RCA
Cables: Ricefound Manila
Phone: 88 48 69

IRRI designs and develops a selection of agricultural equipment which is subsequently manufactured in a number of Asian countries; drawings and other technical information are provided free, subject to a simple agreement. Although their designs are primarily agricultural, a number of energy-related developments are under way, including research into simple windpumps, steam engines, a kerosene fuelled batch dryer, pumps and water lifting equipment, etc.

INTERNATIONAL INSTITUTE OF TROPICAL AGRICULTURE
PMB 5320
Ibadan
Nigeria
Phone: Ibadan 23741
Lagos 33931

Practical research and development work on systems for reducing the energy inputs into tropical agriculture, while taking into account the needs of the tropical small farmer as well as larger farming systems.

INTERNATIONAL SOLAR ENERGY SOCIETY (ISES)
Headquarters:
P.O. Box 52
Parkville
Melbourne
Victoria
Australia 3052
Phone: 03 347 491

U.K.-ISES
c/o The Royal Institution
19 Albemarle Street
London W1X 3HA
U.K.
Phone: (01) 493 6601

ISES has branches in numerous countries and organises seminars and conferences on technical aspects of using solar energy and publishes various publications which are available to ISES members at a considerably reduced rate. This is the primary organisation promoting solar energy on an international scale, and its publications include contributions from most of the more prominent workers in this field.

k

KAJIMA INSTITUTE OF CONSTRUCTION TECHNOLOGY
19-1 Tobitakyu 2 Chome
Chofu-Shi
Tokyo 182
Japan
Cables: Kajima Tokyo
Phone: (0424) 85 1111

This group is doing research into the operations of a solar heated heat pump. This is a dual system that absorbs heat from solar radiation via an outdoor solar panel during winter, and radiates heat from the building via the same panel, during the summer nights. The unit has not yet come into production as research into it has not been completed. It is reported to have a C.O.P. of 400-500% with the compressor absorbing a power input of 2.6kW. It is likely to be commercially available once development is completed.

KINGSTON REYNOLDS THOM & ALLARDICE LTD
44 Wakefield Street
Auckland
New Zealand
Telex: NZ 21385
Cables: Kingsdice
Phone: 30189

Consultants in development of geothermal resources, reconnaissance, geophysical, geochemical investigation, exploratory drilling management, feasibility studies for development of resource. Operations in South East Asia.

m

MIDDLETON ASSOCIATES
6 Crescent Road
Suite 28
Toronto
Canada M4W 1T3
Phone: (416) 961 5136

A firm of Consultants who provide supporting service to organisations interested in establishing themselves in renewable energy development or implementation. Will locate designers, developers and people with specialised technical expertise in resource assessment, planning, engineering and execution.

n

NATIONAL ACADEMY OF SCIENCE/NATIONAL RESEARCH COUNCIL
Board on Science and Technology for International Development
Commission on International Relations
2101 Constitution Avenue
Washington D.C. 20418
U.S.A.

Publishes various booklets on development, including *More Water for Arid Lands,* and *Energy for Rural Development,* which are overviews of the current 'state of the art' in small scale energy and water lifting. These are available free from the above address.

NATIONAL CENTRE FOR ALTERNATIVE TECHNOLOGY
Llwyngwen Quarry
Machynlleth
Powys
Wales
U.K.
Phone: 0654 2400

This is an important experimental centre where a variety of commercially-developed and experimental "ambient energy" systems can be inspected by the public. NCAT also publish useful Do-it-Yourself plans and leaflets on equipment they have developed which do not gloss over some of the problems that were discovered and explains solutions that were tried. This organisation has an extensive book list of publications from other sources, e.g. *Practical and Philosophic Approaches to Alternative Technology, Ecological and Environmental Issues, Growing and Cooking Food, Crafts,* etc. They have a large collection of useful information sheets available at 5p each, plus 15% postage and packing. Their own plans include a *5W Wind Generator, Water Pump, Pumping Windmill, Solar Water Heater, Sail Windmill (Cretan)* — most of these cost £0.20 plus 15% postage. Write, enclosing SAE or Reply Coupons for details of air-mail postage and complete list, (they are a non-profit charity and request a SAE with correspondence). Some of their plans are stocked by IT Publications.

NATIONAL RESEARCH CENTRE
Sh. El-Tahrir
Dokki
Cairo
Egypt

Cables: Research Cairo
Phone: 982433

This major government-run establishment has an active solar energy research centre running research and development programmes on a variety of aspects of solar energy conversion. The NRC is also active in the field of wind energy research and a number of other energy related fields.

NATIONAL RESEARCH DEVELOPMENT CORPORATION OF INDIA
61 Ring Road
Lajpatnagar III
New Delhi 24
India

Telex: 031 3214
Cables: Natredevco
Phone: 625608

This group is not a commercial organisation but develops designs for equipment for local manufacture in the country. The corporation deals with advances in all fields of technology and is available to people in other countries as an information source for any specific problems.

THE NATURAL ENERGY CENTRE
161 Clarence Street
Kingston upon Thames
Surrey KT1 1QT
England

Telex: 21887 MONCO G
Phone: 01 549 5888/9

This company markets commercially available energy equipment utilising renewable energy resources and actively promotes the use of renewable resources with a newsletter and various useful publications. The Centre in effect supplies hardware, designs and installs and acts as consultants and it also houses the Natural Energy Association which publishes books, holds seminars and provides advice for its members.

The following principal hardware is available from this source:-

a. solar collectors. Solarex, Solaray, Maxsum, Liebi, solar air conditioner, plus various other kits and items.

b. wind generators. Elektro, Dunlite, Wesco, Noah and others.

c. wind pumps. Oasis, Sparco.

d. water turbines. Ossberger.

Also available are inverters, wood and waste-burning stoves, heat pumps, electronic energy monitors, methane digesters, instrumentation, etc., etc.

They offer a useful publication *How to Use Natural Energy* (price £1.75 net) which gives a good broad overview of the various systems available and catalogues products available from the Natural Energy Centre.

o

O'BROCK WINDMILL SALES
Route 1 12 St
North Benton
Ohio 44449
U.S.A.

Suppliers of windmills and spare parts for windmills including Heller-Aller, Dempster, Aermotor, and also Davey hydraulic rams.

p

PRAIRIE SUN AND WIND COMPANY
4408 62nd Street
Lubbock
Texas 79414
U.S.A.

Phone: 806 795 1412

This company offers a wind power consultancy service and it is preparing designs for various wind machines. They also have agencies for a number of windmill manufacturers including Aeropower, Heller Aller "Baker" Windpumps, Wincharger (wind generator and selection of engine powered generating sets plus "loose" generators) from Dyna Technology's Winco Division, Dwyer and Maximum wind measuring instruments, Surrette lead-acid storage batteries plus various other similar products.

r

THE ROYAL SCIENTIFIC SOCIETY
P.O. Box 6945
Amman
Jordan

The Mechanical Engineering Department of the RSS is active in solar energy, with a particular interest in desalination systems. They publish a newsletter and research reports.

RURAL INDUSTRIES INNOVATION CENTRE (RIIC)
P.O. Box 138
Kanye
Botswana

Telex: 435 BD
Phone: Gaberone 2942

A research organisation specialising in development technology which is active in a number of fields such as small industries, improved low cost housing construction, as well as small-scale energy conversion. Experimental bio-gas, mud stoves, solar stills and wind-pumps and generators are being developed.

s

SKYTHERM PROCESSES AND ENGINEERING Phone: (213) 389 2300
2424 Wilshire Boulevard
Los Angeles
California 90057
U.S.A.

This consultancy group specialise in natural heating and cooling of buildings using their patented Skytherm water blanket system developed by Harold Hay. They act as consultants to architects, engineers and governments to provide temperature control in buildings.

SOFRETES
(Société Francaise d'Etudes Thermiques et d'Energie Solaire)
B.P. 163
45203 Montargis
France

A pioneering institution working in the field of solar power heat engines (see entry for Renault Moteurs Development) and on a variety of other solar energy conversion systems.

SOLAR ENERGY COMPANY Phone: 603 424 4000
Deerwood Drive
Merrimack
New Hampshire 03054
U.S.A.

Developers of simple concepts in the energy field embracing solar energy, waste energy and wood energy systems.

SOLAR USAGE NOW (SUN) Phone: 419 937 2226
Box 306
450 E Tiffin Street
Bascom
Ohio 44809
U.S.A.

This company stocks a very comprehensive range of solar energy equipment and components. A 120 page catalogue is available for US$2.00 and includes a wealth of clearly detailed entries too numerous to list here. The catalogue also includes a wide selection of books and journals on alternative energy and on energy saving that can be ordered from SUN.

STEERING COMMITTEE ON WIND-ENERGY FOR DEVELOPING COUNTRIES (SWD) Cables: Dehave Amersfoort
P.O. Box 85
Amersfoort
The Netherlands

This group exists to try and stimulate the use of wind energy in developing countries. Its membership is drawn mainly from Dutch technical universities and an active research and development programme to develop suitable equipment is in hand both in the Netherlands and in a variety of overseas countries. This group can offer training in wind energy related specialisations and also carries out consultancy work in this field. A number of publications are available on technical aspects of windmill design; one copy of each publication is available free of charge to research institutes in third world countries and many of these would be invaluable to serious students or prospective researchers in this field. Typical titles available include *Wind Energy Utilisation in Sri Lanka — Potentialities and Constraints, Construction Manual for a Cretan Windmill, Rotor Design for Horizontal Axis Windmills,* etc. Virtually all publications are available in English, and some are in French.

SUNPOWER INC.
48 West Union Street
Athens
Ohio 45701
U.S.A.

This group have produced prototype stirling engines using solid fuel or solar power as the heat source.

Currently under development are:

a. Solar powered stirling engine system of 1kW electrical output.

b. Solid fuel powered stirling engine of 500W electrical output.

c. Solar powered water pumps.

W.T. Beale, a noted pioneer in developing free piston Stirling engines is associated with this work.

SUN POWER SYSTEMS Phone: 813 366 3050
1121 Lewis Avenue
Sarasota
Florida 33577
U.S.A.

Specialists in low temperature heat engines using freon vapour as a working fluid. Developing various prototypes.

SURPLUS CENTER
1000 W O St
P.O. Box 82209
Lincoln
NB 68501
U.S.A.

Stock a wide range of bits and pieces useful for those involved in building their own energy conversion systems.

t

TECHNOLOGY CONSULTANCY CENTRE Phone: Kumasi 5351 5360
University of Science and Technology
University Post Office
Kumasi
Ghana

A centre with extensive experience in the setting up

and provision of support to small businesses and industries in Ghana.

TOOL (Technische Ontwikkeling Ontwikkelings Landen) Phone: (020) (92) 6892
Mauriskade 61a
Amsterdam
The Netherlands

TOOL is a joint organisation formed by groups from seven Dutch universities, which operates a technical enquiry service on problems of rural development. They also publish a periodical in Dutch called *Vraagbaak,* and various information leaflets and plans in English and Dutch, for example *Simple Methods of Soap Manufacture; The Salawepump*: how to build a simple hand pump.

TROPICAL PRODUCTS INSTITUTE Phone: 01 242 5412
56-62 Gray's Inn Road
London WC1X 8LU
U.K.

TPI gives information on renewable resources (plant and animal products, solar energy, etc.,) of less developed countries and provides expert advice on the scientific, technological and economic problems arising post harvest. TPI also carries out research, provides consultants, offers specialist training and publishes numerous technical and economic reports.

U

UNITED STIRLING (SWEDEN) AB & CO. Telex: Stirlin S 32379
Fack
201 10 Malmo 1
Sweden

This group have developed prototype stirling engines of 40kW, 75kW and 150kW which it is hoping to market in 1982. The expected efficiency in vehicles is hoped to be in excess of 35% and the 75kW unit shown weighs 350kg while the 40kW unit weights 180kg.

V

VITA (VOLUNTEERS IN TECHNICAL ASSISTANCE) Phone: (301) 277 7000
3706 Rhode Island Avenue
Mt. Rainier
Maryland 20822
U.S.A.

Publishers of a range of leaflets and books on most aspects of rural development. Most publications are written for application by technical assistance people in developing countries.

VITA Technical Bulletins: (selection on energy related topics)

1kw River Generator	US$1.00
Wood Burning Oven	US$1.00
Solar Water Heater	US$1.00
Double Drum Sawdust Stove	US$1.00
Hydraulic Ram Pump	US$3.00
Low Cost Development of Small Water Power Sites	US$2.95
Hydraylic Ram Pump	US$1.95
Low Cost Windmill for Developing Nations	US$2.95
Design Manual for Water Wheels	US$4.50
Tanzanian Windmill	US$3.00
Savonius Rotor Construction	US$3.25
Pedal Power	US$4.95
Evaluation of Solar Cookers	US$5.95

A few available in other languages.

VOLUNTEERS IN ASIA Phone: (425) 497 3228
Appropriate Technology Project
Box 4543
Stanford
CA 94305
U.S.A.

This is a group which consists of volunteers who are associated with three American universities and who work on AT projects in co-operation with institutes and organisations in Asia. Publishers of the *Appropriate Technology Source Book.*

W

WINDWORKS Phone. 663 4408
P.O. Box 329
Route 3
Mukwonago
Wisconsin 53149
U.S.A.

Windworks offer consultancy services in the fields of wind energy, load management, etc., and produce various publications and plans as follows:

25ft diameter Sail Windmill with 42ft tower set of plans	$25.00
Wind Energy Bibliography	$ 3.00
Educational Wind Energy Display Chart	$ 3.25

They also stock various other publications on wind energy from other sources and they are agents for 'Gemini' inverters.

WEST WIND COMPANY Phone: 728 9940
Wilson Mesa
Telluride
Colorado 81435
U.S.A.

This company offers information and consultancy on wind energy systems, plus various components including a temperature differential electronic controller for solar water heating systems, a voltage controller for battery charging systems, and an automatic load sensing switch. They are developing various inverters and permanent magnet alternators for use with wind generators.

WORLD BANK Cables: Intbafrad, Washington D.C.
1818 11 Street Phone: (202) 477 1234
Washington D.C. 20433
U.S.A.

Has departments specialising in economic assessments of LDC energy prospects, which carries out individual studies of LDC's energy needs, and finances LDC energy projects (normally at government level only).

WORLDWATCH INSTITUTE Phone: (202) 452 1999
1776 Mass. Avenue NW
Washington D.C. 20036
U.S.A.

This institute is concerned with renewable energy sources for developing countries and has a special interest in the "Firewood Crisis". Publications available.

THE WHOLE EARTH TRUCK STORE Phone: 415 323 0313
558 Santa Cruz Avenue
Menlo Park
CA 94025
U.S.A.

This is primarily a book store specialising in alternative energy, ecological, self-sufficiency and related publications. They also operate a world-wide mail order service which sells hard-ware as well as books.

11

List of agents and distributors by country

The information given here is as supplied by the manufacturers.

Abu Dhabi

CATERPILLAR OVERSEAS SA Mohamed Abdulrahman Al-Bahar P.O. Box 441 Abu Dhabi	Telex: 259 AH Cable: Moatasim Phone: 22387
KOHLER INTERNATIONAL LIMITED Sadiq Brothers P.O. Box 169 Abu Dhabi	
KLOCKNER-HUMBOLDT-DEUTZ AG Bruno Company Trading Electric & Mechanic P.O. Box 4264 Abu Dhabi	Telex: AH 2464 Cable: Bruno Phone: 42363
LUCAS SERVICE OVERSEAS LIMITED Automotive Parts Company P.O. Box 2647 Abu Dhabi	Phone: 22373
NISSAN DIESEL MOTOR COMPANY LIMITED Mohammed Bin Massood & Sons P.O. Box 322 Abu Dhabi	
PERKINS ENGINES LIMITED Technical Services & Supply Company (Arabian Gulf) Limited P.O. Box 277 Abu Dhabi	Cable: Techserve Phone: 22331-22332
JOHN ROBSON (SHIPLEY) LIMITED Abdulla Bin Ahmed Zarouni & Bros. P.O. Box 1020 Al-Ain Abu Dhabi	
VOLVO PENTA Mohammad Bin Masaood & Sons P.O. Box 322 Abu Dhabi	Telex: Mastor AH 249 Phone: 41167, 41236

Afghanistan

DETROIT DIESEL ALLISON Peiwar Motors Incorporated Khushhal Khattack Road P.O. Box 243 Kabul	Cables: Pemotor Phone: 23857
DORMAN DIESELS LIMITED Afghanistan Commercial Company Limited P.O. Box No. 3044 Charahi-Sedarat Shahri Nau Kabul Kabul	Cables: Comerco Phone: 20081
KIRLOSKAR OIL ENGINES LIMITED Dwarkadas Kalyanji P.O. Box 39 Kabul	Cables: Raj Phone: 21915/16
LEYLAND INTERNATIONAL Indamer Motor Division P.O. Box 202 Kabul	Cables: Albaldini

LUCAS SERVICE OVERSEAS LIMITED
Shaffi Automobile Company Limited
P.O. Box 142
Outside Lahori Gate
Kabul

Cables: Shaffi Auto
Phone: 20364

PERKINS ENGINES LIMITED
The Afghan Motor Service and Parts Company
Zenda-Banon Workshops
P.O. Box 86
Kabul

Cables: Service Kabul
Phone: Kabul 20920, 20925

PETTERS LIMITED
Anglo Afghan Trade Centre
P.A. Box 363
Jade Sadar Mohd Hashim Khan
Kabul

Cables: Anglaf
Phone: 21193

Algeria

DETROIT DIESEL ALLISON
Detroit Diesel Allison International-Europe
c/o General Motors France
Louis Roche
55 à 58 Avenue
Gennevilliers
92231
France

Telex: Genlmot 62050
Cables: Parautexap Gennevilliers
Phone: 790-70-00 Gennevilliers

DORMAN DIESELS LIMITED
(Oran Province only)
A Magnol & Cie
Boite Postal 8020
(Oran Saint Michel)
5 rue Sidi-Ferruch
Oran

Cables: Magnol Oran
Phone: 412-95, 449-21

KLOCKNER-HUMBOLDT-DEUTZ AG
BCA
Bureau Consultatif de la Société
Klockner-Humboldt-Deutz AG
5 Impasse de la Revolution
B.P. El Mouradia 181
El Mouradia
Alger

Telex: 52835
Cables: Deutzmotor
Phone: 604970, 602475/76

LEYLAND INTERNATIONAL
Sonacome
B.P. No. 15
Rouiba
Algiers

Telex: Sondcv 52773 Alger
Cables: Sonacome
Phone: 664160/64
664180/84

LUCAS SERVICE OVERSEAS LIMITED
DIAL (Diesel Algerie)
47 Boulevard Rouchai Boualem
Algiers

Telex: 52356 DIAI
Cables: Dial-Alger
Phone: 662055, 662057

PERKINS ENGINES LIMITED
Diesel Algerie Dial
47 Boulevard Capitaine Rouchai Boualem
Alger

Telex: 52356 DIAI
Cables: Dial-Alger
Phone: 662055, 662057

PETBOW LIMITED
Sovesta (Soc. de Services Techniques)
Vilmain
20 Chemin des Sept Merveilles
Algiers

PETTERS LIMITED
Sonacome D.C.E.I.
P.O. Box 72
Avenue De L'Ain (Cinq Maisons),
El Harrach
Alger

Telex: 52.938
Phone: 76.64.06/10

American Samoa

KOHLER INTERNATIONAL LIMITED
G.H.C. Reid & Company Limited
P.O. Box 1269
Pago Pago
96799

Andorra

LEYLAND INTERNATIONAL
Garatge Central S.A.
Les Escaldes
Principat

(Agent of Tugs & Lighters Ltd. Trinidad)
P.O. Box 120

Angola

ALLIS CHALMERS
Maquinas E. Equipamentos Technicos De Angola S.A.R.L. (META)
Caixa Postale 3175-C
Luanda

Telex: 3386 META AN
Cables: Meta
Phone: 6701 through 6705

BRIGGS & STRATTON CORPORATION
Agro Tecnica, Lda,
P.O. Box 10.100
Luanda

Cables: Agrotechnica
Phone: 25671/2

CATERPILLAR OVERSEAS SA
Sorel S.a.r.l.
P.O. Box 408
Luanda

Telex: 3229
Cables: Sorel Luanda
Phone: 7-22-81/2/3/4/5

DETROIT DIESEL ALLISON
Casa Americana Commercial S.A.R.L.
P.O. Box 1208C
Luanda

Telex: 998-3039 Caci An
Cables: Americana
Phone: 35190 through 35194

DORMAN DIESELS LIMITED
Friauto S.A.R.L.
Divisao de Maquinas
Rua Paiva Couceiro Nos 373/375
C Postal 1255
Luanda

Telex: Friauto 3229 Sorel An
Cables: Friauto
Phone: Angola 80214, 80191/2

KLOCKNER-HUMBOLDT-DEUTZ AG
P.O. Box 6157
Luanda

Telex: 3245
Cables: Vega
Phone: 32481

LEYLAND INTERNATIONAL
Uniauto-Organizacoes Tecnicas S.A.R.L.
Caixa Postal 1236
Luanda

Telex: 3053 Uniao-An
Cables: Uniao
Phone: 27631/5

LUCAS SERVICE OVERSEAS LIMITED
Motores De Angola S.A.R.L.
Caixa Postal 2622C
Luanda

Cables: Motorang
Phone: 81900/81901

PERKINS ENGINES LIMITED
Motores De Angola (S.A.R.L.)
Caixa Postal 2622-C
Luanda

Cables: Motorang
Phone: 23577/81900

RUGGERINI MOTORI S.p.A.
Agro-Tecnica LDA
P.O. Box 10100
Rua Direita de Luanda 114-116
Luanda

VOLVO PENTA
Samil
Caixa Postal 5065
Luanda

Telex: 3074 Tecnil AN
Cables: Samil
Phone: 70174/5/6/7/8
70180/1/2

Anguilla

PETTERS LIMITED — Cables: Expan
Expan Company Limited
Anguilla

Antigua

DETROIT DIESEL ALLISON — Telex: WB 113; Cables: Mendesteph; Phone: 21224
Stephen R. Mendes (Antigua) Limited
(Agent of Tugs & Lighters Ltd. Trinidad)
P.O. Box 120
The Colonial House
18 Thames Street
St. Johns

LEYLAND INTERNATIONAL — Cables: Bennet Antigua
George W. Bennett, Bryson & Company
P.O. Box 162
St. Johns

PETTERS LIMITED — Cables: Mendesteph
Stephen R. Mendes Limited
P.O. Box 120
Colonial House
18 Thames Street
St. Johns

Argentina

ALLIS CHALMERS — Telex: 122064 AR INMAQ; Cables: Insudmaq; Phone: 32-3661/32-3669
Compania Subamericana De Industria Y Comercio S.A.
25 de Mayo 460
Casilla de Correo 5083
Buenos Aires

BRIGGS & STRATTON CORPORATION — Cables: Autocamion; Phone: 811-0041
Autocam S.A., C.I. F.e.I.,
Rivadavia 3872
Buenos Aires

CATERPILLAR OVERSEAS SA — Telex: 012-1739; Cables: Macrosa Baires; Phone: 792-0021 through 29
Macrosa Crothers Maquinarias S.A.C.I.F.
Av. Fondo de a Legua 1232
Martinex (Partido San Isidro)
Buenos Aires

DETROIT DIESEL ALLISON — Telex: 012-1609 Buenos Aires; Cables: Bairautoex; Phone: 755-6949
c/o General Motors Argentina S.A.
Osvaldo Cruz 2900
1293 Capital Federal

DORMAN DIESELS LIMITED — Telex: 122621 Ar Raimp; Cables: Raimport Baires; Phone: 37-8115
C.I.T.I.R. S.A.
Casilla Correo
41-Suc 46
Buenos Aires

FORD MOTOR COMPANY LIMITED — Telex: 012 1175; Phone: 392 4112
Ford Motor Argentina S.A.
Casilla Correo 696
Correo Central
Buenos Aires

L. GARDNER & SONS LIMITED — Phone: 33-1325, 33-9044
Banham Hermanos y Compania S.A.
Calle Peru 362
Buenos Aires

KATOLIGHT LIMITED
Ing. Iglesias S.R.L.
Blas Parere 6979
Santa Fe

KLOCKNER-HUMBOLDT-DEUTZ AG — Telex: SAMM 121038 AR; Cables: Motores; Phone: 309472
Sociedad Argentina de Maquinas y Motores S.A.C.I.
Belgrano 685 – 5°
RA Buenos Aires

LEYLAND INTERNATIONAL — Telex: 390-12-1887; Cables: Prudante
A.G. Pruden & Company S.R.L.
Calle Bouchard 680 (R.90)
Buenos Aires

LUCAS SERVICE OVERSEAS LIMITED — Cables: Lucaserve Lomas Mirador Baires; Phone: 740-9046, 68-9815/9086
Lucas Service Argentina S.R.L.
Avenida San Martin 4200
Casilla Correo 17,
Lomas del Mirador
Partido de la Matanza
PCA
Buenos Aires

PERKINS ENGINES LIMITED — Telex: 121162; Cables: Perkinsa; Phone: 33-8231/36
Perkins Argentina S.A.I.C.
Bolivar 368
Buenos Aires

PETTERS LIMITED — Phone: 38 0210
Alplomet Motors S.A., (Offices)
Venezuela 1292
Buenos Aires

VOLVO PENTA — Telex: Public Booth 3909 Buenos Aires; Cables: Sudvolvo; Phone: 38-4135, 38-4160
Volvo Sudamericana S.A.C.I.
Moreno 1142
Buenos Aires

WHITE ENGINES INCORPORATED — Telex: 012-1861; Phone: 396-9851 (Int'l Div)
Buxton
Calle Florida 683 – 10th Floor
Buenos Aires

Australia

ALLIS CHALMERS — Telex: AA20604; Cable: Detroitco; Phone: 638-0133
Diesel Engine & Transmission Company
2 South Street
P.O. Box 176
Rydalmere
2116 New South Wales

ALLIS CHALMERS — Telex: 92443; Cables: Westutt Perth Western Australia; Phone: 79-1616/19 1103
Wesfarmers Tutt Bryant PTY. Limited F/A
174-184 Railway Parade
P.O. Box 21
Bassendean
6054 Western Australia

BRIGGS & STRATTON CORPORATION — Telex: 20238; Cables: Perkaust; Phone: 648-4088
A.N.I. Perkins
A division of the ANI Corporation Limited
16 Parramatta Road
P.O. Box 117
Lidcombe
New South Wales 2141

CATERPILLAR OVERSEAS SA
Caterpillar of Australia Ltd.,
P.O. Box 35
Niddrie
Melbourne
Victoria 3042

DETROIT DIESEL ALLISON — Telex: 30168, or 30212 Melbourne; Cables: GMP & IP Melbourne; Phone: 03-792-0111
c/o General Motors Holdens PTY. Limited
Princes Highway
P.O. Box 163
Dandenong
Victoria 3175

DORMAN DIESELS LIMITED
GEC Diesels Australia Limited
1 Winterton Road
P.O. Box 135
Clayton 3168
Victoria
Telex: AA 30994
Cables: Eedaust Melbourne
Phone: 554-8344

FORD MOTOR COMPANY LIMITED
Ford Motor Co. of Australia Limited
Parts and Service Division
Private Mail Bag 2
Campbellfield
Victoria 3061
Telex: 30624
Phone: 359 0211

L. GARDNER & SONS LIMITED
Knox Schlapp PTY. Limited
Corner Milford and Banks Street
East Victoria Park
Western Australia 6101
Telex: 92704
Phone: 61-6822

L. GARDNER & SONS LIMITED
Knox Schlapp PTY. Limited
135-139 McEvoy Street
P.O. Box 67
Alexandria 2015
New South Wales
Telex: 20483
Phone: 699-8333

KATOLIGHT LIMITED
Westinghouse Brake and Signal Company
Australia PTY Rectifier Division
P.O. Box 4
Newport
Victoria 3015

KIRLOSKAR OIL ENGINES LIMITED
Winsland & Co. PTY. Limited
P.O. Box 195
Bentley 6102
Western Australia 6106
Telex: AA 27370 Barpump
Phone: 686166

K.A.B. Projects PTY. Limited
322 Pacific Highway
Hornsby
New South Wales 2077
Telex: AA 27370
Cables: Veebee Sydney
Phone: 47 1863
47 3307
47 3329

KLOCKNER-HUMBOLDT-DEUTZ AG
Deutz Australia PTY. Limited
P.O. Box 241
Cochranes Road
Moorabbin 3189
Victoria
Telex: Deutzmo Melbourne AA 324
Cables: Deutzmotor Melbourne
Phone: 957688 (6 lines)

KLOCKNER-HUMBOLDT-DEUTZ
Tutt Bryant Limited
2-8 South Street
Rydalmere
New South Wales 2116
Telex: Leotutt 20140 Sydney
Cables: Tutbry Sydney
Phone: 638-0133

LEYLAND INTERNATIONAL
Leyland Motor Corporation of Australia Ltd.
(Head Office)
P.O. Box 346
332/342 Oxford Street
Bondi Junction
New South Wales 2022
Telex: 20222
Cables: Leyaust Sydney
Phone: 387 0233

LUCAS SERVICE OVERSEAS LIMITED
Lucas Industries Australia Limited
1156 Nepean Highway
Cheltenham
Victoria 3192
Telex: 30.453
Cables: TLX 30.453 Lucasaust Melbourne
Phone: 93.0311

Correspondence to
Lucas Industries Australia Limited
Private Bag 3
Cheltenham
Victoria 3192

NISSAN DIESEL MOTOR CO. LIMITED
U.D. Australia Pty. Ltd.
P.O. Box 91
Rosebery
New South Wales 2018

PERKINS ENGINES LIMITED
Perkins Engines Australia PTY. Limited
P.O. Box 156
Dandenong 3175
Victoria
Telex: 30816
Cables: Perkoil Melbourne
Phone: 792-0431

PETTERS LIMITED
A.N.I. Australia PTY. Limited
A.N.I. Perkins Division
P.O. Box 117
16 Parramatta Road
Lidcombe
Sydney
New South Wales 2141
Telex: 20238
Cables: Perkaust, Sydney
Phone: 648-4088

RUGGERINI MOTORI S.p.A.
Engine Powered Equipment
P.O. Box 29
Carnegie
Victoria
Telex: 32817 Volkvic

VOLVO PENTA
Volvo Penta (Australia) Pty. Limited
40 Hilly Street
Mortlake
New South Wales 2137
Telex: 21414
Cables: Volvaust
Phone: 736 2666

Volvo Australia Pty. Limited
24 Viking Drive
Dacra 4076
Brisbane
Queensland
Telex: 40403
Phone: 722144

Volvo Australia Pty. Limited
Cherry Lane
Brooklyn 3025
Melbourne
Victoria
Telex: 32231
Phone: 3992933

VOLVO PENTA
Morton Industries
80/88 Port Road
Hindmarsh
Adelaide
South Australia
Cable: Mortonsaw
Phone: 465971

Sandmac Engineering
Hay Street
P.O. Box 63
Subiaco WA 6008
Perth
Western Australia
Telex: 92498
Cable: Sansu Perty W.A.
Phone: 813733

WHITE ENGINES INCORPORATED
B.W.S. Properties PTY. Limited
21 Meeks Road
Marrickville 2204
Cable: Hobaust Sydney
Phone: 51-5013
51-1896

Mailing address
P.O. Box 62
Marrickville 2204

Austria

BRIGGS & STRATTON CORPORATION
A & M Bucek
Krausegasse 8
Postfach 11
A 1112 Vienna XI
Cables: Bucekmotor
Phone: 741101

CATERPILLAR OVERSEAS SA
Eisner Baumaschinen GmbH
Postfach 100
Strasse No. 1,
Objekt 27
Industriezentrum No. Sud
A-2351 Wr. Neudorf
Vienna
Telex: 07-9213
Cables: Eisnerbaumages
Phone: (2236) 83-5-45

DORMAN DIESELS LIMITED
Ing Franz Schmachtl K.G.
4021 Linz-Donau 11,
Ehrentletzbergerstr 2
Postfach 362
Telex: (0) 21373
Cables: Schmachtl-Linz
Phone: 21974 Linz

FORD MOTOR COMPANY LIMITED
Ford Motor Company (Austria) K.C.
Rainerstrasse 27
P.O. Box 2
5021 Salzburg
Telex: 63201
Phone: 74551

KATOLIGHT LIMITED
Ferdinand Berger
Stadplatz 50 A-4960
Schwanenstadt

KLOCKNER-HUMBOLDT-DEUTZ AG
F.M. Tarbuck & Company
St. Ulrichstr 19
A - 6840 Gotzis
Telex: 5229126
Phone: 2371

KOHLER INTERNATIONAL LIMITED
Prochaska & Cia GmbH
9 Strasse
Sudbanhol-Frachten-bahnos
1100-Vienna

LEYLAND INTERNATIONAL
British Leyland Austria K.G.
Wasserfeldstrasse 15
A 5020 Salzburg
Telex: 063479

PERKINS ENGINES LIMITED
Indukont Technische Handelsges M.B.H.
Kolingasse 19
A-1092 Wien
Fach 200
Telex: 07-4022
Cables: Industriekontor
Phone: 34 74 76

PETTERS LIMITED
Karl I. Hoffman K.G.
A 4600 Wels
Boschstrasse 44
P.O. Box 247
Telex: 025 – 422
Phone: (07242) 7422, 4247

RUGGERINI MOTORI S.p.A.
H. Leithmayer K.G.
A. Wenzel Pragergasse 7-9
A-3002 Purkersdorf
Telex: 01-3441

VOLVO PENTA
Schweden-Bootsmotor-Import Gesellschaft mbH
Donaustrasse 22
1195 Wien
Telex: 75961
Phone: 371679

Azores

CATERPILLAR OVERSEAS SA
Sociedade Tecnica de Equipamentos e Tractors S.A.R.L.
Apartado 1351
Lisbon
Portugal
Telex: 12778
Cables: Stretra Sacavem
Phone: 251-1011

LEYLAND INTERNATIONAL
Varela & Cia. LDA
Ponta Delgada
St. Miguel

PERKINS ENGINES LIMITED
Leacock & Cia LDA
Caixa Postal 416
Funchal
Madeira
Rua Major Reis Gomes 13
Cables: Leacock
Phone: 22101

VOLVO PENTA
Motores e Automoveis Jotacor Lda
Rua do Mercado 17
Ponta Delgada-S
Miguel-Azores
Cables: Jotacar
Phone: 22696

Bahamas

BRIGGS AND STRATTON CORPORATION
Bahama Engines Limited
P.O. Box 6170
Nassau

CATERPILLAR OVERSEAS SA
Bahamas Tractor & Equipment Limited
P.O. Box N-3238
Nassau
Telex: NS125
Cables: Bahtrac
Phone: (809) 323-5701

DORMAN DIESELS LIMITED
Bahamas Tractor & Equipment Limited
P.O. Box N-3238
Nassau
Cables: Bahtrac Nassau
Phone: (809) 323-5701

LEYLAND INTERNATIONAL
Nassau Motor Company Limited
P.O. Box N-8165
Shirley and Dowdeswell Streets
Nassau
Cables: Nasmo

LUCAS SERVICE OVERSEAS LIMITED
Bay Street Garage Limited
Dowdeswell Street
P.O. Box N3916
Nassau
Cables: Bay Street Garage
Phone: 80932, 22604

PERKINS ENGINES LIMITED
Earthmovers Limited
Wulff Road
P.O. Box N-4894
Nassau
Cables: Earthmovers
Phone: 3-4343

PETTERS LIMITED
Earthmovers Limited
P.O. Box 4894
Nassau
Telex: 1675
Cables: Earthmovers Nassau
Phone: Nassau 3-4343

Bahrain

CATERPILLAR OVERSEAS SA
Ahmed & Essa Al-Jalahema
P.O. Box 5357
Manama
Telex: 299GJ
Cables: Bahar
Phone: 713-607

L. GARDNER & SONS LIMITED
Abdulla Yusuf Fakhro
P.O. Box 39
Lulu Road
Manama
Telex: 8211. GAC CH
Phone: 53531

KATOLIGHT LIMITED
Mohamed Jalal & Sons Co. Limited
P.O. Box 747
Bahrain

KIRLOSKAR OIL ENGINES LIMITED
Salman A. Kaiksow
P.O. Box 80
Bahrain
Telex: 8349 TYLHTL GJ
Cables: Kaiksow
Phone: 3735

KLOCKNER-HUMBOLDT-DEUTZ AG
Gray MacKenzie u.Co.Ltd.
P.O. Box 210
Bahrain
Telex: GJ 8212
Cables: Gray
Phone: 50804

LEYLAND INTERNATIONAL
Ahmed & Abdulla Fakhro
P.O. Box 39
Bahrain
Cables: Fakhro
Phone: 5908, 3531

LUCAS SERVICE OVERSEAS LIMITED
Gray, Mackenzie & Co. Limited
P.O. Box 210
Bahrain
Telex: Gray BHN GJ 8212
Cables: Gray
Phone: 53537

NISSAN DIESEL MOTOR CO. LIMITED
Y.K. Almoayyed & Sons
P.O. Box 143
Manama

PERKINS ENGINES LIMITED
Yusuf Bin Ahmed Kanoo
P.O. Box 45
Bahrain
Telex: 6215 GJ
Cables: Kanoo
Phone: 4081-9

PETTERS LIMITED
Khalil M.K. Alsharif
Bahrain Workshop
P.O. Box 404
Manama
Cables: Bahworks
Phone: 51730

JOHN ROBSON (SHIPLEY) LIMITED
Y.K. Almoayyed & Sons
P.O. Box 143
Bahrain

Bangladesh

BRIGGS & STRATTON CORPORATION
Auto Equipment
58, Motijheel Commercial Area
Dacca 2
Cable: Autoquip
Phone: 251708

CATERPILLAR OVERSEAS SA
Greenland Engineers & Tractors Co. Ltd.
7 Shantibagh
P.O. Box 541
Dacca 17
Cables: Mechanical

DORMAN DIESELS LIMITED
Hives Engineering Limited
P.O. Box 865
American Life Insurance Building
4th Floor (Pent House)
18-20 Motijheel Commercial Area
Dacca 2
Cables: Hives
Phone: 246183, 246184, 252867

KIRLOSKAR OIL ENGINES LIMITED
Dacca Fibres Limited
125/A Motijheel Commercial Area
Dacca 2
Telex: 843 Dacfib
Cables: Dacfib
Phone: 282191/3

KLOCKNER-HUMBOLDT-DEUTZ AG
Deutz-Bangladesh Limited
31, Bangabandhu Avenue
P.O. Box 249
Dacca 2
Telex: Res. 302905
Cables: Deutzmotor
Phone: 280959, 250279

LEYLAND INTERNATIONAL
Macneill & Kilburn Limited
PSL Building
(Hotel Inter Continental)
P.O. Box 36
Dacca 2
Cables: Macneills
Phone: 255434/5/6

LUCAS SERVICE OVERSEAS LIMITED
Levatus (Bangladesh) Limited
Globe Chamber
104 Motijheel Commercial Area
Dacca
Cables: Levatus
Phone: 257226 254868

PERKINS ENGINES LIMITED
Shahnawaz (Bangladesh) Limited
9/C Motijheel
Dacca 2
Cables: Autoshez
Phone: 245482, 246011, 246012

PETBOW LIMITED
Blackwood Hodge (Bangladesh) Limited
P.O. Box 453
Dacca 2

PETTERS LIMITED
M.E.T.C.O.
(Machinery & Equipment Trading Co.)
P.O. Box 541
7 Shantibagh
Dacca 17
Phone: 25 1217, 25 3497

WHITE ENGINES INCORPORATED
S.M.Bazlool Huq & Company
Hossain Chamber (2nd Floor)
105, Agrabad Commercial Area
P.O. Box 350
Chittagong
Cables: Bazlool Co Chittagong
Phone: 83838

Barbados

CATERPILLAR OVERSEAS SA
USI West Indies Limited
Bosvigo
Eagle Hall
P.O. Box 178
Bridgetown
Telex: WB258
Cables: Usiwil Bridgetown
Phone: 6-5072

KOHLER INTERNATIONAL LIMITED
Central Foundry Ltd
P.O. Box 240
Pierhead Lane
Bridgetown

LEYLAND INTERNATIONAL
Dowding Estates & Trading Co. Ltd.
1372-1376 Bay Street
Bridgetown
Cables: Detco Barbados
Phone: 64697

LUCAS SERVICE OVERSEAS LIMITED
Harrisons Electrical Co. Limited
Lucas House
Tudor Street
P.O. Box 304
Bridgetown
Telex: 301
Cables: Harrison Bridgetown

PERKINS ENGINES LIMITED
Alstons Marketing Company Limited
P.O. Box 24 (Fontabelle)
Bridgetown
Cables: Wiljohn Barbados
Phone: 93143 93432

PETTERS LIMITED
Central Foundry Limited
P.O. Box 240
Bridgetown
Cables: Cenfoundry
Phone: 64365

Belgium

ALLIS CHALMERS
Societe Anonyme Hilaire Van Der Haeghe E
Boomsesteenweg 174
B2610 Wilrijk
Telex: (3) 566
Cables: VDH WIL b
Phone: 031/38.10.80

BRIGGS & STRATTON CORPORATION
S.A. Mertens & Straet
50-54 Rue du Printemps
1050 Brussels
Telex: 25754
Cables: Ametradexo
Phone: (02) 6486445 (1OL)

CATERPILLAR OVERSEAS SA
Treco — Tractor & Equipment Company S.A.
Steenweg op Brussels 340
B-1900 Overijse
Telex: 23386 BRU
Cables: Trecosa Brussels
Phone: (02) 687-60-20

DORMAN DIESELS LIMITED
Constructiewerkhuis
E Van Wingen PVBA
Terneuzenlaan 23
9000 Gent
Telex: 11 071
Phone: (91) 51 12 33

FORD MOTOR COMPANY LIMITED
Ford Motor Company (Belgium) S.S.
P.O. Box 27
B2030 Antwerp
Telex: 31460
Phone: 410080

L. GARDNER & SONS LIMITED
Automobiles Miesse S.A.
42-44 Rue des Goujons
Brussels
Phone: 02/522.03.45

KLOCKNER-HUMBOLDT-DEUTZ AG
S.A. Magirus-Deutz N.V.
12-14 Place de la Gare
Berchem-Ste-Agathe
B1080 Brussels
Telex: Deutz Bru 21597
Cables: Magirus-Deutz
Phone: 4656566 4657178

LEYLAND INTERNATIONAL
British Leyland (Belgium) S.A.
1 Eggestraat
B-2800 Mechelen
Telex: 25159 Britbe or 22487 Flywheel
Phone: 45833

NISSAN DIESEL MOTOR COMPANY LIMITED
Matermaco, S.A.
710-718 Ch de Louvain
Brussels

PERKINS ENGINES LIMITED
Ets Hunter & Co. P.V.B.A.
St. Bernardsesteenweg 858/864
B2710 Hoboken
Antwerp
Telex: 31 958 Perkun B
Cables: Perkhunt
Phone: 031-27 39 70

RUGGERINI MOTORI Sp.A
Marine Motors Agencies
Verbindingsdok W.K. 12A
2000 Antwerp
Telex: 31488 Marimo B

WHITE ENGINES INCORPORATED
Marine Motors Agencies
Verbindingsdok W.K. 12a
Antwerp
Telex: Marinmotor Antwerp 31488
Cable: Marinmotors
Phone: (02) 15-73-53

Belize

CATERPILLAR OVERSEAS SA
R.H. Eyles & Sons Limited
North Front Street
P.O. Box 8
Belize City
Cables: Eyles Belize
Phone: 2369

DORMAN DIESELS LIMITED
The Belize Electricity Board
P.O. Box 327
Belize City
Cables: Electricity
Phone: 3404 and 2141

LEYLAND INTERNATIONAL
Belize Distributors Limited
77 North Front Street
P.O. Box 145
Belize City
Cables: Distribution
Phone: 2500, 2501, 2514

LUCAS SERVICE OVERSEAS LIMITED
Belize Distributors Ltd
77 North Front Street
P.O. Box 145
Belize
Cables: Distribution Belize
Phone: 2500, 2501, 2514

PERKINS ENGINES LIMITED
Hofius Limited
P.O. Box 226
19 Albert Street
Belize City
Cables: Hofius
Phone: 2231/2

Benin

CATERPILLAR OVERSEAS SA
Soger A.O.
B.P. 33
Cotonou
Telex: 5232
Phone: 31 33 59

LEYLAND INTERNATIONAL
Cie Francaise De l'Afrique Occidentale
Agence Centrale
P.O. Box 908
Cotonou
Cables: Senafrica Cotonou

PERKINS ENGINES LIMITED
Sogre-Ao
B.P. 33
Akpakpa
Telex: Ergenep 232
Cables: Gastonegre Cotonou
Phone: 23-77/23-80/33-18

Bermuda

BRIGGS & STRATTON CORPORATION
Masters Limited
Front Street
Hamilton
Telex: 3430
Cables: Mastand
Phone: 5-4321

KOHLER INTERNATIONAL LIMITED
The Bermuda Motor & Cycle Accessories Co. Ltd.
P.O. Box 1005
Dundonald Street West
Hamilton

LEYLAND INTERNATIONAL
Pearman Watlington & Company
P.O. Box 840
Hamilton
Telex: 3246
Cables: Triton
Phone: 809-29-53232

LUCAS SERVICE OVERSEAS LIMITED
The Bermuda Motor & Cycle Accessories Co. Ltd.
The Lucas House
P.O. Box 1005
Hamilton
Cables: Accessories
Phone: 50736

PERKINS ENGINES LIMITED
The Bermuda Motor & Cycle Accessories Co. Ltd.
P.O. Box 1005
The Lucas House
Hamilton
Cables: Accessories
Phone: 1-3186

PETTERS LIMITED
Burland Conyers & Marirea Limited
P.O. Box 292
St. John's Road
Pembroke
Cables: Burlandco
Phone: 2-2067

Bhutan

KIRLOSKAR OIL ENGINES LIMITED
Tashi Commercial Corporation
P.O. Phuntsholing
Cables: Tashi
Phone: 63

Bolivia

BRIGGS & STRATTON CORPORATION
Hansa Ltda
Cajon Postal No. 1402
La Paz
Telex: BX 5261
Cables: Hansa
Phone: 54426

CATERPILLAR OVERSEAS SA
International Machinery Co. (Bolivia). S.A.
Casilla 852
La Paz
Telex: 5227
Cables: Intermaco
Phone: 40972, 53787

DORMAN DIESELS LIMITED
MacDonald & Co. (Bolivia) S.A.
Casilla 879
La Paz
Telex: BX 5262
Cables: Donal
Phone: 54 380-1-2

KATOLIGHT LIMITED
Cablebol
P.O. Box 394
Cochabamba

KLOCKNER-HUMBOLDT-DEUTZ AG
Hansa Ltda.
Mercado
esquina Yanacocha
Cajon Postal 1402
La Paz
Telex: BX 5261
Cables: Hansa
Phone: 54426

KOHLER INTERNATIONAL LIMITED
Tecna Limitada
Casilla 4360
Avenida Montes 437, Esq
Batallon Ilimany
La Paz

LEYLAND INTERNATIONAL
Linale & Weiss SA.
Avenida Montes 603/605
Castilla de Correo 216
La Paz
Cables: Linale

LUCAS SERVICE OVERSEAS LIMITED
Hansa Ltda
Cajon Postal 1402
Calle Mercado Esquina Yenococha
La Plaz
Telex: 5261
Cables: Hansa
Phone: 54425, 54445, 53661

NISSAN DIESEL MOTOR CO. LIMITED
Compania Comercial Industrial Bolivia
P.O. Box 954
La Paz

PERKINS ENGINES LTD
S.A.C.I. (Sucs) De Simon F. Bedoya
Ingavi 1058
La Paz
Telex: 3560009-5311
Cables: Saci
Phone: 55114-53706-55075

PETTERS LIMITED
Sociedad Anonima Comercial Industrial
Calle Ingavi 1058
Casilla 346
La Paz
Telex: ITT 3560009
Cables: Saci
Phone: 55075, 55797, 53706

WHITE ENGINES INCORPORATED
Martin y Cia S.A.
Casilla 1467
La Paz
Telex: Martin BX 5269
Cables: Martin

Botswana

CATERPILLAR OVERSEAS S.A.
Botswana Earthmoving Machinery Co. (Pty) Ltd.
Dumela Industrial Sites
P.O. Box 137
Francistown
Telex: 212BD
Cables: Tractors
Phone: 547

KATOLIGHT LIMITED
K.G. Council Ltd
P.O. Box 879
Gabarone

Brazil

BRIGGS & STRATTON CORPORATION
Borghoff S/A
Rua Riachuelo 243
Caixa Postal 619
Rio de Janeiro
Cables: Borgmagneto
Phone: 242-3720

CATERPILLAR BRAZIL S.A.
Av. des Nacoes Unidas 1516
Mail address: Caixa Postal 8239 01.000
Sao Paulo SP
Telex: 391-21132
Cables: Caterpilar SP
Phone: 247-1011/5919/1100

DORMAN DIESELS LIMITED
Cobrel Maquip S.A.
Comercio e Engenharia
Avenida Rio Branco 138
7th Floor Bozano Simonsen Building
Rio de Janeiro
ZC-21-CE P20-000
Telex: (38) 31505 Boznes
Phone: Rio de J 2241422

FORD MOTOR COMPANY LIMITED
Ford Willys Do Brazil S.A.
Caiya Postal 8610
Sao Paulo
Phone: 63-5151

KLOCKNER-HUMBOLDT-DEUTZ AG
Otto Deutz S.A.
Motores e Tratores
Mailing address
Caixa Postal 841
09700 Sao Bernardo do Campo
Telex: 114179 Deut BR
Cables: Ottodeutz

KOHLER INTERNATIONAL LIMITED
Eximport-Importacao e Exportacao Ltda
P.O. Box 76
Rua Marechal Deodoro 75-A Sala 10
Manaus
Amazonas

LEYLAND INTERNATIONAL
Philippe Daou S.A.
Rua Marcilio Dias 134
Caixa Postal 254
Manaus

LUCAS SERVICE OVERSEAS LIMITED
CAV do Brasil S.A. Ind. e. Com.
Rodovia Raposo Tavares
Km 30
Cotia Sao Paulo
Telex: 1121412
Cables: Lucasbrasil and Vanteri
Phone: 211-7576

PERKINS ENGINES LIMITED
Motores Perkins S.A.
Caixa Postal 30 028 c
Sao Paulo
CEP 09700
Estado de Sao Paulo
Telex: 114013
Cables: Perkoil
Phone: 448-1499

PETTERS LIMITED
Moto Importadora
P.O. Box 94
Free Zone
Manaus
Amazonas
Telex: 922188 Moto BR
Cables: Motoimportadora
Phone: 25060 27009

VOLVO PENTA
Carbrasmar S.A. Indústria e Comécio
Av. Brasil 14936
Caixa Postal 1449
Rio De Janeiro 91
Telex: Public Booth 031903 for Carbraswares
Cables: Carbrasmares
Phone: 30-9830 30-9955

British Virgin Islands

PERKINS ENGINES LIMITED
Diesel Marine Supply (Tortola)
Limited
P.O. Box 61
Phone: 42268

Brunei

CATERPILLAR OVERSEAS SA
Tractors Malaysia Berhad
Jalan Setia DI-Raja
P.O. Box 268
Kuala Belait

LEYLAND INTERNATIONAL
Champion Motors (Singapore)
Pte. Ltd.
P.O. Box 627
31 Leng Kee Road
Singapore 3
Telex: RS 400 (for Champmotors)
Cables: Champmotor
Phone: 28361-9

LUCAS SERVICE OVERSEAS SA
Far East Motors (Pt.) Ltd
54 Orchard Road
P.O. Box 2697
Singapore 9
Telex: 21772 Wearnes
Cables: Ignition
Phone: 328341/8

Bulgaria

DORMAN DIESELS LIMITED
Guest Industrials Limited
Engineering Division
Tubs Hill House
London Road
Sevenoaks
Kent
U.K.
Telex: 95531 Guestelex SVKS
Cables: Guestind Sevenoaks
Phone: Sevenoaks 59311

PETTERS LIMITED
Interpred Pirin
Room 401
2 Stamboliiski Street
Sofia
Telex: 284 BG
Phone: 87.14.41

Burundi

CATERPILLAR OVERSEAS SA
Chanic
B.P. 930
Bujumbura
Cables: Chanusa
Phone: 3284

LEYLAND INTERNATIONAL
Compagnie De L'Afrique Orientale
'Old East' S.A.R.L.
Bp. 330
Bujumbura 37
Cables: Broche
Phone: 2692/3

Burma

ALLIS CHALMERS
Myanma Export Import Corporation
(Agency Division)
77/91 Sule Pagoda Road
P.O. Box 404
Rangoon
Telex: Ubmeic BM 2005
Cables: Myanagent
Phone: 13420, 18164

CATERPILLAR OVERSEAS SA
Caterpillar Far East Ltd
P.O. Box 3069
Hong Kong
Telex: HX3305 CFEL
Cables: Catfareast HKG

DETROIT DIESEL ALLISON
Blackwood Hodge Group
Services Ltd.
Hunsbury Hill Avenue
Northampton
NN4 9QT
U.K.
Telex: 31476
Cables: Suntract
Phone: 0604-61111

DORMAN DIESELS LIMITED
Myanma Export Import Corporation
(Agency Division)
77/79 Sule Pagoda Road
P.O. Box 404
Rangoon
Telex: Ubmeic BM 2005
Cables: Myanagent Rangoon
Phone: 13420, 18164

KLOCKNER-HUMBOLDT-DEUTZ AG
Myanma Export Import
Corporation
(Agency Division)
77-91 Sule Pagoda Road
Rangoon
Telex: Ubmeic BM 2005
Cables: Myanagent Rangoon
Phone: 13420, 18164

LEYLAND INTERNATIONAL
Autocars (Burma) Limited
55 Barr Street
P.O. Box 257
Rangoon
Cables: Autocars

LUCAS SERVICE OVERSEAS
LIMITED
Myanma Export Import Corporation
(Agency Division)
77/91 Sule Pagoda Road
Rangoon
Telex: 2017
Cables: Myanagent
Phone: 11258

PERKINS ENGINES LIMITED
Myanma Export & Import
Corporation
(Agency Division)
P.O. Box 404
77/91 Sule Pagoda Road
Rangoon
Telex: VBMEIC BM 2005
Cables: Myanagent
Phone: Rangoon 13429, 18164

WHITE ENGINES
INCORPORATED
Dodge & Seymour
XM World Trade Incorporated
Ona World Trade Center
Suite 4501
New York 10048
U.S.A.
Telex: 232886
Cables: Eximco Newyork
Phone: (212) 432-5200

Cameroon

CATERPILLAR OVERSEAS SA
S.H.O. Cameroun Departement
Tractafric
B.P. 4017
Douala
Telex: 5217 KN
Cables: Tractafric
Phone: 42-40-83

DETROIT DIESEL ALLISON
Division Matforce Diesel
C.I.A.C.A.M.
P.O. Box 4025
Douala
Telex: 842-650043 Paris France
Cables: Nigerfrin Paris 8 France
Phone: 256-72-00 Paris France

DORMAN DIESELS LIMITED
Paterson Zochonis & Co. Ltd.
P.O. Box 4009
Avenue R Poincare
Douala
Telex: 5207 A/B Zochonis DLA
Cables: Zochonis
Phone: 42-18-78

KLOCKNER-HUMBOLDT-DEUTZ AG
Valcke Freres
Untervertretung: Dacam
Route du Docteur Jamot
B.P. 4028
Douala
Telex: 5237
Phone: 42507

LEYLAND INTERNATIONAL
Paterson Zochonis & Co. Ltd.
B.P. 4009
Douala
Cables: Zochonis Douala

LUCAS SERVICE OVERSEAS
LIMITED
S.O.C.O.M.I.
P.O. Box 4025
Douala
Telex: 5320
Phone: 42.46.66.42.31.81

PERKINS ENGINES LIMITED
Socomi
Rue Jarnot
B.P. 609
Douala

Telex: 5272
Cables: Somatin
Phone: 42.35.40

Canada

BOMBARDIER-ROTAX GmbH
Bombardier Limited
Valcourt

JOE 2LO

Telex: 5832550 – Purchasing Dept. 5832551 – Research Dept.
Phone: (514) 532-2211

CATERPILLAR OVERSEAS SA
Caterpillar of Canada Limited
1550 Caterpillar Road
Mississauga
Ontario

Telex: 06-961149/961151 (from U.K.)
Phone: (416) 279-9901

DORMAN DIESELS LIMITED
Ruston Diesels Limited
2 Paxman Road
Etobicoke
Ontario
M9C 1B6

Telex: 06-967542 Spares Depot 06-967721
Cables: Ruston Toronto
Phone: Toronto 622 3060/3 (Area Code 416)

FORD MOTOR COMPANY LIMITED
Ford Motor Company of Canada Limited
Industrial Products Dept.
National Parts Dept.
8000 Dixie Road
Bramalea
Ontario

Telex: 0229161
Phone: 459 2210

L. GARDNER & SONS LIMITED
Western Diesel Sales & Manufacturing Limited
25 Victoria Drive
Vancouver B.C.

KLOCKNER-HUMBOLDT-DEUTZ AG
Deutz Diesel (Canada) Ltd.
90 Montee de Liesse
Montreal
Quebec H4T 1N4

Telex: 5-826776
Cables: Airdiesel MTL
Phone: 5-826776

LEYLAND INTERNATIONAL
British Leyland Motors Canada Limited
P.O. Box 5033
4445 Fairview Street
Burlington
Ontario L7R 4A3

Telex: 061 8678
Cables: Leymotcan Burlington
Phone: (416) 632-3040

LUCAS SERVICE OVERSEAS LIMITED
Lucas Industries Canada
Automotive Equipment Division
280 Yorkland Boulevard
Willowdale
Ontario M2J 1R9

Telex: Lucas Tor 06-966552
Cables: Lucaspares Toronto
Phone: 416-491-3520

PERKINS ENGINES LIMITED
Perkins Engines Canada Ltd
7 Meridian Road
Rexdale
Ontario

Telex: 0221225
Cables: Perkoil Toronto
Phone: 677/4960

PETTERS LIMITED
Scona Electric Limited
9950-75th Avenue
Edmonton
Alberta

Phone: 403-433-7779

VOLVO PENTA
Atlas Polar Company Limited
60 Northline Road
P.O. Box 160
Toronto 374
Ontario

Telex: 0696-3580
Cables: Atlaspolar
Phone: 757-5131

Canary Islands

DETROIT DIESEL ALLISON
Miquel Ortega S.A.
Carlos Mauricio Blandy
44-48 Las Palmas de Gran Canaria

Telex: 22456 Madrid Spain
Cables: Mitega Madrid Spain
Phone: 20.13.35

DORMAN DIESELS LIMITED
Blandy Brothers Y Cia S.A.
Apartado No. 176
Santa Cruz de Tenerife

Telex: Ten 92341 Blande
Cables: Blandybros
Phone: 2264.40/4

KLOCKNER-HUMBOLDT-DEUTZ AG
Talleres Venequeras
C/Gordillo, s/nQ
Espigón del Castillo
Puerto de la Luz

Phone: 263615 261858

LEYLAND INTERNATIONAL
Blandy Brothers & Co. S.A.
Leon Y Castillo 535
P.O. Box 24
Las Palmas

Cables: Blandy Bros. Las Palmas
Phone: 26.33.40-4

PERKINS ENGINES LIMITED
Blandy Brothers Y Cia S.A.
Carretera del Rosario
Km.5 TACO
Santo Cruz de Tenerife

Telex: 92341 Blan-E
Cables: Blandybros
Phone: 226440/5

RUGGERINI MOTORI SpA
Inesa S.A.
Imeldo Seris 51 alto
Santa Cruz de Tenerife

VOLVO PENTA
Macmor S.A.
Edificio Doramas
Avda de Escaleritas Num 70
Las Palmas de Gran Canaria

Cables: Macmor
Phone: 25 30 44, 25 09 55 25 43 02, 25 32 89.

Cape Verde

CATERPILLAR OVERSEAS SA
Sociedade Tecnica de Equipamentos e Tractors S.A.R.L.
Apartado 1351
Lisbon
Portugal

Telex: 12778
Cables: Stetra Sacavem
Phone: 251-1011

LEYLAND INTERNATIONAL
C. De Vasconcelos (Herdeides) Ltda
Caixa Postal 4
Praia
Cape Verde Islands

Cables: Tamu Praia

Central African Republic

CATERPILLAR OVERSEAS SA
S.H.O. Centrafrique Departement Tractafric
B.P. 959
Bangui

Telex: 5252 (014)
Cables: Tractafric
Phone: 2420

LEYLAND INTERNATIONAL
SIMA Service Autos
B.P. 820
Bangui

PERKINS ENGINES LTD.
Somecaf
B.P. 294
Bangui

Telex: Penteco 5228
Phone: 27.63

Chad

CATERPILLAR OVERSEAS SA
S.H.O. Tchad
Département Tractafric
B.P. 450
Ndjamena

Telex: 5214KD
Cables: Tractafric Ndjamena
Phone: 24-51

DETROIT DIESEL ALLISON
Nouvelle Societe Commerciale du Kouilou Niari
P.O. Box 476
Fort Lamay
Cables: Unasedec
Phone: 2800

LEYLAND INTERNATIONAL
SCOA Service Autos
B.P. 474
Fort Lamy

LUCAS SERVICE OVERSEAS LIMITED
Tchadis Auto
B.P. 474 M'Djamena
Telex: 5241
Phone: 2664

PERKINS ENGINES LTD.
N.S.C.K.N.
P.O. Box 476
Fort Lamy
Telex: 5210
Cables: Sedec
Phone: 28 07

Chile

ALLIS CHALMERS
SALFA S.A.C.I.
Avenue Rondizzoni 2130
Casilla 1089
Santiago
Telex: 3520063
Cables: Salfa
Phone: 93031, 93032, 93033

BOMBARDIER-ROTAX GmbH
Fabrica de Cerraduras y Candados POLI
Casilla 129
Buin

BRIGGS & STRATTON CORPORATION
Lanz y Cia Ltda
Casilla 16389 – Correo 9
Santiago
Telex: 40637 CL
Cables: Lanz
Phone: 74 06 73

CATERPILLAR OVERSEAS SA
Gildemeister S.A.C.
Agustinas – Esquina Amunategui
Casilla 99-D
Santiago
Telex: 392, 40588
Cables: Gildemeist
Phone: 82525

DETROIT DIESEL ALLISON
Jaras y Cia Ltda.
Sucesores de Raul Jaras Barros y Cia. Ltda.
Av. Baron de Juras Reales 5250
P.O. Box 10114
Santiago
Telex: Jarasca SG0334 or RCA 332-334
Cables: Jarasca
Phone: 777-807

DORMAN DIESELS LIMITED
Cia Minera y Comercial Sali Hochschild
Casilla 3127
Santiago
Telex: Public Booth Santiago 260 – for Sali Hochschild Section Equipos
Cables: Hochschild
Phone: 713118

KLOCKNER-HUMBOLDT-DEUTZ AG
Estudiantes 150
(Loreta Altura 400)
Portugal 160
Casilla Correo 2116
Santiago
Telex: Sgo.369
Cables: Socoa
Phone: 770475

KOHLER INTERNATIONAL
Distribuidora Industrial Y Agricola Del Sur A.
P.O. Box 1625
San Antonio 1625
Santiago

LEYLAND INTERNATIONAL
Javier Echeverria Y cia Saci
Avda San Francisco 121
Santiago
Cables: Echev

LUCAS SERVICE OVERSEAS S.A.
Servicio Lucas-Blandford S.A.
General Mackenna 1238
P.O. Box 4D
Santiago
Cables: Blandford
Phone: 711420

PERKINS ENGINES LTD.
Distribuidora Perkins Chilena Ltda
Avenida Espana 69
Casilla 14931
Santiago
Telex: Cummchile SG0 260 for Perkchile
Cables: Perkchile
Phone: 82113/4/5

PETTERS LIMITED
E. Thurston & Cia Ltda
Avenida Gral Buines 204
Casilla 9032
Santiago
Telex: SG0 260 Edplumpton
Cables: Edplumpton
Phone: 66476

VOLVO PENTA
Divolvo S.A.
Echaurren 11 esq. Alameda
Casilla 14085
Santiago
Telex: SGO-260 (with address)
Cables: Divolvo
Phone: 68530, 82174, 80900

WHITE ENGINES INCORPORATED
Alberto Grunwald S.
Casilla 9931
Santiago
Phone: 94331

China

DORMAN DIESELS LIMITED
Biddle Sawyer & Co. Limited
32 St Mary at Hill
London EC3R 8DH
U.K.
Telex: 887971 Bidrourplat
Cables: Bidsawya London
Phone: 01-623 9333

L. GARDNER & SONS LIMITED
Dodwell & Co. Limited
P.O. Box 36
Hong Kong
Telex: Dodco HX 3204
Phone: H237011

LEYLAND INTERNATIONAL
Biddle Sawyer & Co. Limited
Plantation House
Mincing Lane
London EC3
U.K.
Telex: 28266/28221
Cables: Bidsawya
Phone: 01-623 9333

Colombia

ALLIS CHALMERS
Colombiana del Caribe S.A. (Colcaribe)
Carrera 43 No. 12.8.61
Apartado Aereo 80317
Bogota
Telex: 044-676 Holunie
Phone: 47-11-12, 47-74-28, 47-25-72

BRIGGS & STRATTON CORPORATION
Larsen, Vargas & Cia Ltda
Apartado Aereo 5199
Av.5 No.9-30
Bogota
Cables: Larco
Phone: 419-481

CATERPILLAR OVERSEAS SA
General Electric de Colombia S.A.
Km 7 Carretera a Bosa
Apartado Aereo 3644 y 6799
Bogota
Telex: 044-704
Cables: Gecolsa
Phone: 384020, 384032, 304452, 704

DETROIT DIESEL ALLISON
Motores S.A.
Avenida Las Americas 39A-95
Apartado Aereo 80040
Zona Postal No.6
Bogota
Telex: 396-044887
Cables: Motores
Phone: 693066

DORMAN DIESELS LIMITED
Tracey & Compania S.A.
Apartado Aereo 3597
Calle 13 No. 23-60
Bogota
Telex: Bog 44675
Cables: Tracey
Phone: 43.94.72

FORD MOTOR CO. LIMITED
Mr. Stewart
Calle 19 No.7-30
Room 807
Bogota 1

KATOLIGHT LIMITED
Electrointerandina Ltda
Apartado Aereo 1237
Bogota

KLOCKNER-HUMBOLDT-DEUTZ AG — Telex: 44785; Cables: Pluto; Phone: 414202, 416025
Held & Cia S.A.
Edificio de la Bolsa
Calle 14 No. 8-79
Oficinas 718
Bogota

KOHLER INTERNATIONAL
Boza & Cia Ltda
Carrera 13 No. 16-66
Bogota

LEYLAND INTERNATIONAL — Cables: Praco
Praco Limitada
Calle 13 No. 32-36
Bogota

LUCAS SERVICE OVERSEAS LIMITED — Cables: Eldiesel; Phone: 25150, 23284
Laboratorios Electro-Diesel Ltda
Calle 47 No. 43-70
Apartado Aereo 21-26 (Airmail)
Barranquilla

PERKINS ENGINES LIMITED — Cables: Ivor; Phone: 12-072, 15019, 11810, 11
Casa Inglesa Ltda
Via 40 No.53-59
Apartado Aereo No. 13 (Airmail)
Barranquilla

PETTERS LIMITED — Cables: Tecdiesel; Phone: 567662
C.J. Jacobsthal
Resident Regional Executive
Apdo Aereo 91105 Chico
Bogota

RUGGERINI MOTORI SpA
Industrias Forestales S.A.
Apartado Aereo 53039
Bogota

WHITE ENGINES INCORPORATED — Telex: Holo Co 44676; Cables: Corema; Phone: 69-56-03
Corporacion de Repuestos y
Maquinaria Ltds
(Corema Ltda)
Carrera 30 No. 22C-19
Bogota

Congo

CATERPILLAR OVERSEAS SA — Telex: 8217 (014); Cables: Tractafric; Phone: 2867, 2869
S.H.O. Congo
Département Tractafric
B.P. 697
Pointe Noire

DETROIT DIESEL ALLISON — Cables: SCKN; Phone: 28-01
Société Commerciale Du
Kouilou Niari-Congo
P.O. Box 34
Brazzaville

KLOCKNER-HUMBOLDT-DEUTZ AG — Telex: 8201; Cables: Davum; Phone: 2593/4
Davum Outre-Mer
B.P. 691
Pointe Noire

LEYLAND INTERNATIONAL — Cables: Senafrica
Cie Francaise De L'Afrique
Occidentale
Agence Centrale
B.P. 247
Brazzaville

LUCAS SERVICE OVERSEAS LIMITED — Telex: 403; Cables: Sedec; Phone: 28-01, 28-05
Matforce Diesel (Div. of SCKN)
B.P. 146
Pointe Noire

PERKINS ENGINES LIMITED — Telex: 5254 KG; Phone: 29.09
S.A. Pierre Gonthier
Méchanique Générale
B.P. 205
Brazzaville

Cook Islands

LEYLAND INTERNATIONAL — Telex: Citco RG 2013; Cables: Donald
The Cook Islands Trading
Corporation Ltd.
P.O. Box 92
Raratonga

Costa Rica

ALLIS CHALMERS — Telex: 2244; Cables: Tractosa San Jose; Phone: 22.48.94, 22.49.94
Tractomotriz S.A.
Avenida 10-Calle 36 2244
Apartado 10255
San Jose

BRIGGS & STRATTON CORPORATION — Cables: Zitro; Phone: 22.43.07
Ortiz & Cia Ltda
Apartado 4194
San Jose

CATERPILLAR OVERSEAS SA — Telex: C.R.110; Cables: Matra; Phone: 21.00.01
Machinery & Tractors Ltd.
La Uruca
P.O. Box 426
San Jose

DETROIT DIESEL ALLISON — Telex: 3032244; Cables: Tractosa; Phone: 224894
Tractomotriz S.A.
Apartado No. 10255
San Jose

DORMAN DIESELS LIMITED — Cables: Aceinca; Phone: 61.01.28
Aceinca Ltda
Apartado 203
Puntarenas

KATOLIGHT LIMITED
Equipos Nieto S.A.
Apartado 1353
San Jose

KLOCKNER-HUMBOLDT-DEUTZ AG — Cables: Amrhein; Phone: 229188
Franz Amrhein & Co. Limited
FACO
Apartado 1766
CR – San Jose

KOHLER INTERNATIONAL LIMITED — Cables: Kohlerint; Phone: 25.94.10, 24.13.71
Apartado 66
San Pedro Montes De Oca
San Jose

LEYLAND INTERNATIONAL — Cables: Autoro
Auto Europeos S.A.
P.O. Box 10030
Paseo Colon v Calle 38
San Jose

LUCAS SERVICE OVERSEAS LIMITED
Electro-Diesel Ltda
Avenida Central
Calle 19/21
P.O. Box 4112
San Jose
Cables: Eldiesel
Phone: 223518, 223618

NISSAN DIESEL MOTOR CO. LIMITED
Autopista General Canas
Apartado 10223
San Jose

PERKINS ENGINES LIMITED
Almacen Electra S.A.
P.O. Box 730
San Jose
Telex: CR-2143
Cables: Alesa
Phone: 21-43-92

PETTERS LIMITED
Ferreteria Miguel
Macaya & Cia Ltda
Apartado 268
San Jose
Cables: Macaya
Phone: 22.20.20/21

WHITE ENGINES INCORPORATED
Alberto Arco S.A.
Apartado 296
Calle 3 Norte No.150
San Jose
Telex: CR-181
Cable: Larce
Phone: 22.45.55

Crete (Greece)

BOMBARDIER-ROTAX GmbH
Eleftheris Hatzidakis
1 Daedalus Street – 3rd Floor
P.O. Box 146
Iraklion
Phone: 223-357/284-420

CATERPILLAR OVERSEAS SA
Avras S.A.
P.O. Box 1250
Omonia
Athens
Telex: 21-4661/4662
Cables: Avratrac
Phone: 571-6611

Cuba

LEYLAND INTERNATIONAL
c/o The Cuban Commercial Office
54 Conduit Street
London W.1.
U.K.

Cyprus

ALLIS CHALMERS
General Engineering Company Limited
Ledra Street
P.O. Box 1208
Nicosia
Telex: 2434
Cables: Genco
Phone: 3034

CATERPILLAR OVERSEAS SA
Cyprus Trading Corporation Limited
8 Arnalda Street
P.O. Box 1083
Nicosia
Telex: 2415
Cables: Cytraco
Phone: 6.20.21/22

DETROIT DIESEL ALLISON
Kaisis Engineering Ltd.
M.Georgalla Street
P.O. Box 4708
Nicosia
Telex: 2519 Kaigroup
Cables: Kaiplant
Phone: 44348

DORMAN DIESELS LIMITED
Ledra Engineering Co. Limited
Pindarus Street
No. 5L MNX
P.O. Box 2255
Nicosia
Telex: 2363 Promitheus
Cables: Leco
Phone: 74180

L. GARDNER & SONS LIMITED
G.C. Pierides Limited
P.O. Box 1011
Nicosia
Phone: 72024/5/6

INDIAN NATIONAL DIESEL ENGINE CO. LTD
Sarvas C. Charalambous
P.O. Box 1473
Nicosia

KATOLIGHT LIMITED
Fahed Khoury
Isfa Supplies Ltd.
P.O. Box 4913
Nicosia

KLOCKNER-HUMBOLDT–DEUTZ AG
Char. Pilakoutas & Sons Limited
Alexandria Road
P.O. Box 1168
Nicosia
Telex: 2245
Cables: Pilakoutas
Phone: 76312/3/4

LEYLAND INTERNATIONAL
D.J. Demades & Sons Limited
P.O. Box 1104
Nicosia
Cyprus
Telex: Nicosia 2463
Cables: Didemades
Phone: 32111/2/3/4

PERKINS ENGINES LIMITED
Char. Pilakoutas & Sons Limited
P.O. Box 1168
Larnaka Street
Nicosia
Telex: 2245 Nicosia
Cables: Pilakoutas
Phone: 76312/3/4

PETBOW LIMITED
Middle East Sales Promotion
P.O. Box 262
Larnaca

PETTERS LIMITED
General Engineering Co. Limited
P.O. Box 208
Ledra Street 171-9
Nicosia
Telex: 2434
Cables: Genco
Phone: 3034

RUGGERINI MOTORI SpA
Agragaria Limited
P.O. Box 348
Limassol

VOLVO PENTA
The Amazon Enterprise Limited
57 Archbishop Makarios 111 Avenue
P.O. Box 173
Larnaca
Cables: Amazon
Phone: 3478
4027

Czechoslovakia

CATERPILLAR OVERSEAS SA
Phoenix Praha A.S.
Pod Plynojemem
18000 Prague 8
Liben
Telex: 122591
Cables: Phoenix
Phone: 83.87.90

PETTERS LIMITED
Intersim Limited
Topolova 14
Prague 10
Zahradni Mesto
Telex: 121963, 121971
Phone: 753-067, 753-068

Dahomey

LUCAS SERVICE OVERSEAS LIMITED
John Walkden
B.P. 24
Contonou
Telex: 5214
Phone: 31 30 06

Denmark

BRIGGS & STRATTON CORPORATION
Ketner Teknik
23 Fabriksparken
DK-2600 Glostrup
Telex: 22211
Cables: Motorketner
Phone: (02) 451122

CATERPILLAR OVERSEAS SA
Enmaco A/S
363 Park Alle
P.O. Box 138
2600 Glostrup
Copenhagen
Telex: 16795
Cables: Earthmover
Phone: (01) 45-22-11

DETROIT DIESEL ALLISON
c/o General Motors International
Borgmester Christiansens Gade 40
DK-2450 Copenhagen
Telex: 16727
Cables: Copeautoex
Phone: 01.302211

DORMAN DIESELS LIMITED
A/S Dorman Diesels (Salg og Service)
Frydensbergvej 8
DK 3660 Stenloese
Telex: 42538 Dorman Dk
Cables: Dorman
Phone: 03.171455

FORD MOTOR COMPANY LIMITED
Ford Motor Company A/S
Sluseholmen 1
2450 Copenhagen S.V.
Telex: 22218
Phone: 310510

L. GARDNER & SONS LTD.
Dansk Gardner Diesel A/S
Naerum Hovedgade 1
2850 Naerum
Telex: 37337
Phone: (03) 19 22 88

KLOCKNER-HUMBOLDT-DEUTZ AG
Ingeniørfirmaet
Viggo Bendz A/S
Roskildevej 519-523
DK - 2600 Copenhagen — Glostrup
Telex: 22985
Cables: Viggobendz
Phone: 964222

KOHLER INTERNATIONAL
Danpower APS
Houmannsgade 53
8700 Horsens

LEYLAND INTERNATIONAL
Dansk Oversoisk Motor Industri A/S.
Sondre Ringvej 35
2600 Glostrup
Copenhagen
Telex: 33125
Cables: Domimotors
Phone: 961410

PERKINS ENGINES LTD.
Fred Rasmussen-Odense
Pjentedamsgade 21
Post Box 368
DK 5100 Odense
Telex: 59866 Freras DK
Cables: Fredras
Phone: (09) 112216

PETTERS LIMITED
A/S Nordan
25 Vallensbaekvej
DK-2600 Glostrup
Copenhagen
Telex: 33218
Cables: Nordan
Phone: (02) 45.88.66

RUGGERINI MOTORI S.p.A.
Fred Rasmussen
Pjentedamsgade 21
P.O. Box 368
DK 5100 Odense
Telex: 59866 Freras DK
Cables: Fredras
Phone: (09) 112216

VOLVO PENTA
Firma Johs Thornam
Industrihuset
Kalvebod Brygge 20
1560 Copenhagen V
Telex: 27491
Cables: Thornam
Phone: (01) 14 07 82

WHITE ENGINES INCORPORATED
Fred Rasmussen
Pjentendamsgade 21
Postigiro 8042
5100 Odense
Telex: 59886
Cables: Fredras
Phone: (09 112216

Djibouti

CATERPILLAR OVERSEAS SA
Anciens Comptoirs Ries
B.P. 2106
Djibouti
Telex: 223
Cables: Ries
Phone: 2467/2455

LEYLAND INTERNATIONAL
Societe Commerciale de La Mer Rouge
(SOCOMER)
B.P. 676
Djibouti
Cables: Socomer

PERKINS ENGINES LIMITED
T.F.A.I.
Anciens Comptoirs Ries
B.P. 2106
Djibouti
Telex: 823FS
Cables: Ries
Phone: 2467/2455

Dominica

DETROIT DIESEL ALLISON
A.C. Shillingford & Co. Limited
Agent of Tugs & Lighters Ltd., Trinidad
Cross Street
Roseau
Telex: 387284 Trinidad
Cables: Tuglighter Port-of-Spain Trinidad
Phone: 2481 Dominica

LEYLAND INTERNATIONAL
H.H.V. Whitchurch & Co. Limited
P.O. Box 71
Roseau
Telex: Whitdom D0614
Cables: Whitchurch Roseau
Phone: 2181

LUCAS SERVICE OVERSEAS LIMITED
Smith & Co. Limited
P.O. Box 9
Roseau
Cables: Smithco Roseau

PERKINS ENGINES LTD
P. Mallalieu Limited
P.O. Box 70
Roseau

PETTERS LIMITED
A.C. Shillingford & Company
P.O. Box 123
Roseau
Cables: Shillingford,
Phone: 2481

Dominican Republic

ALLIS CHALMERS
Atlas Commercial Co.
Apartado 107
Calle 30 De Marzo 22
Santo Domingo R.D.
Telex: 3460024 Atlas Com Co.
Phone: 682-6230

BRIGGS & STRATTON CORPORATION
Transco S.A.
Apartado 1205
Santo Domingo
Cables: Transco
Phone: 682-7262/689-9576

CATERPILLAR OVERSEAS SA
Implementos y Maquinarias C por A
Carretera Duarte Kilometre 5
Apartado 171
Santo Domingo
Telex: RCA 4183 IMCA
Phone: 566-5171
Cables: Imca

DETROIT DIESEL ALLISON
Quisqueya Motor C por A
P.O. Box 54
Santo Domingo
Telex: 3460106
Cables: Quiquemato
Phone: 9-7021

DORMAN DIESELS LIMITED
Atlas Commercial Co C Por A
Apartado 107
Santo Domingo
Telex: 3460024
Cables: Atlas Comco
Phone: 2-6230

KLOCKNER-HUMBOLDT-DEUTZ AG
Jaime Mendex Sucs C por A
Apartado 1127
Zona Postal 2
Avenida San Martin 98
Santo Domingo R.D.

Telex: ITT 3460377
Cables: Jamendex
Phone: 5658861

KOHLER INTERNATIONAL
Ferreteria Americana C por A
Apartado 1181
Avenida San Martin 175
Santo Domingo

LEYLAND INTERNATIONAL
Cia Anglo Americana C Por A
Avenida San Martin No.18
P.O. Box 856
Santo Domingo

Cables: Anamco

LUCAS SERVICE OVERSEAS LIMITED
Mercantil Antillana C por A
30 de Marzo No. 110
Apartado 791
Santo Domingo

Telex: 3460229
Cables: Mercantil
Phone: 689-8166/7

NISSAN DIESEL MOTOR CO. LIMITED
Motorambar S.A.
Avenue Abraham Lincoln 206
Apartado 1381
Santa Domingo

PERKINS ENGINES LIMITED
Mercantil Antillana C por A
Apartado 791
30 De Marzo No.110

Telex: 3460229
Phone: 689-8166/7
565-2391
565-2655

PETTERS LIMITED
Importadora Tropical C por A
Avenue San Martin v Leopold Navarro
Apartado 750
Santo Domingo

Telex: ITT 3460253
REA 326198
Cables: Tropical
Phone: 689-8353, 682-7704, 682-002

WHITE ENGINES INCORPORATED
Atlas Commercial Co. C por A
P.O. Box 107
Calle 30 De Marzo 22
Santo Domingo

Cables: Atlascomco
Phone: 2-2921 (2)

Dubai

BRIGGS & STRATTON CORPORATION
Mona Trading Company
P.O. Box 5012
Dubai

Telex: 5727
Cables: Monatco
Phone: 26361

CATERPILLAR OVERSEAS SA
Mohamed Abdulrahman Al-Bahar
P.O. Box 1170
Deira

Telex: 5445 DB
Cables: Bahar Dubai
Phone: 60255

DETROIT DIESEL ALLISON
General Navigation and Commerce Company
Zabeel Palace Road
P.O. Box 1911
Dubai

Telex: 958-5504 Genaco Du
Cables: Genavco
Phone: 31065

DORMAN DIESELS LIMITED
Abdulla Haji Kamber Awazi
P.O. Box 42
Dubai

Telex: DB 531
Cables: Kamber
Phone: 21303

INDIAN NATIONAL DIESEL ENGINE CO. LIMITED
Ahmed Ramjan Juma
P.O. Box 2574
Dubai

KATOLIGHT
Ohamanmal Keshavdas
P.O. Box 2068
Dubai

KIRLOSKAR OIL ENGINES LIMITED
Saeed & Mohammed Al-Naboodah
P.O. Box 1200
Deira

Telex: 5511 SMA DB
Cables: Alnaboodah

KLOCKNER-HUMBOLDT-DEUTZ AG
Arabian Gulf Mechanical Center Ltd.
P.O. Box 1145
Sharyah

Telex: 8086 Batha SH
Phone: 23378

LEYLAND INTERNATIONAL
Ali Bin Adbulla Al Owais
P.O. Box 4
Dubai

Telex: DB 450
Cables: Aliowais
Phone: 4556 4555

LUCAS SERVICE OVERSEAS LIMITED
P.O. Box 275
Dubai

Telex: 5527 Bhatia DB
Cables: Bhatiabros
Phone: 23245

PERKINS ENGINES LIMITED
Yusuf Bin Ahmed Kanoo
P.O. Box 290
Dubai

Telex: 5451 DB
Cables: Kanoo
Phone: 24861/21462

PETTERS LIMITED
Nasser Abdullatif
P.O. Box 1219
Dubai

Telex: 5540 Nasser DB
Cables: Nasser
Phone: 21374

RUGGERINI MOTORI S.p.A.
Al Amodi Stores & Company
P.O. Box 1274-5090
Deira

Telex: 6859 Amoodi DB

Ecuador

ALLIS CHALMERS
Delta-Delfini & Cia Ltda
9 de Octubre 2303
Apartado 4893
Guayaquil

Telex: 043203 Delta Ed
Cables: Delta
Phone: 361737

BRIGGS & STRATTON CORPORATION
Ivan Bohman Compania Ltda
Casilla 1317
Guayaquil

Telex: 3205
Cable: Boman
Phone: 307060

CATERPILLAR OVERSEAS SA
Importadora Industrial Agricola S.A.
Avda. Otto Arosemena G.Km3
P.O. Box 562
Guayaquil

Telex: 3215 Rosal Ed
Cables: Rosal
Phone: 384-700

DETROIT DIESEL ALLISON
Fincom-Servicio C.A.
Avenida Rocafuerte 752
P.O. Box 1217
Guayaquil

Telex: 393043355
Cables: Fincomser
Phone: 304367/7

DORMAN DIESELS LIMITED
Mecanos S.A.C.
Casilla 5608
Guayaquil

Telex: 3264 Mecano Ed
Cables: Mecano
Phone: 16626

KATOLIGHT LIMITED
Cotrac
Cuero y Caicedo 224
Quito

KLOCKNER-HUMBOLDT-DEUTZ AG
Electro Ecuatoriana S.A.
Avenida Colon 1011
Apartado 1123
Quito
Telex: 22115 Electr. ED
Cables: Electro
Phone: 525800

KOHLER INTERNATIONAL
Arthur Fried Cia Ltda
P.O. Box 2680
Calle Mejia 347
Quito

LEYLAND INTERNATIONAL
Morisaenz S.A.C.
Apartado 625
Panamericana
Norte La 'Y'
Quito
Telex: 2158 Morisa ED
Cables: Morisaenz

LUCAS SERVICE OVERSEAS LTD
Gustavo A Lopez
Apartado 3369
Guayaquil
Cables: Dunlopez
Phone: 39-0004 639-5133

NISSAN DIESEL MOTOR CO. LTD.
MYTSA Maquinas y Tractores S.A.
Panamericana Norte Km 2.6
Casilla 3941
Quito

PERKINS ENGINES LTD
Organizacion Commercial Vallejo
Araujo S.A.
P.O. Box 717
Avenida 9 de Octubre 721-723
Guayaquil
Cables: Vallarjo
Phone: 5-26960

PETTERS LIMITED
Rectificadora Botar S.A.
P.O. Box 3344
Avenida 10 de Agosto 5980
Quito
Telex: 308 02 2176
Cables: Recticiguenal
Phone: 241-544

JOHN ROBSON (SHIPLEY) LIMITED
Superagencias C.A.
P.O. Box 666
Avenida 9 de Octubre 822
Guayaquil

RUGGERINI MOTORI SPA
Superagencies C.A.
P.O. Box 666
Avenida 9 de Octubre 822
Guayaquil
Telex: 308 043222
Super ED

VOLVO PENTA
Fincom Guayaquil SA
Avenida de Las Americas
Casilla de correo 5557
Guayaquil
Telex: 43312
Cable: Fincomquil
Phone: 51 56 12
51 44 10
51 30 72

WHITE ENGINES INCORPORATED
B. Aviles Alfaro & Cia
Attn. White-Hercules Dept
Apartado Postal 354
Guayaquil
Telex: 3531051
Cable: Baviles
Phone: 391-100

Egypt

CATERPILLAR OVERSEAS SA
The Tractor & Engineering Co. Ltd
Agriculture & Earthmoving Dept.
23 Boustan Street
P.O. Box 1400
Cairo
Telex: 2286 Teco UN
Cables: Aziz
Phone: 21704

DETROIT DIESEL ALLISON
Carlin Middle East
1103 Corniche El Nil Street
Apt. 2 Garden City
Cairo
Telex: 927-2366 Extension 085 CME
Cable: Cocarlin
Phone: 27451/2

DORMAN DIESELS LIMITED
Getco
2 Sherif Pacha Street
Lewa Building
Cairo
Telex: 2113 Alfcai-U.N. ATT.006/Mec
Cables: Kamelkof
Phone: 43967/53436

FORD MOTOR CO. LIMITED
Ford Motor Co. (Egypt) S.A.E.
Avenue Victor Emmanuel III
Alexandria

L. GARDNER & SONS LTD
Alexandria Marine Equipment Co.
Mikhail Abadeer Street
Rushdy
Alexandria

INDIAN NATIONAL DIESEL ENGINE CO. LTD
Md. Khalil El Kewefi
28 Talat Harb Street
9th Floor
Cairo

KATOLIGHT LIMITED
Delta Export & Import
20 Salah Salem Street
Alexandria

KLOCKNER-HUMBOLDT-DEUTZ
Ectra Egyptian Consulting Trading Co. Ltd
12A Hassan Sabry Street
Zamalek
Cairo

LEYLAND INTERNATIONAL
M.I.S.R. Car Trading Co.
28 Talaat Harb Street
Cairo
Cables: Automisr

LUCAS SERVICE OVERSEAS LIMITED
Missr Car Trading Company
12 Abdel Khalek Sarwat
Cairo
Telex: 362
Cables: Automiser
Phone: 43655, 45598, 908 19

NISSAN DIESEL MOTOR CO. LTD
General Trade Co.
P.O. Box 720
Cairo

PERKINS ENGINES LIMITED
The Engineering General Co. S.A.A.
P.O. Box 588
9-11 Orabi Street
Cairo
Telex: 2249
Cables: Egici
Phone: 49291/4

PETBOW LIMITED
Tractor & Engineering Co. S.A.E.
23 Sh Pres Abdel-Salam Aref
P.O. Box 1400
Cairo

PETTERS LIMITED
Tractor & Engineering Co. S.A.E.
18 Rue Emad El Dine
P.O. Box 366
Cairo
Cables: Crocodile
Phone: 32724

RUGGERINI MOTORI Spa
S.A. Ashmawy B.Sc.
14 El-Bohtory Street
Kubba Gardens
Cairo

El Salvador

ALLIS CHALMERS
Tecnica Universal Maegli S.A.
(Tecun S.A.)
3a Avenida 3-21
Zona 9
P.O. Box 590
Guatemala City
Telex: 275 Tecun GU
Cables: Tecunsa (Tecunsael-San Salvador, El Salvador)
Phone: 65783/4/5/6/7

BRIGGS & STRATTON CORPORATION
Almacenes Saprissa
Alameda Roosevelt 2419
San Salvador
Cables: Maquinaria
Phone: 23-3178, 23-3367

CATERPILLAR OVERSEAS SA
Compania General de Equipos S.A. de C.V.
Klm.5 Carretera a Sta. Tecla
Apartado (06) 1000
San Salvador
Telex: Cogesa-20074 Area Code 301
Cables: Cogesa
Phone: 23-23-23

DETROIT DIESEL ALLISON
Compania Importadora De Maquinaria S.A.
Alameda Roosevelt y 53 Avenida Norte
San Salvador
Telex: 301-20126
Cables: Cidema
Phone: 23-1144

DORMAN DIESELS LIMITED
Promarca Alvarado y Cia
Apartado 506
San Salvador
Cables: Promarca
Phone: 21-9081

KATOLIGHT LIMITED
Goldtree Lieves S.A. de C.V.
Apartado Postal (06) 195
San Salvador

KLOCKNER-HUMBOLDT-DEUTZ AG
Representaciones Caribia S.A.
Septima Avenida Norte Nr.222
Apartado 268
San Salvador
Cables: Caribia
Phone: 219547

KOHLER INTERNATIONAL
Distribution Tecnica S.A.
7A Calle Poniente 527
San Salvador

LEYLAND INTERNATIONAL
Auto Commercial Grange
Alameda Roosevelt y 43 Avenida Norte
San Salvador
Cables: Grane

LUCAS SERVICE OVERSEAS LIMITED
Automotriz Sabater S.A.
Alameda Roosevelt 35 Avenida Sur
San Salvador
Cables: Ponsaba
Phone: 231600

NISSAN DIESEL MOTOR CO. LIMITED
Distribuidors Mejia Delgado S.A.
25 Avenida Norte y 3ra
Calle Poniente
San Salvador

PERKINS ENGINES LTD.
Salvador Machinery Company S.A.
Apartado Postal 125
Boulevard Ejercito Nacional
San Salvador
Cables: Simac
Phone: 21-8410

PETTERS LIMITED
Siman S.A.
Alameda Rossevelt 3114
Apartado Postal 161
San Salvador
Telex: 20165 Simansa
Cables: Simansa
Phone: 23-6333

WHITE ENGINES INCORPORATED
Almacenes Saprissa
Alameda Roosevelt 2419
San Salvador
Cables: Maquinaria
Phone: 23-3178, 23-3367

Equatorial Guinea

CATERPILLAR OVERSEAS SA
Finanzauto S.A.
Plaza de las Cortes 6
Madrid 14
Spain
Telex: 27752
Cables: Finanzauto Madrid
Phone: (91) 448-2700

LEYLAND OVERSEAS SA
Casa Amilivia Ltda
Santa Isabel
Fernando Poo
Cables: Akonibe

Ethiopia

CATERPILLAR OVERSEAS SA
Ries Engineering Share Co.
P.O. Box 1116
Addis Ababa
Telex: 151133
Cables: Riestract
Phone: 21082

DETROIT DIESEL ALLISON
The Motor & Engineering Co. of Ethiopia Ltd.
P.O. Box 1767
Addis Ababa
Telex: 976-21019
Cables: Moenco
Phone: 45708

DORMAN DIESELS LIMITED
Petram Company
Negash Oda Building
Addis Ababa
Cables: Petram
Phone: 121557/117162

KLOCKNER-HUMBOLDT-DEUTZ AG
Seferian & Co. (Ethiopia) Ltd. S.C.
Near Mexico Square
P.O. Box 64
Addis Ababa
Telex: 21070 Seferian
Cables: Seferco
Phone: 488100/444719 152197

LEYLAND INTERNATIONAL
Mitchell Cotts & Co. (Ethiopia) Ltd.
P.O. Box 527
Addis Ababa
Cables: Mitcotts
Phone: 47160-4

LUCAS SERVICE OVERSEAS LIMITED
Auto Accessories & Components Dept.
Mitchell Cotts & Co (Ethiopia) Ltd.
P.O. Box 527
Addis Ababa
Telex: 21036
Cables: Mitcotts
Phone: 44716015

PERKINS ENGINES LTD.
Ries Engineering Share Company
P.O. Box 1116
Addis Ababa
Telex: 21082
Cables: Riestrac
Phone: 51133

RUGGERINI MOTORI SpA
Giuseppe Selicato
P.O. Box 100
9 Jubilee Street
Asmara

Falkland Islands

LEYLAND INTERNATIONAL
Estate Loius Williams
Port Stanley
Cables: Williams

LUCAS SERVICE OVERSEAS LIMITED
The Falkland Islands Trading Co. Ltd
Stanley

PETTERS LIMITED
Falkland Islands Co. Limited
Port Stanley

Faroe Islands

LEYLAND INTERNATIONAL
Dansk Oversoisk Motor Industri As.
Sondre Ringvej 35
2600 Glostrup
Copenhagen
Denmark
Telex: 22252
Cables: Domimotors Copenhagen
Phone: 961410

PETTERS LIMITED
William Holm-Jacobsen
Torshavn
Telex: 81242 WHJ
Phone: 2360

VOLVO PENTA
Firma Alfred Johannesen
Tingshúsvegur 44-46
P.O. Box 101
Torshavn
Cables: Alf
Phone: 1735

Fiji Islands

BRIGGS & STRATTON CORPORATION
Coral Island Motors
G.P.O. Box 48
Walu Bay, Suva
Phone: 311655

CATERPILLAR OVERSEAS SA
Carpenters Tractor and Equipment
154 Queens Road
Private Mail Bag
Suva
Cables: Carptrac, Suva

DETROIT DIESEL ALLISON
Clyde Engineering (Pacific) Ltd
6 Lami Street
P.O. Box 1068
Lami
Suva
Telex: NZ2377
Cables: Clypac Suva Fiji
Phone: 361382

LEYLAND INTERNATIONAL
Motor Corporation (Fiji) Ltd.
71 Moala Street
P.O. Box 418
Suva
Phone: 383144

Suva Motors Ltd.
P.O. Box 34
Lami P.O. Box 1068
Suva

LUCAS SERVICE OVERSEAS LIMITED
Burns Philp (South Sea) Co. Ltd.
P.O. Box 355
Suva
Telex: FJ2127
Cables: Burnsouth

NISSAN DIESEL MOTOR CO. LIMITED
Suva Motors
P.O. Box 34
Suva

PERKINS ENGINES LTD
Clyde Engineering (Pacific) Ltd.
P.O. Box 1068
6 Lami Street
Lami
Suva
Phone: 361382

PETTERS LIMITED
Suva Motors Limited
P.O. Box 34
Victoria Parade
Suva
Cables: Motors
Phone: 22634

JOHN ROBSON (SHIPLEY) LTD
Wainiyaku Ltd.
Vuna
Taveuni

Finland

BRIGGS & STRATTON CORPORATION
Oy Promotor Ab
Ormustie 14
00700 Helsinki 70
Telex: 122205
Cables: Promotor
Phone: 354144

CATERPILLAR OVERSEAS SA
Wilhuri Oy Witraktor
01530 Helsinki-Lento
Telex: 12-618
Cables: Witraktor Helsinki
Phone: 826-311

DORMAN DIESELS LIMITED
Dieselkeskus Oy
Fredrikinkatu 43B
00120 Helsinki 12
Telex: 121382
Cables: Dieselkeskus
Phone: 648426

FORD MOTOR CO. LIMITED
O/Y Ford A/B
Henry Fordinkatu 6
Postilokero 46
Helsinki
Telex: 12743
Phone: 12161

KLOCKNER-HUMBOLDT-DEUTZ AG
Oy Grönblom AB
Postfach 370
Mekaanikonkatu 6
00101 Helsinki 10
Telex: 12542
Cables: Grönbloms
Phone: 80-7554411

LEYLAND INTERNATIONAL
Oy Suomen Autoteollisuus AB
Henkilöautot
Lauttasaarentie 52
P.O. Box 129
00201 Helsinki 20
Telex: 121473 Sisul SF
Cables: Voimavaaunu
Phone: 673283

PERKINS ENGINES LIMITED
Oy Hans Palsbo AB
Pulttitie 20
SF 00810 Helsinki 81
Telex: 12434 Palsb SF
Cables: Palsbos
Phone: (90) 782100

PETTERS LIMITED
Oy Mercantile AB
P.O. Box 29
Mannerheimintie 12B
Helsinki
Telex: 12-416
Cables: Mercantile
Phone: 60-981

RUGGERINI MOTORI SpA
Enon Teollisuus
Koivumaentie 14
SF-00680 Helsinki 68
Telex: 123213 Masi SF

VOLVO PENTA
Oy Volvo-Auto AB
Marinavdelningen
Batbyggarevägen 9
SF-00210 Helsingfors 21
Telex: 12736
Cables: Volvoauto
Phone: 711 311

WHITE ENGINES INCORPORATED
Oy Suomen Autoteollisus AB
Fleminginkatu 27
P.O. Box 10307
Helsinki 10
Telex: 121176 Sisue-SF
Cables: Autoteollisus
Phone: 70101

France

BRIGGS & STRATTON CORPORATION
2 & 4 Rue Guy Moquet
95100 Argenteuil
Telex: 695 109

CATERPILLAR OVERSEAS SA
Hy. Bergerat, Monnoyeur SA
6 Rue Christophe Colomb
75008 Paris
Telex: 660-911
Cables: Tibi
Phone: 723.61.32-723.61.34

DETROIT DIESEL ALLISON INTERNATIONAL EUROPE
c/o General Motors France
56 a 58 Ave. Louis Roche
92231 Gennevilliers
Telex: Genlmot 62050
Cables: Parautexap
Phone: 790-70-00

DORMAN DIESELS LIMITED
Ets E Anduze & Cie
47 Rue Servan
75 Paris 11 e
Telex: Obertin 22259
Phone: (805) 90-76, 90-77

FORD MOTOR CO. LTD.
Ford (France) SA
344 Avenue Napoleon Bonaparte
92 Rueil Malmaison
B.P. 90
Telex: 69963
Phone: (01) 977 0505

L. GARDNER & SONS LIMITED
Navigair
5 Avenue Du 11 Novembre
Antibes-06
Phone: 34.70.26

KIRLOSKAR OIL ENGINES LIMITED
Société Grossol
14 Rue Chaptal
92 Lewallois Perret
B.P. 104
Paris
Telex: 620207
Phone: 757 82 90

LEYLAND INTERNATIONAL
British Leyland France SA
Rue Ambroise Croizat
Z1 Argenteuil
B.P. 32
F-95 Argenteuil
Telex: 60875 60866
Phone: 982-09-22

PERKINS ENGINES LIMITED
Moteurs Perkins SA
55 Boulevard Ornano
93203 Saint Denis
Telex: 620 251 Saint Denis
Cables: Perkoil Paris
Phone: 243-04-40

PETTERS LIMITED
N. Rey
Resident Regional Executive
1 Rue Schlumberger
92430 Marnes la Coquette
Phone: 9262937

RUGGERINI MOTORI SpA
Nauder
P.O. Box 46
22 Avenue Barthelemy Thimonnier
69300 Caluire
Telex: 310843 Noderalp Calui

VOLVO PENTA
Volvo Penta France SA
Chemin de la Nouvelle France
78130 Les Mureaux
Telex: 60722
Phone: 474.72.01

WHITE ENGINES
Bell International
10 Rue des Oziers
95 — Pierrelaye
Telex: Belinter 69573 — F
Phone: 464-24-13

French Polynesia

JOHN ROBSON (SHIPLEY) LIMITED
RC Papeete No.32B
B.P. Box 628
Papeete
Polynesie Francaise

Gabon

CATERPILLAR OVERSEAS SA
S.H.O. Gabon
Département Industrial
Tractafric
P.O. Box 2147
Libreville
Telex: 5210 (014)
Cables: Tractafric
Phone: 2070/2309

DETROIT DIESEL ALLISON
Division Matforce Diesel
Hatton & Cookson Ltd.
P.O. Box 10075
Libreville
Telex: Hatton 5224 GO
Phone: 2331

KLOCKNER-HUMBOLDT-DEUTZ AG
Valcke Frères
Z.J. d'Oloumi
P.O. Box 53
Bd. Jules Rémy Isembé
Libreville
Telex: 5234
Phone: 22104, 22383

LEYLAND INTERNATIONAL
Cie Francaise de L'Afrique Occidentale
Agence Centrale
P.O. Box 1181
Libreville

LUCAS SERVICE OVERSEAS LIMITED
Matforce Diesel (Div. of Hatton & Cookson Ltd.)
P.O. Box 1075
Libreville
Telex: 5216
Cables: Hatton
Phone: 2303

NISSAN DIESEL MOTOR COMPANY LIMITED
Hatton et Cookson
B.P. 4052
Libreville

PERKINS ENGINES LIMITED
Diesel Gabon
B.P. 205
Libreville
Telex: 6223 Diesel-Gabon
Cables: Diesel Gabon
Phone: 21.18

PETTERS LIMITED
Hamelle-Afrique S.A.
B.P. 31
Route de l'Aviation
Libreville
Telex: Hamelaf 5292
Cables: Hamelaf
Phone: 325.38

Gambia

LEYLAND INTERNATIONAL
Cie Francaise de L'Afrique Occidentale
P.O. Box 297
Bathurst
Telex: GV 213
Cables: Senafrica
Phone: 473

PERKINS ENGINES LTD.
S. Madi Ltd.
11 Russell Street
Bathurst
Telex: Madi GV 209
Cables: Madi
Phone: 372/373/282/559

PETTERS LIMITED
Cie Francaise de L'Afrique Occidentale
Technical Dept.
P.O. Box 297
Banjul
Cables: Senafrica
Phone: 472

Gaza

DORMAN DIESELS LIMITED
Mourtaga Engineering & Trading Co.
Omer El-Moukhtar Street
P.O. Box 4
Gaza
Cables: Mourgaza
Phone: 1473/522

L. GARDNER & SONS LIMITED
Saamir Commercial Agencies
P.O. Box 23
Gaza
Telex: 341938 Saamr II
Cables: Saamr
Phone: 177

PETTERS LIMITED
Saamir Commercial Agencies
P.O. Box 23
Omer El Mouktar Street
Gaza
Telex: 341938 Saamr II
Cables: Saamr
Phone: 177

Germany

BRIGGS & STRATTON CORPORATION
Ernst Hahn
Postfach 1520
Ringstrasse 12-18
Fellbach 7012
Telex: 7254431
Cables: Hahn D
Phone: 0711/5003402, 5003274

CATERPILLAR OVERSEAS SA
Zeppelin-Metallwerke GmbH
Zeppelinstrasse 1-5
Postfach 2003
8046 Garching bei Munchen
Telex: 5215-821/606/607
Cables: Zeppelinmetall
Phone: (089) 32-00-01

DORMAN DIESELS LIMITED
Elna GmbH
2084 Rellingen Bei Hamburg
Siemensstrasse 35
Telex: 02-189123
Cables: Elektronautik
Phone: (04101) 301-1

FORD MOTOR CO. LIMITED
Ford Werke AG
Parts Operations
5 Koeln 71
Industrie Strasse
Postfach 714444
Telex: 885410
Phone: (0221) 7191

KATOLIGHT LIMITED
Kaptan
6 Frankfurt/Main
Kaiser Str 70/IV
Frankfurt

KOHLER INTERNATIONAL LIMITED
J. Wizemann & Co.
7-9 Quellenstrasse
P.O. Box 501260
D-7000 Stuttgart

LEYLAND INTERNATIONAL
A. Bruggemann & Co. GmbH
P.O. Box 1940
Harffstrasse 53
4000 Düsseldorf
Telex: 1939

PERKINS ENGINES LIMITED
Perkins Motoren GmbH
8752 Kleinostheim
Postfach 12
Telex: 4188869
Phone: 06027/8081

PETTERS LIMITED
Siegfried Schumacher GmbH
5252 Runderoth/RHI.D
Muhlenberg 6
Telex: 88476
Phone: 02263-5081/2

RUGGERINI MOTORI SpA
Goebler-Hirthmotoren
Postfach 20
Max Eyth Strasse 10
7141 Benningen/Neckar

VOLVO PENTA
Volvo Penta Deutschland GmbH
Redderkoppel 5
Postfach 9069
2300 Kiel-Friedrichsort
Telex: 02-92764
Phone: 0431/39096-9

WHITE ENGINES INCORPORATED
Ern Motorenteile K.-G.
Schinkelstrabe 46-48
4-Dusseldorf
Telex: 0858-2618
Cables: Ern Dusseldorf
Phone: (0211) 35-35-36

Ghana

ALLIS CHALMERS
Paterson, Simons & Co. (Ghana) Ltd
Korle Lagoon
Guggersburg Road
P.O. Box 480
Accra
Telex: 883771
Cables: Paterson London (England)

CATERPILLAR OVERSEAS SA
Tractor & Equipment
Division of the UAC of Ghana Ltd.
P.O. Box 5207
Accra-North
Cables: Machtrac
Phone: 21900

DETROIT DIESEL ALLISON
Blackwood Hodge (Ghana) Ltd.
Ring Road
P.O. Box 126
Accra
Telex: 974-2147
Cables: Suntract
Phone: 21255 Accra

DORMAN DIESELS LIMITED
The Ghana Consolidated Machinery & Trading Co. Ltd.,
P.O. Box 3400
Swanmill
Accra
Telex: 2008 Accra Unamerch
Cables: Machsuper
Phone: 63921/2, 64822

L. GARDNER & SONS LTD.
IKQ Motors & Co. Ltd.,
P.O. Box 227
Accra
Phone: 26278

INDIAN NATIONAL DIESEL ENGINE CO. LTD.
The C.F.A.O. Service Transit
P.O. Box 154
Tema

KIRLOSKAR OIL ENGINES LIMITED
Agricultural Engineers Limited
P.O. Box 3707
Ring Road West
Accra
Cables: Agrico
Phone: 28260/28292

KLOCKNER-HUMBOLDT-DEUTZ AG
Firma Rowedder, Delmenhorst
Uber Firma Reiss & Co.,
D 757-4 Clement Papatio Avenue
P.O. Box 3074
Accra
Telex: 2040
Phone: 65105-8

KOHLER INTERNATIONAL LIMITED
Interassociates Ghana Ltd.
Ring Road Central
P.O. Box 5238-5269
Accra

LEYLAND INTERNATIONAL
Leyland Motors (Ghana) Ltd.
P.O. Box 2969
Accra
Cables: Leymotors
Phone: 28806

LUCAS SERVICE OVERSEAS
Lucas House
P.O. Box 5731
Ring Road
Accra North
Cables: Unalucas
Phone: 28774 28789

NISSAN DIESEL MOTOR CO. LIMITED
Trans Africa Eng. & Motor Industry (Ghana) Ltd.
P.O. Box 7269
Accra

PETTERS LIMITED
Pasico (Ghana) Ltd.
P.O. Box 480
Accra
Cables: Pasico
Phone: 64658

RUGGERINI MOTORI SpA
Japan Motors
P.O. Box 2516
Accra

Gibraltar

DETROIT DIESEL ALLISON
H. Sheppard & Co. Ltd.
P.O. Box 130
Waterport
Cables: Marina Cibraltar
Phone: 2183

L. GARDNER & SONS LIMITED
H. Sheppard & Co. Ltd
P.O. Box 130
Waterport
Phone: 2183

LEYLAND INTERNATIONAL
A.M. Capurro & Sons Ltd.,
20 Line Wall Road
P.O. Box 130
Gibraltar
Telex: GK 258
Cables: Capurro Gibraltar

Grand Cayman

PETTERS LIMITED
Kirkconnel Bros. Ltd.
P.O. Box 72
Telex: CP 232
Cables: Kirk B
Phone: 9-2521

Greece

ALLIS CHALMERS
Technical and Commercial
of Greece SA
61 Academias Street
P.O. Box 727
Athens
Telex: 21-5760 Cris GR
Cables: Tecellas
Phone: 625629

BOMBARDIER-ROTAX GmbH
P.N. Evmirides
12, Rue Paparigopoulou
Thessaloniki
Phone: 204934

BRIGGS & STRATTON
CORPORATION
Catopodis Triantaphyllos &
Co. Inc.,
63 Stournara Street
Athens 102
Cables: Autoserve
Phone: 541-511, 541-512

CATERPILLAR OVERSEAS SA
Avras SA
Athinon and Kiffisou Aves
P.O. Box 1250
Omonia
Telex: 21-4661/4662
Cables: Avratrac
Phone: 571-6611/17
571-1401

DETROIT DIESEL ALLISON
Technica S.Malcotsis SA
14 Marni Street
Athens 103
Mail: 50 Papastratou Street
Piraeus
Telex: 863-215362 TSM GR
Cables: Mapeca, Athens
Phone: 831-577, 816-759
Athens

DORMAN DIESELS LIMITED
General Enterprises SA
43A, 3rd Septemvriou Street
Athens 103
Telex: 219395 GPGE GR
Phone: 822 6645

FORD MOTOR CO. LIMITED
Ford of Europe Inc.,
Athens Area Office
118c Kifissia Avenue
Athens 614

L. GARDNER & SONS
LIMITED
Dion Drossos & Co. Ltd.,
28 Carolou Street
Athens 107
Telex: 215058
Phone: 530-565

KATOLIGHT LIMITED
Cable Engineering Ltd
Piraeus 246
Athens (310)

KLOCKNER-HUMBOLDT-
DEUTZ AG
Magirus-Deutz Hellas A.E.
Leoforos Kifissou 94
P.O. Box 629
Athen-Egaleo
Telex: 1-5356 Athen
Cables: Ottomoto
Phone: 5616960-66

KOHLER INTERNATIONAL
LIMITED
Andrew Apostolopoulos Co.
Successors
3-1 Paparrigopoulou Street
Athens 124

LEYLAND INTERNATIONAL
KENTAVROS SA
131 Tera Odos
Athens
Telex: 216089
Cables: SO Fosti
Phone: 3467201-209

NISSAN DIESEL MOTOR CO. LTD
NIC J. Theocarakis S.A.
169 Athinon Ave.
Athens

PETTERS LIMITED
The Alson Engineering Co.,
Leonidhopoulos Works
12b K.Mavromichali Street
Piraeus 24
Telex: 212835 Leo GR
Cables: Xenmas
Phone: (21) 411 3817/8

RUGGERINI MOTORISpA
A.B.E. R.H.
P.O. Box 17
Chalkis

VOLVO PENTA
Saracakis Brothers SA
71 Leoforos Athinon
P.O. Box 410
Athens 301
Telex: 215420, 215988
Cables: Saracal
Phone: 0030-21-365320,
367011

WHITE ENGINES
INCORPORATED
Ergex Ltd.,
35 Deligiori Street
Athens 107
Cables: Telergex
Phone: 535.946

Greenland

VOLVO PENTA
Godthab Autoservice
Rasmussen & Knudsen
Box 24
3900 Gadthab
Phone: 1263

Grenada

DETROIT DIESEL ALLISON
Grenada Yacht Services
Agent of Tugs & Lighters Ltd.
Trinidad
P.O. Box 183
St. George's
Telex: 387284 Trinidad
Cables: Tuglighter, Port-of-Spain, Trinidad
Phone: 2508 St. George's

LEYLAND INTERNATIONAL
Jonas Browne & Hubbard
(Grenada) Ltd.
P.O. Box 25
St. George's
Cables: Hubbard Grenada
Phone: 2087 (2 lines)

LUCAS SERVICE OVERSEAS
LIMITED
Gomez & Western Service Limited
Lowthers Lane
P.O. Box 373
St. George's
Cables: Gowest, St. George's
Phone: 3022

PETTERS LIMITED
W.E. Julien & Co. Ltd
Young Street
P.O. Box 76
St. George's
Cables: Julien Granada

Guadeloupe

CATERPILLAR OVERSEAS SA
Yves Massel and Cie
Route de Raizet
Box 210
97156 Pointe-A-Pitre
Telex: 791GA
Cables: Massely Pointeapitre
Phone: 82-15-36

KLOCKNER-HUMBOLDT-
DEUTZ
Valcke Frères
Untervertretung:
ARG Puech — Socomi
B.P. 896
Morne Vergain
Point-A-Pitre
Telex: 029869
Phone: 82-52-59

LEYLAND INTERNATIONAL
Establissements Succes
Maurice Deher-Lesaint
34 Rue Bebian
Pointe-A-Pitre
Cables: Mauher
Phone: 82-00-99

LUCAS SERVICE OVERSEAS
LIMITED
L. Loret et Cie
P.O. Box 610
113 Rue de Nozieres
Pointe-A-Pitre
Telex: 029745GL
Cables: Deslo
Phone: 82-07-18, 82-38-20,

PETTERS LIMITED
Establissements Philippe Vivies
Zone Industrielle de Jarry
B.P. 187
Pointe-A-Pitre

Telex: Agmoli 029719/GL
Cables: Fivi
Phone: 82-11-17, 82-10-28

Guam

ALLIS CHALMERS
F.L. Moylan Company
Moylan Building
Marine Drive-Agana
Guam

Telex: 721118 RCA
Cables: Flmco
Phone: 772-6738

BRIGGS & STRATTON CORPORATION
Mid-Pac Far East Inc.
P.O. Box 7420
Tamuning

Telex: M-PAC-GM RCA 721-207
Phone: 646-5447

DETROIT DIESEL ALLISON
Mid Pac Far East Inc.
P.O. Box 7420
Tamuning 96911

Telex: M-Pac-GM RCA 721207
Phone: 746-5447

KOHLER INTERNATIONAL LIMITED
Atkins Kroll (Guam) Ltd.
P.O. Box 428
Tamuning
Guam 96910

LEYLAND INTERNATIONAL
Overseas Enterprises Limited
P.O. Box 6096
Tamuning
Guam 96911

Cables: Paulsung
Telex: 246-6638

Guatemala

ALLIS CHALMERS
Tecnica Universal Maegli
SA (TECUN SA)
3a Avenida 3-18
Zona 9
P.O. Box 590
Guatemala City

Cables: Tecunsa (Tecunsael-San Salvador, El Salvador)
Phone: 65783, 4, 5, 6, 7.

BRIGGS & STRATTON
Almacenes Concordia
18 Calle 6-85
Zona 1
Guatemala

Cables: Concordia
Phone: 26-345, 24-534

CATERPILLAR OVERSEAS SA
Mayatrac SA
Kilometroll
Carretera a Amatitlan
Apartado Postal 1793
Guatemala City

Telex: 273 Matrac CU
Cables: Mayatrac
Phone: 481061, 2, 3, 4, 5.

DETROIT DIESEL ALLISON
CIDEAgro
10A Avenue 30-57
Zone 5
Guatemala City

Telex: 305-254 – Cidea – GU
Cables: Cidea
Phone: 61531

DORMAN DIESELS LIMITED
Felix Montes y Cia SA
8a Calle 3-27
Zona 1
Guatemala CA

Telex: 4195 femco GU
Cables: Remontes
Phone: 20136

FORD MOTOR COMPANY LIMITED
Calerias Espana
7a Avenida 11-95
Oficino 3
Zone 9
Guatemala City

KATOLIGHT
Galmar Importaciones y Exportaciones
Central Comercial
Zona 4
Guatemala City

KLOCKNER-HUMBOLDT-DEUTZ AG
Auto Marina SA
Calzada
Raul Aguilar Batres 23-82
Zona 11
Guatemala D.C.

Telex: A Guatelgu para Automasa
Cables: Automasa
Phone: 460008, 480397, 460383
332, 334, 335

KOHLER INTERNATIONAL LIMITED
Felix Montes & Cia Ltda
8a Calle 3-38
Guatemala City

LEYLAND INTERNATIONAL
Almacenes 'Concordia'
Calle 18 de Septembre 6-85
Zona 1
Guatemala City

Cables: Concordia

LUCAS SERVICE OVERSEAS LIMITED
Figueroa y Cia Ltd.
La Calle 1-51
Zona 9
Guatemala City

Telex: 305358
Cables: Falcoa
Phone: 61025

NISSAN DIESEL MOTOR CO. LIMITED
Auto Mercantil S.A.
Apartado Postal 2643
Caizada Aguilar Batres 23-16
Zona 11
Guatemala City

PETTERS LIMITED
Almacen de Maquinaria Topke
Via 4 No.5-52
P.O. Box 678
Zona 4
Guatemala City

Cables: Etopke City
Phone: 64056 & 64190

JOHN ROBSON (SHIPLEY) LTD.
Orion
Maquinaria Y Accessorios Industriales
5a Calle 3-21
Zona 9
Apartado Postal 919
Guatemala C.A.

WHITE ENGINES INCORPORATED
Equipos Mecanicos de Guatemala SA
Apartado Postal 472
Guatemala C.A.

Cables: GIMAC
Phone: 63326, 63859

Guiana (French)

CATERPILLAR OVERSEAS SA
Yves Massel & Cie
B.P. 171
Cayenne

Telex: 527 FG
Cables: Masselco
Phone: 312948

KLOCKNER-HUMBOLDT-DEUTZ AG
Valcke Frères
Untervertretung: Agence Guyannaise de Mécanique A.G.M.
Angle routes de Cabasson et de la Madeleine
B.P. 350
Cayenne

Telex: AGM 549 FG
Phone: 311469

LEYLAND INTERNATIONAL
Cie Commerciale des Antilles Francaises
F. Tanon & Cie
B.P. 225
Cayenne
Cables: Tanon

PETTERS LIMITED
P.I.D.E.G.
B.P. 170
Le Larivot
Cayenne
Phone: 615

Guinea

CATERPILLAR OVERSEAS SA
Manutention Guinéenne
B.P. 336
Conarky
Phone: 621-42

LEYLAND INTERNATIONAL
Paterson Zochonis Guinea SA
Conakry
Cables: Zochonis

Guinea Bissau

CATERPILLAR OVERSEAS SA
STET
Sociedade Tecnica de Equipamentos e Tractores S.à.r.l.
Apartado 1315
Lisbon
Portugal
Telex: 12778
Cables: Stetra Sacavem
Phone: 251-1011

DETROIT DIESEL ALLISON
Sociedade Tecnica De Equipmentos
Industriais E Acessorios Ltda
P.O. Box 159
Bissau
Cables: Steia
Phone: 2742

LEYLAND INTERNATIONAL
Barbosas & CTA
Rua Das Pedras Negras
35E 41
Lisbon 2
Portugal
Cables: Guicolim Lisbon

PETTERS LIMITED
Sociedade Commercial Untramarina
S.A.R.L.
Apartado 23
Bissau
Cables: Ultra

Guyana

CATERPILLAR OVERSEAS SA
Guyana Tractor & Equipment Company
P.O. Box 604
Georgetown
Telex: GY238
Cables: Guytrac
Phone: 61130, 67181

DETROIT DIESEL ALLISON
Associated Industries Ltd.
Ruimveldt 5
P.O. Box 77
Georgetown
Telex: 312-241
Cables: Ainlim Guyana
Phone: 6729-67295

DORMAN DIESELS LIMITED
Vernon H Gibson Ltd
P.O. Box 424
Georgetown
Telex: CY211 — starting each message with "please pass to Vergil phone 2812 or 5404".
Cables: Vergib Demerara

LEYLAND INTERNATIONAL
Guyana Gajraj Limited
13 Water & Bentinck Streets
P.O. Box 440
Georgetown 2
Cables: Gadwhite
Phone: 2606

LUCAS SERVICE OVERSEAS LIMITED
13-15 Water Street
Georgetown
Telex: GY212
Cables: Boost
Phone: 66171

PETTERS LIMITED
Bookers Stores Limited
49-53 Water Street
Georgetown
Telex: GY212
Cables: Campcult Demerara
Phone: Central 1136/66171

Haiti

CATERPILLAR OVERSEAS SA
Haytian Tractor & Equipment Company SA
Ave. Haile Selassie
P.O. Box 1318
Port-Au-Prince
Telex: 3490074
Cables: Haytractor,
Phone: 2-1750, 2-3837

DETROIT DIESEL ALLISON
Haitian Marine & Industrial Co.
Rue Lowverture
Gonaives
Telex:
Cables: Haimarine
Phone: 3-3226

KOHLER INTERNATIONAL LIMITED
Charles Fequirer & Cie
B.P. 398
84 Rue Dantes Destouches
Corner of Rue Du Center
Port-Au-Prince

LEYLAND INTERNATIONAL
Auto SA
360 Boulevard J.J. Dessalines
P.O. Box 147
Port-Au-Prince
Cables: Autosa
Phone: 2772

LUCAS SERVICE OVERSEAS LIMITED
Compagnie Haitienne de Moteurs SA
P.O. Box 162
Port-Au-Prince
Cables: Haimo
Phone: 22230

NISSAN DIESEL MOTOR CO. LIMITED
Autorama S.A.
P.O. Box 1046
Rue de Peuple
Port-Au-Prince

PETTERS LIMITED
Commercial Francisco Oliver
P.O. Box 356
Port-Au-Prince
Cables: Oliver
Phone: 3193

JOHN ROBSON (SHIPLEY) LTD.
Commercial Francisco Oliver
P.O. Box 356
Port-Au-Prince

WHITE ENGINES INCORPORATED
Charles Fequiere & Cie
B.P. 398
84 Rue Dante Destouches
Port-Au-Prince
Telex: 3490094
Cables: Fequiere Port-Au-Prince
Phone: 2-2149

Honduras

ALLIS CHALMERS
Distribuidora De Repuestos SA
3a Avendida 11 y 12 Calles
Comayagiiela
Apartado Postal 46
Tegucigalpa
Telex: 1148 Diresaht
Cables: Diresa, Tegucigaipa (Honduras)
Phone: 2-7492

BRIGGS & STRATTON CORPORATION
Soc. General de Comercio
Empresa Alvarez
Apartado 568
San Pedro
Sula
Cables: Sogema
Phone: 52-13-10

CATERPILLAR OVERSEAS SA
Casa Commercial Mathews, SA
Barrio La Bolsa
Comayaguela, D.C.
P.O. Box 39
Tegucigalpa
Telex: 1109 Cemcol HT
Cables: Cemcol
Phone: 22-3164

DETROIT DIESEL ALLISON
Corporacion Commercial SA de C.V.
La Avenida 1111
Comayaguela D.C.
P.O. Box 207
Tegucigalpa
Cables: Motz
Phone: 22-67-90, 22-51-95

DORMAN DIESELS LIMITED
Distribuidora de Repuestos SA
(Diresa)
Apartado Postal No.46
Tegucigalpa D.C.
Cables: Diresa
Phone: 2-4437, 2-8482

KOHLER INTERNATIONAL LIMITED
S.E.M.P.E. SA
Apartado Postal No.219
Boulevard Kennedy
Tegucigalpa

LEYLAND INTERNATIONAL
Sociedad De Equipos Mecanicos Y
Productos Especializados SA,
De C.V.
(S.E.M.P.E.)
Apartado 219
Tegucigalpa
Cables: 1134 SEMPE — HT

LUCAS SERVICE OVERSEAS LTD.
G.L. Schofield
Apartado 1488
Tegucigalpa C.A.

NISSAN DIESEL MOTOR CO. LIMITED
Sociedad de Equipos Mecaicos
y Productos Especializados S.A.
Apartado 219
Tegucigalpa D.C.

PERKINS ENGINES LIMITED
Agencia Rene Sempe
Apartado Postal 219
Tegucigalpa D.C.
Telex: 1134 Sempe HT
Cables: Sempe
Phone: 2-5101

PETTERS LIMITED
Empresa de Servicios Agricola
SA de C.V.
Apartado 278
7a Avenido S.O. No.62
San Pedro
Sula
Telex: 5522 ESA SA HT
Cables: ESA
Phone: 520025
522185

WHITE ENGINES INCORPORATED
Fomento Internacional
SA de C.V.
P.O. Box 337
Tegucigalpa D.C.
Cable: Fomento
Phone: 2-2433 or 2-8015

Hong Kong

BRIGGS & STRATTON CORPORATION
King Tah Steel Ball &
Bearing Co.
P.O. Box 2120
Kowloon Central Post Office
959 Canton Road
Mongkok
Kowloon
Telex: 75028 HKBRG HX
Cables: HKBRGS
Phone: 3-300231

CATERPILLAR OVERSEAS SA
Caterpillar Far East Limited
P.O. Box 3069
Hong Kong
Telex: HX3305 CFEL
Cables: Catfareast HKG

DETROIT DIESEL ALLISON
American Engineering Corp.
(Hong Kong) Ltd.
1207-1213 Princes Building
Ice House Street
Hong Kong
Telex: 780-83116 Inten HX
Cables: Amengo
Phone: H 239081-239085

DORMAN DIESELS LIMITED
World Wide Engineers Ltd.
Room 503
Landwide Building
118-120 Austin Road
Kowloon
Telex: 84761 HX
Cables: Laichuli
Phone: 3-691985

L. GARDNER & SONS LIMITED
Dodwell & Co. Ltd
P.O. Box 36
Hong Kong

KLOCKNER-HUMBOLDT-
Hongkong United Dockyards Ltd.
Hung Hom
Hong Kong
Telex: 73547 Hudhkh X
Cables: Hudrep
Phone: 33411

KOHLER INTERNATIONAL LIMITED
Industrial Engineers Ltd.
P.O. Box 1838
311 Windsor House
12 Desvoeux Road Central
Hong Kong

LEYLAND INTERNATIONAL
Metro Dodwell Motors Ltd.
42 Floor
Connaught Centre
Hong Kong
Telex: HX 75696
Cables: Metcar
Phone: 5-702381

LUCAS SERVICE OVERSEAS LIMITED
Auto-Electric Ltd
Lucas House
231-235 Gloucester Road
P.O. Box 20641
Hong Kong
Telex: Dodwell 73204
Cables: Lucaserve
Phone: 5-738241-4

NISSAN DIESEL MOTOR CO. LIMITED
Dah Chong Hong, Ltd.
Commercial Vehicles Department
152A-D Prince Edward Road
Kowloon

PERKINS ENGINES LIMITED
M.K. Gilman & Co Ltd.
World Trade Centre
24th Floor
280 Gloucester Road
Causeway Bay
Hong Kong
Telex: 73358 Gilman HK
Phone: 5-793083

PETBOW LIMITED
Reiss Bradley & Co. Ltd.
P.O. Box 78
701-704 Realty Buildings
Hong Kong

PETTERS LIMITED
Autodiesel Trading Corporation
55 Gloucester Road
Hong Kong
Cables: Autodiesel
Phone: 5-272091,
5-276739,
5-278373

RUGGERINI MOTORI SpA
Wai Gin Ltd
P.O. Box 8642
Mongkok Post Office
711-713 Nathan Road —
3rd Floor
Mongkok
Kowloon
Telex: 75878 Waigi

Hungary

CATERPILLAR OVERSEAS SA Universal Co. Ltd. P.O. Box 54 Budapest 1135	Telex: Cables: Phone:	22-5347 Universal 424-514. 882-318
DORMAN DIESELS LTD I. Bier & Son (Overseas) Ltd. Kemp House 152/160 City Road London EC1V 2PE U.K.	Telex: Cables: Phone:	23465 Ionansteel London 01-2537515/6 01-2536173
KLOCKNER-HUMBOLDT-DEUTZ AG Nikex Ungarisches Cserkesz Str. 7 H-1809 Budapest X	Telex: Phone:	22-4971 475-903, 475-140

Iceland

BRIGGS & STRATTON CORPORATION Gunnar Asgeirsson Ltd Sudurlandsbraut 16	Telex: Cables: Phone:	2042 Volver IS 35200
CATERPILLAR OVERSEAS SA Hekla Ltd. Laugavegur 170-172 P.O. Box 5310 Reykjavik	Telex: Cables: Phone:	2018 Hekla 21240-21250
DETROIT DIESEL ALLISON Detroit Diesel Allison International — Europe c/o General Motors Norge A/S P.O. Box 205 2001 Lillestrom Norway	Telex: Phone:	11682 02 713860, 715860 Lillestrom
DORMAN DIESELS LIMITED S. Steffansson Ltd P.O. Box 1006 Grandagardi 5 Reykjavik	Cables: Phone:	Stefansson 1-5579 Reykjavik
L. GARDNER & SON LTD. Jonsson & Juliusson P.O. Box 731 Tryggvagata 4 Reykjavik	Telex: Phone:	2074 25430
KLOCKNER-HUMBOLDT-DEUTZ AG Hlutafelagid Hamar Postholf 1444 Reykjavik	Telex: Cables: Phone:	2070 Hamar 22123-6
LEYLAND INTERNATIONAL P. Stefansson Ltd P.O. Box 5092 Hverfisgata 103 Reykjavik	Telex: Phone:	2151 26911
PERKINS ENGINES LTD. Drattarvelar H.F. Sundurlandsbraut 32 Reykjavik	Telex: Cables: Phone:	2124 Icetractors 86500
PETTERS LIMITED Velar & Skip Ltd. P.O. Box 1006 Gradagardi 5 Reykjavik	Cables: Phone:	Velskip 27544
WHITE ENGINES INCORPORATED Bergur Larusson HF P.O. Box 634 Armula 14 Reykjavik	Cables: Phone:	Blar 81050
VOLVO PENTA Veltif HF Gunnar Asgeirson Ltd Sudurlandsbraut 16 Reykjavik	Telex: Phone: Cables:	42 35200 Volver

India

BRIGGS & STRATTON CORPORATION Aeicorp Pvt Ltd 10 Lall Bazar Street Calcutta 700001	Telex: Cables: Phone:	021-7883 Spares CA Eagerness 23-5120, 23-0879
CATERPILLAR OVERSEAS SA Larsen & Toubro Ltd L & T House Ballard Estate P.O. Box 278 Bombay 400 038	Telex: Cables:	2246 Larsenbro
Tractors India Limited 1 Taratolla Road Garden Reach P.O. Box 323 Calcutta 700 024	Cables:	Diesels
DETROIT DIESEL ALLISON Prem Nath Diesels Private Ltd 7 Scindia House Kasturba Ghandi Marg New Delhi 1	Telex: Cables: Phone:	953-312599 Premotnd Premdiesel 45891
DORMAN DIESELS LIMITED Greaves Cotton & Co Ltd P.O. Box 91 1 Forbes Street Bombay	Telex: Cables: Phone:	2517 Greaves 259771
L. GARDNER & SONS LIMITED W.H. Brady & Co Ltd Brady House 12-14 Veer Nariman Road P.O. Box 26	Phone:	251337
KATOLIGHT E. Krishna Rao & Brothers Visakhopatnam 530 002 Bombay		
KLOCKNER-HUMBOLDT-DEUTZ AG Deutz India Ltd 3-D-2 Court Chambers 35 Sir Vithaldas Thackersey Marg Bombay 400 020	Telex: Cables: Phone:	11-2181 DIL Aircooled 298247
LEYLAND INTERNATIONAL Ashok Leyland Ltd 11/12 North Beach Road Madras 1	Telex: Cables: Phone:	41271 Leyind Madras (Ennore Works) 59340/9 (City Office) 24674
Austin Distributors (Prvt) Ltd 19 Chowringhee Road Calcutta	Cables:	Ausdis
Automobile Products of India Ltd Tractor Division 8 Eastern Avenue Maharani Bagh New Delhi 14		
LUCAS SERVICE OVERSEAS LTD. Lucas Indian Service Ltd 9 Patullo Road Mount Road Madras 600 002	Telex: Cables: Phone:	41-439 Lucasind 812176-9
Aeicorp Private Ltd Mercantile Buildings 2nd Floor 10 Lall Bazaar Street Calcutta 700001		

PERKINS ENGINES LIMITED
Simpson & Co Ltd
P.O. Box 303
202-203 Mount Road
Madras 2
Telex: 41-538
Cables: Simpsons
Phone: Madras 83091

PETBOW LIMITED
William Jacks & Co
(India) Private Ltd
Hamilton House
P.O. Box 335
Ballard Estate
Bombay 1BR

PETTERS LIMITED
Parry & Co Ltd
Dare House
P.O. Box 12
Madras 1
Cables: Distimulo
Phone: 21101/29251

WHITE ENGINES INCORPORATED
Salgaocar Engineers Pvt Ltd
805 New Delhi House
27 Barakhamba Road
New Delhi 1100011
Telex: 3333
Cables: Salepril
Phone: 42408

Indonesia

ALLIS CHALMERS
John D. Hutchinson & Co
(Indonesia) Ltd
Hutchison House
17th Floor
P.O. Box 43
Hong Kong
Telex: 74798
Cables: Alltrakhk
Phone: 5-233081

BRIGGS & STRATTON CORPORATION
P.T. Unimas Motor Wasta
3 Jalan Batu Ceper
P.O. Box 2911 Dkt
Jakarta
Telex: Unimas 46238
Cables: Unimotor
Phone: 56497, 50499

CATERPILLAR OVERSEAS SA
P.T. Trakindo Utama
Kompleks KKO Cilandak
P.O. Box 2282
Jakarta
Telex: 011 44393
Cables: Traktama

DETROIT DIESEL ALLISON
P.T. Garuda Diesel (Ltd)
Jl Letjen Haryono M.T.33
Jakarta
Telex: 796-44100
Cables: Garudadiesel
Phone: 81204, 81097, 84117

DORMAN DIESELS LTD
Berca Indonesia P.T.
1st Floor JL
Cikini Raya 61
P.O. Box 496/JKT
Jakarta
Telex: 2895 Bercacon JKT
Cables: Bercacon
Phone: 40369

KATOLIGHT
P.T. Speed Indonesia Ltd
P.O. Box 2346
Jakarta

KIRLOSKAR OIL ENGINES LTD
Rutan Trading Coy
Djalan Slompretan 16-18
Surabaya
Telex: 3497
Cables: Rutan
Phone: 2841, 3620, 2151, 2502

KLOCKNER-HUMBOLDT-DEUTZ AG
P.T. Lindeteves Indonesia
Jl yos Sudarso
P.O. Box 2651
Sunter 11
Jakarta Utara
Cables: Lindeteves
Phone: 290409, 290934

LEYLAND INTERNATIONAL
P.T. Java Motors
17 Djalan Raya Kramat
Tromolpos 161
Jakarta 1V/4
Cables: Javamotors
Phone: 40693/5

LUCAS SERVICE OVERSEAS LTD
P.T. Fakta Jaya Motor
Tromol Pos 3032/JKT
97 Jalan Lautze
Jakarta
Cables: Factsmot
Phone: 25927

NISSAN DIESEL MOTOR CO. LIMITED
P.T. Imermotors
Djalan Melawai VIII No.10
Kebejoran Baru
Jakarta

PERKINS ENGINES LTD
P.T. Traktor Nusantara
Jalan Gajah Mada 80
Jakarta
Telex: JKT 42844
Cables: Traknusa
Phone: 270919/20062

PETBOW LIMITED
P.T. Porodisa Equipment
Ground Floor
Arthaloko Building
Jalan Jenderal Sudirman
Kav-2
Jakarta

PETTERS LIMITED
P.T. Petindo Permanent V.A. Ltd
Jalan Hayam Wuruk No.AI-3
Pasar Lindeteves
Jakarta
Barat
Cables: Indopermanent
Phone: 22651, 21882

VOLVO PENTA
P.T. Benua United Trading & Engineering Corporation
Jalan Gajah Mada No.209
Jakarta
Telex: 42869
Cables: Betrab
Phone: 21288, 24504

WHITE ENGINES INCORPORATED
P.T. Sumber Mesin Raya
Jl: Gajah Mada No.176-177
P.O. Box 1298/JAK
Jakarta
Barat
Telex: 46111
Cables: Summesra
Phone: 22016 20-417

Iran

ALLIS CHALMERS
Diesel Auto Part SA
247 Abassabad Avenue
P.O. Box 687
Tehran
Telex: 212658 Yasa Ir
Cables: Fouladi
Phone: 851021-8

BRIGGS & STRATTON CORPORATION
Sabet-Pasal Co. Incorporated
Eisenhower Avenue
Jeyhoun Street
Telex: 2519
Cables: Sabetpasal
Phone: 954051-53

CATERPILLAR OVERSEAS SA
Mashinhaye Rahsazi Co Ltd
Av. Saadi 168
P.O. Box 3390
Tehran
Telex: 212357
Cables: Marcoli
Phone: (021) 314001/10

DORMAN DIESELS LTD
Partt Motor Co
Saadi Avenue 588
Tehran
Cables: Parttmotor
Phone: 303207/9

L. GARDNER & SONS LTD
H.G. Mollahzadeh & Bros
Avenue Arvand
P.O. Box 49
Abadan
Phone: Abadan 3778

KATOLIGHT
Karteman Company
No.90 Ramsar Street
Shahreza Avenue
Tehran

KIRLOSKAR OIL ENGINES LTD — Telex: 212261 IMTC IR
Iran Margo Trading Co. Ltd — Cables: Iranmargo
P.O. Box 373 — Phone: 921082/83/84
216-222 Eisenhower Avenue
Tehran

KLOCKNER-HUMBOLDT-DEUTZ AG — Telex: 213239 Varn Ir; Phone: 899585
Varnco Co Ltd
Avenue Sepahbad Zahedi
P.O. Box 41-2166
No.6 Naser Street
Shahdad Building
Tehran 14155

LEYLAND INTERNATIONAL — Cables: Leymotors
British Leyland Motor Corporation Ltd
P.O. Box 11/1545
11 Avenue Sepand
Avenue Iranshahr
Tehran

LUCAS SERVICE OVERSEAS LTD — Telex: 212075
Tundar Distribution & Service Co. — Cables: Irantundarco
P.O. Box 8-6189 — Phone: 547688-9
Ghazvine Avenue No 1682
Tehran

NISSAN DIESEL MOTOR CO. LIMITED
Romco Co. Limited
Ave. Iranshahr North
Kouchen Homa No.13
Tehran

PERKINS ENGINES LTD — Cables: Ferromatex
Ferromatex Technical & Trading Corporation — Phone: 310085-89
Aftab Shargh Building
Avenue Ferdowsi
Tehran

PETBOW LIMITED
Iran Electrical Engineering
249 Avenue Shahreza
Tehran

PETTERS LIMITED — Telex: 212511 Mahyar THN
The Mahyar Industrial Company — Cables: Mahyar
Mahyar Building — Phone: 217093
634 Saadi Avenue
Tehran

JOHN ROBSON (SHIPLEY) LTD
Sherkat Tazamoni Faridoon
Zartoshty va Baradar
Zartoshty Building
Kooche Momtaz Saadi Avenue
Tehran

RUGGERINI MOTORI SpA
Farland Company Ltd
642 Saadi Avenue
Tehran

WHITE ENGINES INCORPORATED — Telex: IEC IR 21-2607
Iran Equipment Co — Cables: Taymour
17 Mellat Avenue — Phone: 311301-311820
Tehran

Iraq

ALLIS CHALMERS — Cables: Diesel
Technical Works W.L.L. — Phone: 90018
Muaskar Al-Rashid
Baghdad

DORMAN DIESELS LIMITED — Cables: Rustengine Baghdad
The Iraqi Diesel Engine Company — Phone: 94799
P.O. Box 2211
Alwiyah

KATOLIGHT
Nabeel Jewaideh
26/3/26 Muasker El Rashid Road
Baghdad

KLOCKNER-HUMBOLDT-DEUTZ AG — Cables: Faryal; Phone: 99215
Faris Trading & Contracting
Alexander Iskander Building
Bustan Kubbah Sa'addon Street
P.O. Box 5681
Baghdad

LEYLAND INTERNATIONAL — Cables: Automobile
General Automobile Company — Phone: 89711
Al-Andalus Square
P.O. Box 3270
Saadoun
Baghdad

LUCAS SERVICE OVERSEAS LIMITED — Cables: Automobile; Phone: 95071
Automobile State Enterprise
P.O. Box 3270
Al-Andalus Square
Baghdad

PERKINS ENGINES — Telex: 2342 Auto I.K.
Automobile State Enterprise — Cables: Automobile
P.O. Box 3270 — Phone: 95071
Al-Andalus Square
Baghdad

PETTERS LIMITED
State Machinery Import Company
Sa'Adon Street
Baghdad

JOHN ROBSON (SHIPLEY) LTD
J.J. Moukhtar
Numaan Street 25G/1/197
Baghdad

Ireland

DETROIT DIESEL ALLISON — Telex: 851-3 1329 Wellingborough
Detroit Diesel Allison International – Europe — Cables: Genmopower
Division of General Motors Ltd — Phone: (0933) 71122
London Road
P.O. Box 6
Wellingborough
Northamptonshire NN8 2DL
U.K.

FORD MOTOR CO LIMITED — Telex: 6121
Henry Ford & Son Limited — Phone: 26881
Cork

KLOCKNER-HUMBOLDT-DEUTZ AG — Telex: 5122; Cables: Suntract
Blackwood Hodge Limited — Phone: 503666
Long Mile Road
Dublin 12

NISSAN DIESEL MOTOR CO. LIMITED
William P. Ryan Ltd.
52 Thorncastle Street
Ringsend
Dublin 4

PERKINS ENGINES LTD — Telex: 5320
McNeill (Ireland) Ltd — Phone: 371601
Collinstown Cross
Cloughram
Co. Dublin

PETTERS LIMITED M & G Limited Cooksdown Industrial Estate Belgard Road Tallaght Co. Dublin	Telex: Phone:	4486 (Mags) 511144
VOLVO PENTA Western Marine Ltd Bulloch Shipyard Dalkey Dublin	Telex: Phone:	4839 800321

Israel

BRIGGS & STRATTON CORPORATION Moise Carasso Sons Limited 26 Rival Street P.O. Box 991 Tel Aviv	Telex: Cables: Phone:	32-114 Mocarasso 33241-6
CATERPILLAR OVERSEAS SA The Israel Tractors & Equipment Co. Ltd New Industrial Area 8 Hamanor Street P.O. Box 214 Holon	Telex: Cables: Phone:	32447 Israelquip 856-125
DETROIT DIESEL ALLISON David Argamany Limited 36 Yitzhak Sade Street P.O. Box 14161 Tel Aviv	Telex: Cables: Phone:	922-03-2470 Coin IL Agradiesel 03-30172
JLO Resses Limited 12 Harakeveth Street P.O. Box 941 Tel Aviv	Telex: Cables: Phone:	34118/9bxtv il for 5141 resses Resses 625 634
KATOLIGHT C.N.E.C. Telecoms Limited P.O.B. 1707 Tel Aviv		
LEYLAND INTERNATIONAL Israel Motor Agency Limited 65 Derech Petach Tikva P.O. Box 20029 Tel Aviv	Telex: Cables:	03-2283 Isomotor
RUGGERINI MOTORI SpA N. Feldman & Son Limited P.O. Box 33181 71 Giborey Israel Street Tel Aviv	Telex:	32276 Felta
VOLVO PENTA Mayer's Cars and Trucks Co. Limited 23 Carelbach Street P.O. Box 16164 Tel Aviv	Telex: Cables: Phone:	32230 Carsmayer 269191
WHITE ENGINES INCORPORATED Arditi Limited P.O. Box 1033 3 Levontin Street Tel Aviv	Cables: Phone:	Arditil 61118/2, 621533

Italy

ALLIS CHALMERS Cav. Tonino Cerioli + C.S.A.S. Via Bentini 9 40013 Castelmaggiore Bologna	Telex: Cables: Phone:	51209 Cerioli Ceriolo Castel- maggiore 700442/3/4
Via Isonzo 34 Rome	Telex: Cables: Phone:	Reitaly 62037 Powerhouse 8440351
BOMBARDIER-ROTAX GmbH Cantiere Jet Craft Quartiere ind. Mirablla 1-20081 Abbiategrass (MI)	Phone:	94.92.17
BRIGGS & STRATTON CORPORATION R.A.M.A. Snc Via Agnoletti (Zona Annonaria) Reggio Emilia 42100	Telex: Cables: Phone:	53396 Rama R.E. (0522) 71852- 72631
CATERPILLAR OVERSEAS SA Compagnia Generale Trattori S.p.A. Direzione Generale Via San Vittore 37 Milan	Telex: Cables: Phone:	39073 Cogetrattori (01) 49-94
Macchine Agricole Industriali Automezzi MAIA S.p.A. Via Nomentana 995 00137 Rome	Telex: Cables: Phone:	61463 Maiarom 820-241/4 824-841/4 823-901
DETROIT DIESEL ALLISON Detroit Diesel Allison International — Europe c/o General Motors Suisse SA Salzhausstrasse 21 2501 Biel-Bienne Switzerland	Telex: Cables: Phone:	34217 Gms-Ch Gms-Ch Bienne 032-215111 Bienne
DORMAN DIESELS LIMITED Officine Meccaniche Ing Contaldi (OMIC) Strada Padana Superiore 307 20090 Vimodrone (Milano)	Telex: Cables: Phone:	35193 Comavim Contaldi 2500441/2/3/4/5
FORD MOTOR COMPANY LIMITED Ford Italiana S.p.A. Parts Distribution Centre Via Appia Nuova 1L57 00178 Rome	Telex: Phone:	62106 (06) 799 3015
L. GARDNER & SONS LIMITED Cantiere Navale Dell'Argentario s.r.l. 58019 Porto S. Stefano (Grosseto)	Phone:	(05 64) 81.29,75, 81.41,15
KLOCKNER-HUMBOLDT-DEUTZ AG Industrie Meccaniche Lombarde S.p.A. via Garcia Lorca 25 22050 Lomagna (Como)	Telex: Cables: Phone:	37606 Imellom Imelombarde 587291
LEYLAND INTERNATIONAL B.L. Italia Via Paolo di Dono Angolo Via Baldovinetti 00144 Rome	Telex: Phone:	79284 (Leylita) 546811
PERKINS ENGINES LIMITED Motori Perkins S.p.A. 22100 Como-Camerlata Via Pasquale Paoli 9/A	Telex: Cables: Phone:	38063 Perkoil Camerlata 504885
PETTERS LIMITED Volpi & Bottoli Via Altobello Ferrara 26034 Piadena (Cr)	Telex: Phone:	30421 Combanca Mastora (0375) 98128/ 98434
VOLVO PENTA Volvo Penta Italia S.r.l. Via Lomellina 18 Casella Postale 30 1-20094 Buccinasco-Milano	Telex: Cables: Phone:	35323 Volvopenta 02-447 97 05 02-447 80 57

Ivory Coast

ALLIS CHALMERS
Compagnie Africaine D'Equipement Industrial
(C.A.E.I.)
30 Rue Thomas Edison
P.O. Box 836
Abidjan
Telex: 768
Cable: Afequip
Phone: 357039

CATERPILLAR OVERSEAS SA
Manutention Africaine
P.O. Box 1299
Abidjan
Telex: 675
Cables: Mea
Phone: 37-33-66
37-33-86/7/8/9/
90/91

DETROIT DIESEL ALLISON
Blackwood Hodge (Cote d'Ivoire)
S.a.r.l.
25 Boulevard Angoulvant
P.O. Box 14066
Abidjan
Telex: 969-Abidjan 2207
Cable: Suntract
Phone: 324083

KOHLER INTERNATIONAL LIMITED
Ets. Peyrissac Cote D'Ivoire
P.O. Box 1272
Abidjan

LEYLAND INTERNATIONAL
Cie Francaise de l'Afrique Occidentale
Agence Centrale
P.O. Box 2114
Abidjan
Telex: 777
Cables: Senafrica
Phone: 226.51, 262.26

LUCAS SERVICE OVERSEAS LIMITED
Matforce Diesel (Div. of CFCIT)
6 Rue Thomas Edison
Zone 4
P.O. Box 1844
Abidjan
Telex: Tecofra
640435
Phone: 35 58 69

NISSAN DIESEL MOTOR CO. LIMITED
Comafrique
B.P. 20817
Abidjan

PERKINS ENGINES
Carena
P.O. Box 453
Abidjan
Telex: Carena 323
Cables: Carena
Phone: 22.22.27, 32.13.36

PETTERS LIMITED
Hamelle-Afrique SA
Km3 Autoroute Sud
P.O. Box 1326
Abidjan
Telex: Hamelaf 401
Cables: Hamelaf
Phone: 35.67.43, 63.28

Jamaica

ALLIS CHALMERS
Kingston Industrial Agencies Limited
381 Spanish Town Road
P.O. Box 80
Kingston 11
Telex: 01042 – Indage
Cables: Indage
Phone: 933-7121, 933-7123

BRIGGS & STRATTON CORPORATION
Will's Battery Co. Limited
66 Slipe Road
Kingston 5
Cables: Battery
Phone: 936-5921

CATERPILLAR OVERSEAS SA
Jamaica Tractor & Equipment Company
379 Spanish Town Road
P.O. Box 213
Kingston
Telex: 2135 Alprojam
Cables: Jamtrac
Phone: 923-9251

DETROIT DIESEL ALLISON
Industrial Equipment Company
22 Bell Road
P.O. Box 22
Kingston 11
Telex: 381-2213 Haultrax
Cables: Powerload
Phone: (92) 36519, 36529

DORMAN DIESELS LIMITED
Caribbean Industrial Equipment Limited
7 South Avenue
Rest Pen
Kingston 10
Cables: Indquip
Phone: 926-1240

KATOLIGHT
Specialist Agencies Limited
P.O. Box 657
Spanish Town

KLOCKNER-HUMBOLDT-DEUTZ AG
The Technical Supply Co. Limited
44 Hagley Park Road
P.O. Box 75
Kingston 10
Telex: 2271 Servocomp JA
Cables: Kemtex
Phone: 9367744

KOHLER INTERNATIONAL LIMITED
Abdulla C. Marzouca Limited
157 Orange Street
Kingston

LEYLAND INTERNATIONAL
Cars & Commercials Limited
67/69 Harbour Street
Kingston
Cables: Carancom
Phone: 23121, 26821

LUCAS SERVICE OVERSEAS LIMITED
John Crook Limited
Lucas House
86-88 Tower Street
P.O. Box 21
Kingston
Cables: Lucasjam
JCA
Phone: 5227/8/9

PERKINS ENGINES LIMITED
Reginald Aitken Limited
437 Spanish Town Road
P.O. Box 280
Kingston 11
Telex: 2170 Marco
Cables: Marco
Phone: 933/6737/9
933/6730, 933-8538, 933-8411

PETTERS LIMITED
Will's Battery Co. Limited
P.O. Box 440
66 Slipe Road
Kingston 5
Cables: Battery
Phone: 936-5921

Japan

ALLIS CHALMERS
Denyo Co. Limited
No.2, 4-Chome
Nakano-Ku
Kamitakada
Tokyo
Telex: 232-2936 Denyo J
Cables: Denweltyo
Phone: 03-552-1201

BRIGGS & STRATTON CORPORATION
Yanase & Co. Limited
Trading Division
Minato-ku
Tokyo
Telex: 78102422551
Cables: Yanaseco
Phone: 452-4311

CATERPILLAR OVERSEAS SA
Caterpillar Mitsubishi Limited
3700 Tana
Sagamihara-Shi
Kanagawa-Ken 229
Telex: J222877
Cables: Caterbishico
Sagamihara

DETROIT DIESEL ALLISON
Tominga & Co. Limited
Ibasen Building
No.5 Nichome Kobunacho
Nihombashi
Chuc-Ku
Tokyo 103
Telex: 781-22435 Tomco
Cables: Tomidiesel
Phone: 662-1851

DORMAN DIESELS LIMITED
Kyotkuto Boeki Kaisha Limited
Second Industrial Machinery Dept
CPO Box 330
Tokyo 100-91
Telex: 22440 Kyokubo
Cables: Mercantile
Phone: 244-3511

FORD MOTOR CO. LIMITED
Ford Motor Company of Japan Limited
Checker Building
5-29 Akasaka 8 Chome
Minato Ky
Tokyo 107
Cables: Fomoja

KLOCKNER-HUMBOLDT-DEUTZ AG
Mitsui Deutz Diesel Engine Co. Limited
Daini Toyo Kaiji Building
3rd Floor
4-8-24 Shinbashi
Minato-ku
Tokyo
Telex: 02842-109 mdeutz j
Cable: Demitsui Deutz
Phone: 4331666

KOHLER INTERNATIONAL LIMITED
Tominaga & Co. Limited
8 Kinugasa-cho
Kita-ku
Osaka

LEYLAND INTERNATIONAL
Shintoyo Motors Company Limited
P.O. Box 581
Hatanodai
Shingawa-ku
Tokyo
Telex: 246/6638
Cables: Shintoprise

LUCAS SERVICE OVERSEAS LIMITED
Nihon-Lucas (Sales & Service) Limited
7-5 Iwamotochu 1-chome
Chiyoda-ku
Tokyo 101
Telex: 265-5888
Cables: Lucastyo
Phone: 863-5187

PERKINS ENGINES LIMITED
Perkins Engines (North Pacific) Ltd.
6th Floor
Reinanzaka Building
14-2 Akasaka 1-chome
Minato-ku
Tokyo 107
Telex: 24292
Phone: (03) 586-7377

PETBOW LIMITED
Mikuni Kikai Kogyo Company Limited
No.2-2 Kajicho
2-chome
Chiyoda-Ku
Tokyo

PETTERS LIMITED
Dodwell & Company Limited
Industrial Department
P.O. Box 297
Kowa Building No 16 Annex
9-20 Akasaka 1-chome
Minato-ku
Tokyo
Telex: International: J22274, J22602
Domestic: 222 2842
Cables: Dodtrade
Phone: (03) 584-2351

VOLVO PENTA
The Penta Japan Limited
Room No. 316
Azabu Heights
5-10, 1-chome
Roppongi
Minato-ku
Tokyo
Telex: J24692
Cables: Volvopentajap
Phone: 03-586-3801/5

Jordan

CATERPILLAR OVERSEAS SA
Jordan Tractor & Equipment Co. Limited
Salt Road
P.O. Box 313
Amman
Telex: 1226
Cables: Jallad
Phone: 383-63/64/65

DETROIT DIESEL ALLISON
Farradj & Company
Jabel El Weibdeh
P.O. Box 974
Amman
Telex: 925-1317 Farradj
Cables: Farradj
Phone: 21662

DORMAN DIESELS LIMITED
Nagib Baki & Sons (Jordan) Ltd.
P.O. Box 572
King Hussein Street
Basharat Building
Amman
Cables: Nagib Baki
Phone: 36121

INDIAN NATIONAL DIESEL ENGINE CO. LIMITED
Ramadhan Trading Company
P.O. Box 12067
Amman

KATOLIGHT
Modern Trading Agency Company
P.O. Box 1377
Amman

KIRLOSKAR OIL ENGINES LIMITED
Nicolas C. Peridakis & Company
P.O. Box 441
King Hussein Street
Amman
Cables: Perico
Phone: 23789

KLOCKNER-HUMBOLDT-DEUTZ AG
Adnan Kassim
P.O. Box 5210
Amman
Telex: 1474 ask jo
Phone: 42378

KOHLER INTERNATIONAL LIMITED
The Commercial & Industrial Co. Limited
Station Street
P.O. Box 379
Amman

LEYLAND INTERNATIONAL
Suleiman Tannous & Sons
P.O. Box 102
Amman
Cables: Tours

LUCAS SERVICE OVERSEAS LIMITED
Suleiman Tannous & Sons Ltd.
Amir Mohammed Street
P.O. Box 102
Amman
Telex: 1324 Sultan Jo
Cables: Tours
Phone: 36180, 36188, 36189, 24768

NISSAN DIESEL MOTOR CO. LIMITED
The Commercial & Industrial Co. Ltd.
P.O. Box 379
Amnan

PETTERS LIMITED
The United Engineering & Mechanical Co.
(Modaco)
P.O. Box 1400
Mahatta Road
Amman
Cables: Modaco
Phone: 55354

PERKINS ENGINES LIMITED
Transjordan Engineering Company
P.O. Box 1
Amman
Telex: 1300
Cables: Tec
Phone: 24198/9, 23354

JOHN ROBSON (SHIPLEY) LIMITED
Nagib Baki & Sons (Jordan) Limited
P.O. Box 572
King Hussein Street
Amman

Kenya

ALLIS CHALMERS
Hokman Bros (East Africa) Ltd
Buckleys Road
P.O. Box 42044
Nairobi
Telex: 22510 Homanea
Cables: Airdrill
Phone: 555388

BRIGGS & STRATTON CORPORATION
Car & General (Kenya) Limited
Cargen House
13 Government Road
P.O. Box 20001
Nairobi
Telex: 22046
Cables: Cargen
Phone: 31570

CATERPILLAR OVERSEAS SA
Construction Equipment
(Div. of Gailey & Roberts Ltd.)
P.O. Box 30331
Nairobi
Cables: Afritrak
Phone: 55-71-88

DETROIT DIESEL ALLISON
Wigglesworth & Co. Kenya Limited
Kingston Road
Industrial Area
P.O. Box 30092
Nairobi
Telex: 963-22241 Pettinato
Cables: Pettinato
Phone: 57022

DORMAN DIESELS LIMITED
Mackenzie (Kenya) Ltd.
Likoni Road
P.O. Box 30010
Nairobi
Telex: 22205
Phone: 555333 (Office)

KIRLOSKAR OIL ENGINES LIMITED
Marina Machineries Co.
(Kenya) Limited
P.O. Box 43266
Nairobi
Cables: Marina
Phone: 22059

KLOCKNER-HUMBOLDT-DEUTZ AG
Achelis (Kenya) Limited
P.O. Box 30378
Kijabe Street
Nairobi
Telex: 22556
Cables: Achelissons
Phone: 559766, 559485

LEYLAND INTERNATIONAL
Leyland Kenya Limited
P.O. Box 18052
Connaught House
Haili Selassie Avenue
Nairobi
Cables: Leyalb

LUCAS SERVICE OVERSEAS LIMITED
Lucas House
Delta Limited
Uhuru Highway
P.O. Box 30519
Nairobi
Telex: 22248
Cables: Delta
Phone: 557422

NISSAN DIESEL MOTOR CO. LIMITED
Ryce Motor Ltd.
P.O. Box 49729
Nairobi

PERKINS ENGINES LIMITED
Century Tractor & Implements
(Kenya) Limited
P.O. Box 30192
Roland Ngala Street
Nairobi
Telex: 22641
Cables: Centrac
Phone: 23035-6, 24940

PETTERS LIMITED
Mackenzie (Kenya) Limited
P.O. Box 30010
Mercantile House
Koinange Street
Nairobi
Telex: 22205
Cables: Mackenzies
Phone: 555333

Khmer Republic

DORMAN DIESELS LIMITED
Societe Commerciale Khmere
(L. Rondon & Co)
P.O. Box 231
61 Moha Vithei Preah-Bat Norodom
Phnom Penh
Cables: Rondon
Phone: 2 4078

LEYLAND INTERNATIONAL
Societe Des Garages Charner
Au Cambodge
215 Vithei Pau Kambo
P.O. Box 274
Phnom Penh
Cables: Fareast St Ĝermain Franc
Phone: 963.48-83, 973.07-15

PETTERS LIMITED
Establissements Moha-Bopea
250 Mona Vithei Preah Monivong
P.O. Box 446
Phnom Penh
Cables: Moha
Phone: 32024

Korea

ALLIS CHALMERS
Woodward & Dickerson
(Korea) Limited
1601 Korean Air Lines Building
118-2K a Namdaemun-Ro
I.P.O. Box 1218
Chung-Ku
Seoul
Telex: 2252
Cables: Woodward
Phone: 24-3551/4

CATERPILLAR OVERSEAS SA
Hae Nin Tractor Co. Limited
Dong Bang Building
18th Floor
250, 2-Ka Taipyong-ro
Chung-Ku
Central P.O. Box 1201
Seoul
Cables: HN Tractor

DETROIT DIESEL ALLISON
Summit Industrial Corporation
Hae Nam Building
No.90 2-KA
Taepyong-No
Chung-Ku
I.P.O. Box 1078
Seoul
Cables: Sumiting
Phone: 28-5338

DORMAN DIESELS LIMITED
Ewkor Trading Co. Limited
901 International Insurance Building
No.8 Yang Dong
I.P.O. Box 1162
Chung-Ku
Seoul
Telex: 2365
Cables: Ewkortc
Phone: 22-2124, 22-5349

KATOLIGHT
Europoean American Corporation
P.O. Box 1467
Seoul

KLOCKNER-HUMBOLDT-DEUTZ AG
Lee-Gee Industrial Corporation
I.P.O. Box 2232
Seoul
Telex: 28548 Sesil K
Cables: Leegee
Phone: 766259

LEYLAND INTERNATIONAL
The Seoul Kyo Tonh Corporation
Limited
185 Setongro
I.P.O. Box 1154
Seoul
Cables: Seoul Kyotong

LUCAS SERVICE OVERSEAS LIMITED
Hyundai Motor Service Company
113-25 4KA Wonhyo Ro
Yongsan Ku
Seoul
Telex: 2338
Phone: 43-3171/6, 43-7901

PERKINS ENGINES LIMITED
Hyundai Motor Co. Limited
Hyundai Building
178 Sejong-Ro
Jongro-Ku
Seoul
Telex: Hdmoco K2391
Phone: 74-8311, 75-6511

PETTERS LIMITED
Shriro (International) Ltd.
9th Floor
Jeban Building
5 Yang Dong
Choong-Ku
Seoul
Telex: Shriroco 23515
Phone: 22.6131/3

WHITE ENGINES INCORPORATED
Pan-A International Co.
Room 1200 Samyun Building
63-2, 2-KA Chungmu-Ro
Joong-Ku
Seoul
Cables: Panabyun Seoul
Phone: 76-1588

Kuwait

BRIGGS & STRATTON CORPORATION
Khalid Yousuf Al-Homaizi Corp.
P.O. Box 252
Kuwait
Telex: 2488
Cables: Homaizico
Phone: 812321

CATERPILLAR OVERSEAS SA
Mohamed Abdulrahman Al-Bahar
Machinery General Office
P.O. Box 148 Safat
Kuwait
Telex: 2302 KT
Cables: Moatasim
Phone: 810-855/56

DETROIT DIESEL ALLISON
Yusuf Ahmad Alghanim & Sons W.L.L.
Construction Equipment and Engine Division
Fourth Ring Road
Rai Industrial Area
Shuwaikh
P.O. Box 223
Kuwait
Telex: 9592069 Autoarabi
Cables: Autoarabi
Phone: 719772 or 730105

L. GARDNER & SONS LIMITED
Al-Sabah & Fakhro Company Limited W.L.L.
Commercial Centre No.3
P.O. Box
Safat 319
Kuwait
Phone: 34138

INDIAN NATIONAL DIESEL ENGINE CO. LIMITED
Abdulazuz Abdullah
Al-Murshed & Brothers
P.O. Box 1874
Kuwait

KATOLIGHT
Yusuf A. Alghanem
P.O. Box 223
Kuwait

KLOCKNER-HUMBOLDT-DEUTZ AG
Ateeqy
Trading & Engineering Enterprises
P.O. Box 699
Kuwait
Telex: 3007 KT
Cables: Ateeq
Phone: 717945/6/7

KOHLER INTERNATIONAL LIMITED
Bader Al Mulla & Brothers
P.O. Box 177 Safat
Kuwait

LEYLAND INTERNATIONAL
Kuwait Automobile & Trading Company
P.O. Box 41
Kuwait
Telex: 3005 Automobile KT
Cables: Automobile
Phone: 34133-34816

LUCAS SERVICE OVERSEAS LIMITED
Alomar Mechanical Engineering Company
P.O. Box 392
Kuwait
Telex: 2152 Omarco Kwt
Cables: Omarco
Phone: 819555/7/8/9

NISSAN DIESEL MOTOR CO. LIMITED
Boodai Trading Co. Ltd.
P.O. Box Safat 1287
Kuwait

PERKINS ENGINES LIMITED
The Trading & Industrial Equipment Co. W.L.L.
P.O. Box Safat 2159
Kuwait
Telex: Success 2224 KWT
Cables: Success
Phone: 819179/819188/819190

PETBOW LIMITED
Morad Yousuf Behbehani
P.O. Box 146
Kuwait

PETTERS LIMITED
Yousif Ibrahim Hizami
P.O. Box 733
Kuwait
Cables: Hizami
Phone: 434645

JOHN ROBSON (SHIPLEY) LIMITED
Al-Humaidan Kuwait Trading & Contracting Est.
Hilali Street
Mohd. N.Al-Hajeri Building
Kuwait

Laos

LEYLAND INTERNATIONAL
Ets. Lath Nhouyvanisvong Enterprises
Commerciales Et Constructions
P.O. Box 33
Vientiane
Cables: Lath-Pakse
Phone: 35

Lebanon

ALLIS CHALMERS
Merican-Curtis International Limited
Al Ghanen Building
Rue Verdun
P.O. Box 11-4986
Beirut
Telex: 20829LE Penyou
Cables: Engtracting
Phone: 353-670

BRIGGS & STRATTON CORPORATION
Michel Andraos & Co., S.A.L.
P.O. Box 447
Beirut
Cables: Micandros
Phone: 220300

CATERPILLAR OVERSEAS SA
M. Ezzat Jallad & Fils
P.O. Box 110208 and 112556
Beirut
Telex: 21624
Cables: Jamla Beirut
Phone: 932-522

DETROIT DIESEL ALLISON
Levant Diesel
Matta Building
Makdesi Street
P.O. Box 11-2867
Beirut
Telex: 20680
Cables: Levandiesel Beirut
Phone: 350180 or 350181

DORMAN DIESELS LIMITED
Nagib Baki & Sons
P.O. Box 11-828
Patriarch Hoyek Street
Khan Antoun Bey
Beirut
Cables: Bakson
Phone: 234618, 271395

L. GARDNER & SONS LIMITED
Saamir Commercial Agencies
P.O. Box 11-2891
Beirut

Telex: 20680 Le Joe Int
Cables: Trancomage
Phone: 295240-238001

INDIAN NATIONAL DIESEL ENGINE CO. LTD.
Abourrousee & Company
P.O. Box 1500
Beirut

KIRLOSKAR OIL ENGINES LIMITED
Abdel Massih Commercial Company
St. Joseph Hospital Street
P.O. Box 80260 Dora
Bourj Hammoud
Beirut

Cables: Amcco
Phone: 262388/264615

KLOCKNER-HUMBOLDT-DEUTZ AG
Emile Acar & Fils
Rue Arz
B.P. 115834
Beirut

Telex: 22218 Rocomy
Cables: Emilacarfils
Phone: 225797, 293774

KOHLER INTERNATIONAL LIMITED
Demetrius Stephanou & Company
P.O. Box 127
Port Street
Fattal Building
Beirut

LEYLAND INTERNATIONAL
Phillippe & Adib Fata
P.O. Box 5309
Beirut

Cables: Fatabros
Phone: 280860, 281533

LUCAS SERVICES OVERSEAS LIMITED
Michel Andraos & Co.,
S.A. Lebanese
Andraos Building
Al Arz Street
P.O. Box 447
Beirut

Cables: Micandros
Phone: 220300

PERKINS ENGINES LIMITED
Abourrousee & Company
P.O. Box 1500
Beirut

Cables: Edrouss
Phone: 227345/235784

RUGGERINI MOTORI SpA
Antoine Hajjar & Fils
P.O. Box 11-1705
Saliba Building
Dora-Beirut

VOLVO PENTA
S. Sadaka & Sons
P.O. Box 4
Zahle

Telex: 21211
Cables: Sadaka
Phone: 823333

Lesotho

CATERPILLAR OVERSEAS S.A.
Barlow's (O.F.S.) Ltd.
Nuffield Street
Hamilton
Bloemfontein 9301
South Africa

Telex: 2-632
Cables: Shipments Bloemfontein
Phone: 82721

Liberia

CATERPILLAR OVERSEAS SA
Liberia Tractor & Equipment Company
United Nations' Drive
P.O. Box 299
Monrovia

Telex: 4282
Cables: Libtraco
Phone: 22279/22057

DETROIT DIESEL ALLISON
Blackwood Hodge (Liberia) Inc.
United Nations Drive
P.O. Box 105
Monrovia

Telex: 937-4223 Suntraclib
Cables: Suntract
Phone: 21039

KATOLIGHT
United States Trading Company
P.O. Box 140
Monrovia

KLOCKNER-HUMBOLDT-DEUTZ AG
United Liberia Rubber Corporation
Randall Street
P.O. Box 51
Monrovia

Telex: 4243
Cables: Unilibru
Phone: 21506, 21708

LEYLAND INTERNATIONAL
Swiss Africa Trading Corporation
P.O. Box 115
Bushrod Island
Monrovia

Telex: 4271
Phone: 21668

PERKINS ENGINES LIMITED
Oost Afrikaansche Compagnie (O.A.C.)
P.O. Box 281
Water Street
Monrovia

Telex: Monrovia 4233
Cables: Jupiter
Phone: 21010, 22655

Libya

ALLIS CHALMERS
Hajmohammed M. Sheibany & Co (SHEBCO)
Sciara Haiti 2
Alitalia Building
P.O. Box 626
Tripoli

CATERPILLAR OVERSEAS SA
General Company for Farm Equipment & Agricultural Necessities (GISMET-EME)
Sidi Mesri
P.O. Box 148
Tripoli

Telex: 20022
Cables: Metrade
Phone: 32520, 41237

DETROIT DIESEL ALLISON
Belashbar GM Power
P.O. Box 2238
Tripoli

Telex: 929.40031 Mobilely

DORMAN DIESELS LIMITED
Mahdi Bettamer & Sons
P.O. Box 66
8 Omar Mukhtar Street
Benghazi

Telex: 20017
Cables: Bettamer
Phone: 3135

KATOLIGHT
Gharabel Trading Agency
P.O. Box 1541
Tripoli

KIRLOSKAR OIL ENGINES LIMITED
Omar Ali Adballa Elhuni & Brothers
P.O. Box 2232
146 Omar Mukhtar Street
Tripoli

Cables: Elhuni
Phone: 36295

KLOCKNER-HUMBOLDT-DEUTZ AG
A. Bramley
P.O. Box 68
Tripoli

Phone: 40588

LEYLAND INTERNATIONAL
Mahari Trading Co. SpA
Sciara Haiti
P.O. Box 2361
Tripoli

Cables: Alram

LUCAS SERVICE OVERSEAS LIMITED
Mahari Trading Company
P.O. Box 2361
Tripoli
Cables: Alram
Phone: 31171/5

NISSAN DIESEL MOTOR CO. LIMITED
The National Company for Trade & Vehicles
P.O. Box 8456
Tripoli

PERKINS ENGINES LIMITED
Libyan Electronics Company
P.O. Box 3680
Sidi El-Masri
Tripoli
Cables: Electronco
Phone: 31840, 39462

PETTERS LIMITED
Libyan Anglo Agencies
(Shebco) Ltd
Darnis Building
Avenue Emile San Lot
P.O. Box 242
Benghazi
Telex: 40047 Laally
Cables: Laal
Phone: 2320-3509

JOHN ROBSON (SHIPLEY) LIMITED
Libyan Development & Investment Company
P.O. Box 3294
Tripoli

RUGGERINI MOTORI SpA
Agricultura Shop
Ramadam Misillati
P.O. Box 2999
Tripoli

VOLVO PENTA
I.L.C.O.
Suani Road Km 2
P.O. Box 60
Tripoli
Cables: Entesar
Phone: 30726, 32108/9

Liechtenstein

CATERPILLAR OVERSEAS SA
Ulrich Ammann Baumachinen AG
Langenthal
Switzerland
Telex: 68446
Cables: Ammannag Langenthal
Phone: (063) 2-27-02 2-51-22

Luxembourg

CATERPILLAR OVERSEAS SA
Bergerat Dutry S.A.
Zone Industrielle
Howald
Luxembourg
Telex: 2543
Cables: Bedutry
Phone: 48-12-21/22

KLOCKNER-HUMBOLDT-DEUTZ AG
Garage Magirus Deutz
Arthur Schmitz
243 route d'Esch
Luxembourg
Telex: 2346
Cables: Garagedeutz-luxemburg
Phone: 487474

Madagascar

CATERPILLAR OVERSEAS SA
Société Henri Fraise Fils & Cie
Route des Hydrocarbures
B.P. 28
Tananarive
Malagasy Republic
Telex: 22218
Cables: Fraisenri
Phone: 227-21/24

Madeira

CATERPILLAR OVERSEAS SA
STET
Sociedade Tecnica de Equipamentos e Tractores S.à.r.l.
Apartado 1351
Lisbon
Portugal
Telex: 12778
Cables: Stetra Sacavem
Phone: 251-1011

DORMAN DIESELS LIMITED
Madeira Engineering Co. Lda
Arsenal de S. Tiago
Rua D. Carlos 1,2A
Caixa Postal 528
Funchal
Cables: Arsenal
Phone: 20191/2

LEYLAND INTERNATIONAL
Leacock & Cia Ltd
Rua Major Reis Gomes 13
Caixa Postal 24
Funchal
Cables: Leacock

PERKINS ENGINES LIMITED
Leacock & Cia Lda
Rua Major Reis Gomes 13
Caixa Postal 416
Funchal
Cables: Leacock
Phone: 22101

VOLVO PENTA
Mendes Gomes & Ca Lda
Rua do Visconde de Anadia 3-4
P.O. Box 556
Funchal
Cables: Svea
Telex: 72144
Phone: 29191/3, 28118

Malagasy

CATERPILLAR OVERSEAS SA
Société Henri Fraise Fils & Cie SA
Route des Hydrocarbures
P.O. Box 28
Tananarive
Telex: 22218
Cables: Fraisenri
Phone: 227-21/24

KLOCKNER-HUMBOLDT-DEUTZ AG
5 Rue Dr Rasamimanana
P.O. Box 1248
Tananarive
Telex: 22216
Cables: Darrimax
Phone: 20611, 23741

LEYLAND INTERNATIONAL
Societe Industrielle & Commerciale de L'Emyrne (SICE)
P.O. Box 1078
Rue Clemenceau
Tananarive
Telex: 216
Cables: Sicegar
Phone: 237-41

LUCAS SERVICE OVERSEAS LIMITED
Ateliers De Mechanique Automobile A.Mec A
Lalana Ravoninahitriniarivo
P.O. Box 392
Tananarive
Telex: 22271
Cables: Ameca
Phone: 230-16

PERKINS ENGINES LIMITED
AMECA
Ateliers De Mecanique Automobile
Routes Des Hydrocarbures
P.O. Box 392
Tananarive
Telex: 222-71 Ameca Tana
Cables: Ameca
Phone: 230-16

PETTERS LIMITED
Ets Modrin & Cie
P.O. Box 23
Tananarive
Cables: Modrin
Phone: 01-47

Malawi

CATERPILLAR OVERSEAS SA
Cesco Ltd.
Stadium Road
P.O. Box 526
Blantyre
Telex: 240
Cables: Cesco
Phone: 30166

DETROIT DIESEL ALLISON
Detroit Diesel Allison International
c/o General Motors South African (Pty) Ltd
Aloes Plant (near Port Elizabeth)
P.O. Box 1137
Port Elizabeth 6000
Telex: 74-7684
Cables: Portautoex
Phone: 6-2384, 6-1131 Port Elizabeth

DORMAN DIESELS LIMITED
Brown & Clapperton Limited
P.O. Box 52
Blantyre
Cables: Beance
Phone: 2352/4

LEYLAND INTERNATIONAL
The Leyland Motor Corporation (Malawi) Ltd
P.O. Box 581
Blantyre
Telex: Blantyre 4138
Cables: Leymotor
Phone: 30477

LUCAS SERVICE OVERSEAS LIMITED
Brown and Clapperton Ltd
P.O. Box 52
Blantyre
Telex: 4243
Cables: Beance
Phone: 34677

PERKINS ENGINES LIMITED
Farming & Engineering Services Limited
Hanover Street
P.O. Box 918
Blantyre
Cables: Dixie
Phone: 2923

PETTERS LIMITED
Farming & Engineering Services Limited
Hanover Street
P.O. Box 918
Blantyre
Cables: Dixie
Phone: 35044

Malaysia

BRIGGS & STRATTON CORPORATION
United Motor Works (M) Sdn. Bhd
Batu Tiga Industrial Estate
P.O. Box 52
Sungei Rengam
Shah Alam
Selangor
Telex: MA 38649
Cable: Unity
Phone: 361911

CATERPILLAR OVERSEAS SA
Tractors Malaysia Berhad
Bangunan Sharidal
Jalan Yong Shook Lin
Section 7
P.O. Box 2465
Petaling Jaya

DETROIT DIESEL ALLISON
General Diesel Malaysia Sdn. Bhd (Main Office)
2 Jalan President Kennedy
Lahat Road
Ipoh
Telex: Genreco RS: 786-21962 Singapore
Cable: Gedes
Phone: 76455

DORMAN DIESELS LIMITED
The General Electric Company of Malaysia
Sendirian Barhad
Diesel Division
P.O. Box 256
Kuala Lumpur
Telex: KL335
Cables: Geclumpur
Phone: 84291/5

FORD MOTOR CO. LIMITED
Ford Motor Co. Pte. Ltd
Malaysia Branch
Sub Lot 11
Jalan Paku 2/6
Kawasan Miel Phase II
Batu Tiga
Selangor
W. Malaysia
Phone: 361900, 361909

INDIAN NATIONAL DIESEL ENGINE CO. LIMITED
Chop Lim Hup Choon
Sibu
Sarawak

KATOLIGHT
NEB
P.O. Box 1003
Kuala Lumpur 22-06

KLOCKNER-HUMBOLDT-DEUTZ AG
Industrial Agricultural Div.
Borneo Motors Sdn. Bhd
Lot 1, Jn 13/2
Petaling Jaya
Telex: Borneo MA 30334
Phone: 51716, 51715, 772629

KOHLER INTERNATIONAL LIMITED
Sandilands Sdn.Bhd
P.O. Box 159
3 Jalan 13/2
Petaling Jaiya Selangor

LEYLAND INTERNATIONAL
Wearne Bros. Malaysia Sdn. Bhd.
P.O. Box 2323
Kuala Lumpur
Telex: Wearne MA 30604

LUCAS SERVICE OVERSEAS LIMITED
Far East Motors
P.O. Box 81
3 Jalan Robertson
Kuala Lumpur
Telex: Wearnes 30604
Cables: Ignition
Phone: 80381/3

NISSAN DIESEL MOTOR CO. LIMITED
Malaysia Tractor & Equipment Sdn. Bhd.
P.O. Box 1206
Sandakan
Sabah

PERKINS ENGINES LIMITED
Perkins Engines Eastern (Malaysia) Sdn. Bhd.
35 Jalan Segambut
Tehgah
Kuala Lumpur 12-07

PETTERS LIMITED
William Jacks & Co. (Malaysia) Sdn. Bhd.
Jalan Bersatu Road 13/4
Petaling Jaya
P.O. Box 286
Kuala Lumpur
Telex: 8430319
Cables: Expanded
Phone: 564121

Mali

CATERPILLAR OVERSEAS SA
Manutention Africaine
P.O. Box 143
Bamako
Telex: 565
Cables: Mea
Phone: 22957, 22549

KLOCKNER-HUMBOLDT-DEUTZ AG
Valcke Frères
Untervertretung: Sté Sofica
P.O. Box 1783
KN 4 — Route de Rufisque
Dakar
Telex: 663 Sofica SG
Phone: 50063, 50537

LEYLAND INTERNATIONAL
Etaperu Soudan
5 Rue Mohamed
P.O. Box 83
Bamako
Cables: Etaperu

PERKINS ENGINES LIMITED
Etabs. Simaga
P.O. Box 27
Segou
Phone: Agence 247.69, Escale 241.47

Malta

DETROIT DIESEL ALLISON
Malta Marine Repairs & Services Ltd
Mariner 1 Hay Wharf
Pieta Creek
Pieta
SA Maison
Cables: Mariner
Phone: 20556

DORMAN DIESELS LIMITED
T. Pace & Co.
Malta Marine Engineering Works
20/24 Zerafa Street
Marsa
Cables: Repairs
Phone: Central 24139

KLOCKNER-HUMBOLDT-DEUTZ AG
Untervertretung I.M.L.
Ronasons International Ltd
P.O. Box 50
113 Britannia Street
Valletta
Cables: 813 Ronaso MT
Phone: 20072, 22956

LEYLAND INTERNATIONAL
Muscat's Garage Ltd
Rue D'Argens
Msida
Telex: MW 232
Cables: Automobile
Phone: 33041

PERKINS ENGINES LIMITED
International Automobiles
20 Republic Street
Valetta
Cables: Interauto
Phone: 30001/2/3

PETTERS LIMITED
Joseph Bezzina & Co. Ltd
44/45 Jetties Wharf
Marse
Cables: Darmzeb
Phone: 139

RUGGERINI MOTORI SpA
Michael Galea
41 Fisher Road
Magat

Martinique

CATERPILLAR OVERSEAS SA
Garage Americain
Ets. Louis Crocquet
38 Avenue Duparquet 38
P.O. Box 579
Fort-De-France
Telex: 02935 MR
Cables: Amgarage
Phone: 71-54-54

DETROIT DIESEL ALLISON
Manufactures Reunies
37 Rue Bertin
Fort-De-France
Cables: Pamaru Martinique
Phone: 35-28

DORMAN DIESELS LIMITED
Guy Vieules
P.O. Box 233
Fort-De-France
Telex: Fort-De-France 665
Phone: 54-69

KLOCKNER-HUMBOLDT-DEUTZ AG
Valcke Freres
Untervertretung:
SMR – SOCOMI
Zone de Jambette Lamentin
P.O. Box 213
Fort-De-France
Telex: 029633 MR – Chamco
Phone: 719363

KOHLER INTERNATIONAL LIMITED
Ets. Rene & Henry Dormoy
P.O. Box 472
Pointe Simon
Fort-De-France

LEYLAND INTERNATIONAL
Givvaa Garaga Austin
Zone Industrielle Voie No.2
Lamentin
Martinique
Telex: 29631 Mr. Traser
Phone: 74.16.28

LUCAS SERVICE OVERSEAS LIMITED
Antilles Mecanique
P.O. Box 497
Zone Industrielle
Lamentin
Fort-De-France
Telex: 029663
Cables: Monplasir
Phone: 741338, 741692

PERKINS ENGINES LIMITED
Ets. Louis Croquet
P.O. Box 579
38 Avenue Duparquet
Fort-De-France
Telex: 635 Amgarage
Cables: Amgarage
Phone: 71-54-54

PETTERS LIMITED
Ets. Rene & Henry Dormoy
P.O. Box 472
Fort-De-France 12.033
Telex: Hermay 029648 MR
Cables: Remoy
Phone: 71.95.16

PERKINS ENGINES LIMITED
Ets. Louis Croquet
P.O. Box 579
38 Avenue Duparquet
Fort-De-France
Telex: 635 Amgarage
Cables: Amgarage
Phone: 71-54-54

Mauritania

CATERPILLAR OVERSEAS SA
Somatrac
P.O. Box 164
Nouakchott
Telex: 571
Phone: 2188

KLOCKNER-HUMBOLDT-DEUTZ AG
Valcke Freres
Untervertretung: ACD
Atelier Chantier de Dakar
P.O. Box 170
Av. Félix Eboué
Dakar
Telex: 420 ACD SG
Phone: 50558

LEYLAND INTERNATIONAL
Cotema
P.O. Box 313
Nouakchott

LUCAS SERVICE OVERSEAS LIMITED
Compagnie Technique Mauritanienne
(Cotema)
P.O. Box 313
Nouakchott
Telex: 88
Cables: Unamotors
Phone: 23-52

PERKINS ENGINES LIMITED
Cotema
Rte du Ksar
P.O. Box 313
Nouakchott
Telex: Cotema NKC 88
Phone: 23.52

PETTERS LIMITED
Hamelle R.I.M.
Route d'Akjoujt
P.O. Box 382
Nouakchott
Telex: Locombe 556
Cables: Hamelaf
Phone: 5.25.38

Mauritius

CATERPILLAR OVERSEAS SA
Blyth Brothers & Co. Ltd
P.O. Box 341
Port Louis
Telex: IW 211
Cables: Ibel Mauritius
Phone: 2-0265

DETROIT DIESEL ALLISON
Doger De Speville & Co. Ltd
P.O. Box 100
Port Louis
Cables: Ybrat Mauritius

DORMAN DIESELS LIMITED
Ireland Fraser & Co. Ltd
P.O. Box 58
Port Louis
Cables: Ireland
Phone: 2-2811

KLOCKNER-HUMBOLDT-DEUTZ AG
MECOM Mechanisation Co. Ltd
13 Eugene Laurent
P.O. Box 694
Port Louis
Telex: 317 IW
Phone: 2-4873/4

LEYLAND INTERNATIONAL
Rogers & Co. Ltd
P.O. Box 60
Port Louis
Telex: IW 221
Cables: Finance
Phone: 1281

LUCAS SERVICE OVERSEAS LIMITED
Rey and Lenferna Ltd
10 Edith Cavell Street
Port Louis
Telex: 297
Cables: Reyfer
Phone: 2-0261

PERKINS ENGINES LIMITED
Hall Geneve Langlois Ltd
P.O. Box 77
Port Louis
Telex: IW 245
Cables: Hagelan
Phone: Port Louis 2-0211/

PETTERS LIMITED
Forges Tardieu Ltd
P.O. Box 20
31 Route Nicolay
Port Louis
Cables: Tardieu Port Louis
Phone: Port Louis 194

WHITE ENGINES INCORPORATED
Harel, Mallac & Co. Ltd
P.O. Box 36
Port Louis
Telex: 219 Port Louis
Cables: Lerca

Mexico

ALLIS CHALMERS
Standard Machinery & Supply Co. S.A.
Boulevard Manuel Avila Comacho No. 112
Apartado Postal 139
Nancalpan de Juarez
Estado de Mexico
Phone: 5-76-35-42
5-76-35-46

BRIGGS & STRATTON CORPORATION
Sociedad Electro Mecanica SA
Manuel Ma. Contreras 25
Apartado Postal 42-164
Mexico 4, D.F.
Telex: 017-71-137
Cable: Selmex
Phone: 5-66-36-00

CATERPILLAR OVERSEAS SA
Maquinaria S.A.
Carretera a Avalos y Calle 2A
Apartado Postal 394
Chihuahua
Telex: Maqsa Chi-034837
Phone: 5-00-49, 5-00-51

CATERPILLAR OVERSEAS SA
Caterpillar Mexicana S.A. de CV
P.O. Box 2781
Monterrey, N.L.
Telex: 038-794 (from U.S.)
Phones: 46-45-32/3/4/5

DETROIT DIESEL ALLISON
Detroit Diesel Allison International – Mexico
c/o General Motors de Mexico, SA de CV
Emerson 432 3rd Floor
Mexico 5, D.F.
Telex: 017-73835
Cables: Genmotomex,
Phone: 531-25-21,
531-94-57

FORD MOTOR COMPANY LIMITED
Ford Motor Company SA
Paseo de lar Reforma 333
Apartado 39 Bis
Mexico IDF
Telex: 7860
Phone: 625 9200

KLOCKNER-HUMBOLDT-DEUTZ AG
Dr. Rio de la Losa 68
Apartado Postal 31-505
MEX – Mexico 7, D.F.
Telex: Otto-Motor/ME
Cables: Ottomotor
Phone: 5787310
1775604

LEYLAND INTERNATIONAL
Refacciones Inglesas De Mexico SA
Dinamarca 64-66
Mexico 6, D.F.
Cables: Refimsa City
Phone: 546-67-27,
535-12-42

LUCAS SERVICE OVERSEAS LIMITED
Electro Diesel de Mexico SA
Arenal 110
Apartado Postal 22-164
Mexico 22, D.F.
Cables: Edemsa
Phone: 5.73.51.00

PERKINS ENGINES LIMITED
Motores Perkins SA
Tiber No 68 ler Piso
Mexico 5, D.F.
Telex: Perkoil Mex
017-71-347
Cables: Perkoilmex
Phone: 528-61-67

PETTERS LIMITED
Cummins de Mexico SA
Norte 35 No.1015
Col. Industrial Vallejo
Mexico City 14, D.F.
Cables: Cummins, Mexico City
Phone: 5-67-37-00

VOLVO PENTA
Motor-Equipos SA
Alemania 14
Mexico 21, D.F.
Telex: 017-7-960
Cables: Motesa
Phone: 49 32 65

WHITE ENGINES INCORPORATED
Sociedad Electro Mecanica SA
Manuel Maria Contreras 25
Apartado Postal 42-164,
Mexico 4, D.F.
Telex: 017-71-137 Selmec Mex
Phone: 566-36-00

Montserrat

LEYLAND INTERNATIONAL
United Motors Limited
Empire House
P.O. Box 224
Plymouth
Montserrat

PETTERS LIMITED
O.R. Kelsick & Co. Ltd
P.O. Box 166
Plymouth
Montserrat
Cables: Orkel
Phone: 2205

Morocco

BRIGGS & STRATTON CORPORATION
Etablissements Frendo
Societe Morocaine De Construction de Materiel Agricole
96 Boulevard de Khouribga
Casablanca
Telex: 21617
Cables: Agricultore-Casa
Phone: 63411-634-12

CATERPILLAR OVERSEAS SA
Société Marocaine des Ets.
P. Parrenin
145 Bd. de la Résistance
Casablanca
Telex: 21733
Cables: Paragri
Phone: 24-06-75

DETROIT DIESEL ALLISON
Detroit Diesel Allison International – Europe
c/o General Motors France
56 a 58 Ave. Louis Roche
92231 Gennevilliers
France
Telex: Genlmot 62050
Cables: Parautexap Gennevilliers
Phone: 790-70-60 Gennevilliers

DORMAN DIESELS LIMITED
Technical Equipments
121 Bd Emile Zola
P.O. Box 612
Casablanca
Telex: 21860 Techequip Casa
Cables: Techequip
Phone: 406-21/22

KATOLIGHT
Sorecotec
19 Rue Hadjamar Riffi
Casablanca

KLOCKNER-HUMBOLDT-DEUTZ AG — Telex: 21690; Cables: Magideutz; Phone: 2415-44, 243022-23, 2416-98-99
Novelle Société
Magideutz SA
4-8 Rue Layris Verger
P.O. Box 24
Casablanca

KOHLER INTERNATIONAL LIMITED
Societe Fenie Brosette
650 Boul Mohammed V
Casablanca

LEYLAND INTERNATIONAL — Phone: 51-16-60
Afric Auto
147 Rue Mustafa El Maani
Casablanca
Mr. Jose Ballesteros Velasco
Borras Y Ballesteros
Calvo Sotelo 60
Ceuta
Spanish Territories
North Africa

LUCAS SERVICE OVERSEAS LIMITED — Telex: 21859 SAA; Cables: Standard; Phone: 2786.23/24
Standard-Auto-Accessoires SA
149 Rue Mohamed Smiha
Casablanca

PERKINS ENGINES LIMITED — Cables: Garagecalpe; Phone: 376-77
Calpe Garage Exploitation SA
23 Rue Victor Hugo
Tangier

Dimateq — Telex: 21907; Phone: 24.48.96/97
83 Boulevard de la Resistance
Casablanca

PETTERS LIMITED — Telex: Hamelaf 23841; Phone: 620.51/2/3/4
Hamelle-Maroc SA
35 Boulevard Hassan-Seghir
P.O. Box 682
Casablanca

Technical Equipments SA — Telex: 21860; Cables: Techequip. R.C. Casa 26.0
P.O. Box 612
121 Boulevard Emile Zola
Casablanca

JOHN ROBSON (SHIPLEY) LIMITED
Technical Equipments
P.O. Box 612
121 Boulevard Emile Zola
Casablanca

WHITE ENGINES INCORPORATED — Telex: 21.815; Cables: Piradio; Phone: 702-61
F. Pignal
21/29 Boulevard Girardot
Casablanca

Mozambique

CATERPILLAR OVERSEAS SA — Telex: 6241 (393); Cables: Remoterra Maputo; Phone: 24187/88
Sociedade Tecnica de Equipamentos Industriais e Agricolas Lda. (STEIA)
P.O. Box 2864
Maputo

DETROIT DIESEL ALLISON INTERNATIONAL — Telex: 74-7684; Cables: Portautoex Port Elizabeth; Phone: 6-2384 or 6-1131 Port Elizabeth
c/o General Motors South African (Pty.) Ltd.
Aloes Plant (near Port Elizabeth)
P.O. Box 1137
Port Elizabeth 6000

DORMAN DIESELS LIMITED — Telex: 6-440FBC MO; Cables: Bridler Can Phumo; Phone: 23031
F. Bridler & Ca Ltda
P.O. Box 65
Can Phumo

L. GARDNER & SONS LIMITED — Phone: 2587-2687
The Beira Engineering Co. Ltd (Marine)
P.O. Box 363
Beira

KATOLIGHT
Sadem
Av. 31 de Janeiro 70
Caixa Postal 2447
Maputo

KLOCKNER-HUMBOLDT-DEUTZ AG — Telex: 6-388 Comel Mo; Cables: Comela; Phone: 27110, 27114
Comel
Consorcio de Maquinas & Electricidade Ltda
P.O. Box 459
Maputo

LEYLAND INTERNATIONAL — Cables: Camiauto; Phone: Camiauto No.6-225
Companhia Distribuidora de Automoveis SARL (Codauto)
Caixa Postale 2510
Maputo

LUCAS SERVICE OVERSEAS LTD — Telex: 6-268; Cables: Agmod; Phone: 733175
Agencias Modernas Ltda
P.O. Box 1066
Avenida de Angola 3016
Maputo

PERKINS ENGINES LIMITED — Telex: 01-268 Penco Mo; Cables: Agmod; Phone: 733175
Agencias Modernas Lda
Caixa Postal No. 1066
Av. de Angola 3016
Maputo

PETTERS LIMITED — Cables: Produce; Phone: 3154
Breyner & Wirth (Beira) Lda
Caixa Postal 293
Rue Costa Serrao 158/166
Beira

RUGGERINI MOTORI SpA
Inagrico Lda
Caixa Postal 1809
Avenida General Machado 1230
Maputo

VOLVO PENTA — Cables: Delvemar; Phone: 26507, 28798
Bonanza
Del Ré & Velloza L da
Av. Paiva Manso 343/347
Caixa Postal 2480
Maputo

Namibia

BRIGGS & STRATTON CORPORATION — Telex: 56-720; Cables: Emzetmotor; Phone: 25401
M & Z Motors & Engineering Ltd.,
P.O. Box 192
Windhoek

CATERPILLAR OVERSEAS SA — Telex: 56-725, 56-875; Cables: Shipments; Phone: 32026
Barlow's S.W.A. Tractor Company
Republiek Road
Windhoek 9100

DETROIT DIESEL ALLISON — Telex: 74-7684; Cables: Portautoex; Phone: 6-2384 or 6-1131
Detroit Diesel Allison International
c/o General Motors South African (Pty.) Ltd.
Aloes Plant (near Port Elizabeth)
P.O. Box 1137
Port Elizabeth 6000
Republic of South Africa

KLOCKNER-HUMBOLDT-DEUTZ AG
Magirus-Deutz (Southern Africa) (Pty.) Ltd.
P.O. Box 365
Wrench Road
Isando Travsvaal 1600
Telex: 86433
Cables: Deutzmotor
Phone: 365381/7

PETTERS LIMITED
Barswa Motors Agric. Dept.
P.O. Box 1708
Windhoek
Telex: 56-871WK
Cables: Barmech
Phone: 2-3261

Nepal

CATERPILLAR OVERSEAS SA
Tractors India Limited
1 Taratolla Road
Garden Reach
P.O. Box 323
Calcutta 700 024
Cables: Diesels

KATOLIGHT
Timber Corporation of Nepal Ltd.
Babar Mahal
Kathmandu

KIRLOSKAR OIL ENGINES LIMITED
Kalpana Trading Co. Ltd.
P.O. Box 83
Kathmandu
Telex: NP 205 AAPU
Cables: Kalpana
Phone: 12-335

LEYLAND INTERNATIONAL
Nepal Construction & Engineering Corporation (Priv.) Ltd.
Dwalkha Chhen
8/324 Pyukha Tole
P.O. Box 156
Kathmandu
Cables: Necoenco

LUCAS SERVICE OVERSEAS LIMITED
Bhajuratna Engineering & Sales (P) Ltd.
P.O. Box 133
Jyoti Bhawan
Kantipath
Kathmandu
Cables: Himaliron
Phone: 11490 and 14327

NISSAN DIESEL MOTOR CO. LIMITED
Nepal Impex (P) Ltd
Ramshah Path
Kathmandu

Netherlands

BOMBARDIER-ROTAX Gmbh
Bonenkamp B.V.
Castellumlaan 2-4-6
De Meern
Phone: 03406-1246/2324

BRIGGS & STRATTON CORPORATION
Motor Snelco B.V.
HJE Wenckebachweg 137-139
Amsterdam 0
Telex: 15272 Mosco
Cables: Motor Snelco
Phone: (020) 932808

CATERPILLAR OVERSEAS SA
Geveke Motoren en Grondverzet B.V.
Spaklerweg 45
Industrieterrein Amstel
Postbus 4091
Amsterdam 0
Telex: 13106
Cables: Globetract
Phone: (020) 94-32-32

DETROIT DIESEL ALLISON INTERNATIONAL – EUROPE
c/o General Motors Continental S.A. Nederland
Parmentierplein 1
P.O. Box 5061
NL-Rotterdam 22
Telex: 21355 or 23018
Cables: Cartruck
Phone: Rotterdam 29-00-00

DORMAN DIESELS LIMITED
BV Ingenieursbureau de Roos Beheer
P.O. Box 1813
Koninginnegracht 92
Den Haag,
Telex: 31215 A/B Roos NL
Cables: Abezet
Phone: (070) 649830

FORD MOTOR CO. LIMITED
N.V. Nederlandsche Ford Automobiel Fabriek
Postbus 795
Amsterdam
Telex: 12303
Phone: 5869111

L. GARDNER & SONS LIMITED
Geveke Elektronica en Automatie N.V.
Nautical Dept.,
P.O. Box 5232
Rotterdam
Telex: 22225
Phone: (010) 29 24 44

KATOLIGHT
MRC Zuid Hollandse Motoren Revisie B.V.
P.O. Box 63008
Rotterdam

KLOCKNER-HUMBOLDT-DEUTZ AG
Deutz Motoren B.V.
Sluisjesdijk 145
Postbus 267
Rotterdam
Telex: 28079
Cables: Deutzmotor
Phone: 298544

LEYLAND INTERNATIONAL
British Leyland Nederland CV
P.O. Box 204
Niewe Gouwe OZ 40
Gouda
Telex: 21553
Cables: Britleyned
Phone: 1820-11122

PERKINS ENGINES LIMITED
Kemper En Van Twist Diesel B.V.
P.O. Box 156
Mijlweg 33
Dordrecht
Telex: 22-124 KVT NL
Cables: Diesel
Phone: 078-30155

PETTERS LIMITED
Petters Continental B.V.
Bunschotenweg 122-126
Rotterdam
Telex: 22344
Cables: Petters
Phone: 010-290333

RUGGERINI MOTORI SpA
Sim Holland B.V.
Antwerpseweg 10
Postbus 60
Gouda
Telex: 26099 Simwo NL

VOLVO PENTA
Nebim Handelmaatscappij B.V.
Zwarteweg 77
P.O. Box 3
Aalsmeer
Telex: 10555
Phone: 02977-24652

WHITE ENGINES INCORPORATED
Petters Continental B.V.
Bunschotenweg 124-126
Rotterdam 3022
Telex: 22344 Petters
Phone: 010-29 03 33

Netherlands Antilles

CATERPILLAR OVERSEAS SA
PBC Machines & Services Inc.
P.O. Box 157
Curacao
Telex: 1159 Pbaiz NA
Cables: Equipment
Phone: 13981, 13866

DETROIT DIESEL ALLISON
N.B. Winkel's Handel Maatschappi
P.O. Box 8
Oranhestad
Cable: Winruba

DORMAN DIESELS LIMITED
Kusters Trading Co. (Curacao) Inc.
P.O. Box 41
Concordia Street No.30
Curacao
Telex: CU6
Cables: Kusters
Phone: 11853-11129

KATOLIGHT
Ferreira Sales and Service
Siouxweg 5
Willemstad,
Curacao

KOHLER INTERNATIONAL LIMITED
S.E.L. Maduro & Sons (Aruba) Inc.
P.O. Box 36
Orenjestad
Aruba

LEYLAND INTERNATIONAL
N.V. Warenhuis Van Der Ree
30 A.L. Boulevard
Aruba
Cables: Vanderree

LUCAS SERVICE OVERSEAS LIMITED
Viana Auto Supply Co. Inc.
P.O. Box 362
San Nicholas
Aruba
Cables: Vasco
Phone: 5120/1

PERKINS ENGINES LIMITED
Wellman Auto Supplies N.V.
P.O. Box 64
Schottegatweg Oost 156-158
Willemstad
Cables: Wellman
Phone: 54012/54354/54289

PETTERS LIMITED
Emile Tackling
P.O. Box 17
Phillipsburg
St. Maarten
Cables: Tackling
Phone: 2327-9

WHITE ENGINES INCORPORATED
Bell Caribbean N.W.
P.O. Box 136
Curacao
Cables: Belicarib
Phone: 93082

Nevis

DETROIT DIESEL ALLISON
S.L. Horsford & Co. Ltd
Agent of Tugs & Lights Ltd.
Trinidad
Basseterre
P.O. Box 45
St. Kitts
Telex: 387284 Trinidad
Cables: Tuglighter
Phone: 2085

New Caledonia

CATERPILLAR OVERSEAS SA
Societe Caledonienne des Tracteurs
Caltrac S.A.
Complexe Edward Pentecost – PK5
Magenta
B.P. C2 Noumea Cedex
Noumea
Cables: Penocean 051 NM

DETROIT DIESEL ALLISON
Cemic S.A.
Societe Commerciale d'Equipement Mecanique
P.O. Box 2059
Noumea
Telex: 714-078
Cables: Cemic
Phone: 739-48

KLOCKNER-HUMBOLDT-DEUTZ AG
Joseph Cheval & Cie
3 Rue de Jean Jaurès
B.P. 100
Noumea
Phone: 73101

LEYLAND INTERNATIONAL
Societe Commerciale d'Equipement Mecanique
CEMIC
Lot 121, Duocos,
Noumea

LUCAS SERVICE OVERSEAS LIMITED
Groupement des Enterprises Berton
B.P. 309
Noumea

NISSAN DIESEL MOTOR CO. LIMITED
Royal Motors
P.O. 981

PERKINS ENGINES LIMITED
Pacific Motors S.A.
24 Rue General Gallieni
P.O. Box C2
Noumea
Telex: Penocean 051 NM
Cables: Penocean
Phone: 757.44 & 757.45

PETTERS LIMITED
ABC Imports
P.B. 617
Noumea
Phone: 20-49

New Hebrides

DETROIT DIESEL ALLISON
D.J. Gubbay & Co. (New Hebrides) Pty. Ltd
P.O. Box 223
Vila
Cables: Trade, Luganville

LEYLAND INTERNATIONAL
Societe Des Automobiles Tracteurs & Materiel Agricole – Nouvelle Hebrides,
(SATMA – NH)
Rue Higginson
Lieu dit Thebakor
Port Villa
New Hebrides
Telex: Satma BP 94
Cables: Pacific Real Estates Villa New

New Zealand

ALLIS CHALMERS
Cable-Price Corporation Limited
C.P.D. House,
108 The Terrace
P.O. Box 10042
Wellington
Telex: NZ3438
Cables: Caprice
Phone: 46-835

BRIGGS & STRATTON CORPORATION
Brown Brothers Ltd.,
P.O. Box 548
Christchurch C.1.
Telex: 4941
Cables: MAZDA
Phone: 65-195

CATERPILLAR OVERSEAS SA
Gough Gough and Hamer Ltd.
24-26 Amyes Road
P.O. Box 16-168
Hornby
Cables: Goughs NZ 4889

DETROIT DIESEL ALLISON
Detroit Diesel Allison International
c/o General Motors New Zealand Ltd.
Railway Avenue
Parts Power & Appliances Dept.
Private Bag
Upper Hutt
Telex: NZ3521
Cables: Genmotor
Phone: 2-7055

DORMAN DIESELS LIMITED
John Burns (Engineering) Ltd
P.O. Box 14445
Carbine Road
Panmure
Auckland 6
Telex: NZ 2481
Cables: Burnseng
Phone: 576-119, 576-604, 575-055

FORD MOTOR CO. LTD.
Ford Motor Company of
New Zealand Limited
Private Bag
Auckland
New Zealand

L. GARDNER & SONS LIMITED — Telex: NZ 4277 — Phone: 60-599
Andrews & Beaven Limited
Private Bag
203-205 Cashel Street
Christchurch

KLOCKNER-HUMBOLDT-DEUTZ AG — Telex: 2340 — Phone: 662119, 665139
Mason Mesco Ltd.
Head Office
P.O. Box 3485
17 Maurice Road
Penrose
Auckland

KOHLER INTERNATIONAL LIMITED
Agency Distributors
(International) Ltd.
P.O. Box 61-133
Apirana Avenue
Otaro
Auckland

LEYLAND INTERNATIONAL — Telex: NZ 3432 — Cables: Mocorho — Phone: 844 039
New Zealand Motor
Corporation Ltd.
P.O. Box 2599
89 Courtenay Place
Wellington 1

LUCAS SERVICE OVERSEAS LIMITED — Cables: Aukas Auckland
Lucas Industries New Zealand Ltd.
309-321 Broadway
Newmarket
Private Bag
Auckland 1

NISSAN DIESEL MOTOR CO. LIMITED
Nissan Motor Distributors (N.Z.) Ltd.,
P.O. Box 61133
Otara
Auckland

PERKINS ENGINES LIMITED — Telex: NZ 3387 — Cables: Clydering — Phone: 698-169
Clyde Engineering Ltd.
P.O. Box 30480
415-427 Hutt Road
Lower-Hutt

PETTERS LIMITED — Telex: NZ 3455 — Cables: Dredger — Phone: 557-728
Richardson McCabe & Co. Ltd.
P.O. Box 792
216/218 Willis Street
Wellington

VOLVO PENTA — Telex: NZ 2991 — Cables: Volpower — Phone: 362-797/8
Scandinavian Motors Limited
115 Beach Road
P.O. Box 1903
Auckland 1

Nicaragua

ALLIS CHALMERS — Telex: 342/0025 — Cables: McGregor — Phone: 6726
Casa Commercial McGregor A.A.
3A Calle S.E. No. 104
Apartado 448
Managua

BRIGGS & STRATTON CORPORATION — Telex: 1336 — Cables: Ludeca — Phone: 40973-77 & 4611
Maquinaria Ludeca S.A.
Apartado 861
Managua

CATERPILLAR OVERSEAS SA — Telex: 5811 — Cables: Nimac — Phone: 3151, 3159
Nicaragua Machinery Company
P.O. Box 469
Managua

DETROIT DIESEL ALLISON — Telex: 3021015 — Cables: Automotive — Phone: 60092
Automotive & Industrial
Equipment Co.,
P.O. Box 323
Managua

DORMAN DIESELS LIMITED — Cables: Cross — Phone: 60114-5
H.F. Cross S.A.
Apartado Postal 1169
Managua D.N.
Nicaragua

KATOLIGHT
Samqui y Asociados
Consultores S.A.
Apartado 1005
Managua

KOHLER INTERNATIONAL LIMITED
Casa Cross S.A.
Apartado 1169
Managua

LEYLAND INTERNATIONAL — Telex: 1237 H.F. Cross — Cables: Cross — Phone: 60114-5
H.F. Cross S.A.
P.O. Box 1169
Managua D.N.

LUCAS SERVICE OVERSEAS LIMITED — Telex: 3421327 — Cables: Cross — Phone: 60114/5
H.F. Cross S.A.
Apartado 1169
Managua

NISSAN DIESEL MOTOR CO. LIMITED
Distribuidora Datsun S.A.
Boulevardo Kennedy
Carretera Norte Km 4½
Apartado Postal 3680
Managua D.N.

PERKINS ENGINES LIMITED — Cables: Cross — Phone: 3485-6-7
H.F. Cross, S.A.
Apartado Postal 1169
Calle 15 de Septembre
Managua D.N.

PETTERS LIMITED — Telex: 1025 McGregor — Cables: McGregor, — Phone: 2-5306-07 — Phone: 2-5306-07
Casa Commercial McGregor S.A.
Apartado 448
Carretera Sur, Kilometre 4,
Managua

Niger

CATERPILLAR OVERSEAS SA — Telex: 5234 — Phone: 72-20-11/12
Manutention Africanine
B.P. 136
Niamey

KLOCKNER-HUMBOLDT-DEUTZ AG — Telex: Nigretia 5227 NI — Phone: 723319
Valcke Frères
Matforce Diesel
B.P. 165
Niamey

LEYLAND INTERNATIONAL — Cables: Torfyrmot
Societe Commerciale De L'Ouest
Africain
Niamey

LUCAS SERVICE OVERSEAS LIMITED — Phone: 33.17, 30.38
Niger Afrique
B.P. 62
Niamey

PERKINS ENGINES LIMITED
Compagnie Transafricaine
Boite Postale 246
Niamey

Telex: 237 Transacit Niamey
Cables: Transacit
Phone: 29.25/6

RUGGERINI MOTORI S.p.A
El Hadji Nagnou Amadou Bonkoukou
B.P. 323
Niamey

Nigeria

ALLIS CHALMERS
Scoatrac – A Division of
Scoa (Nigeria) Ltd.
Dual Carriage Way
Isolo Industrial Estate
Isolo
P.M.B. 1108
Ikeja
Lagos

Telex: 21376 Torfymot
Cables: Torfymot
Phone: 48144-45249

BOMBARDIER-ROTAX GMBH
African Afronic Merchants Co.
P.O. Box 1541
Lagos

BRIGGS & STRATTON CORPORATION
Reiss & Co. (Nigeria) Ltd.
71 Apapa Road
Ebute Metta
P.O. Box 678
Lagos

Telex: 21161
Cables: Reico
Phone: 44348 & 21258 20956

CATERPILLAR OVERSEAS SA
Tractor & Equipment Division of the
UAC of Nigeria Ltd.
P.M.B. 1015
Ebute-Metta
Lagos

DETROIT DIESEL ALLISON
Diesel Sales & Service
(Nigeria) Ltd.
15 Burma Road
P.O. Box 109
Apapa

Telex: 961-21393 Suntra
Cables: Suntract Lagos
Phone: 47107 or 47049

DORMAN DIESELS LIMITED
Machinery & Electrical Equipment
Div. of U.A.C. (Technical) Ltd.
P.M.B. 1015
Taylor Road
Iddo
Ebute Metta
Lagos

Cables: Mequip
Phone: 43310

J.L.O.
Western Nigerian Technical Co. Ltd.
P.M.B. 5148
Ibadan
Dugbe

Telex: 31111
Cables: Wenitra
Phone: 22361/2

KATOLIGHT
D. Nhilene (Pty) Ltd.
23 Capt. Amangala
Port Harcourt

KLOCKNER-HUMBOLDT-DEUTZ AG
Nigerian Technical Comp. Ltd.
Head Office
11-13 Warehouse Road
P.O. Box 356
Apapa
Lagos

Telex: 21476 Niteco
Cables: Niteco
Phone: 45131, 45542, 48074

KOHLER INTERNATIONAL LIMITED
R.T. Briscoe (Nigeria) Ltd
Technical Dept.
P.O. Box 2104 Lagos
22 Creek Road
Apapa Lagos

LEYLAND INTERNATIONAL
Bewac Ltd.
P.M.B. 1016
1 Commercial Road
Apapa

Cables: Bewac

LUCAS SERVICE OVERSEAS LIMITED
Lucas House
Division of UAC of Nigeria Ltd.
P.O. Box 344
19 Creek Road
Apapa

Cables: Unalucas
Phone: 46997

NISSAN DIESEL MOTOR CO. LIMITED
CFAO Motors
P.M.B. 2344
Lagos

PERKINS ENGINES LIMITED
Stokvis (Nigeria) Ltd.
Perkins Department
P.O. Box 136
1 Dawodu Lane
Ebute Metta

Cables: Toolco
Phone: 44609/43391

PETBOW LIMITED
Holt Engineering
24 Creek Road
P.O. Box 217
Apapa

PETTERS LIMITED
Nigerian Motors Industries Ltd.
P.M.B. 1032
26 Wharf Road
Apapa

Cables: Nigermot
Phone: Lagos 46848

VOLVO PENTA
R.T. Briscoe (Nigeria) Ltd.
Technical Department
22 Creek Road
Apapa
P.O. Box 2104
Lagos

Telex: 21249
Cables: Briscoe
Phone: 46470/1, 46479

Norfolk Islands

LEYLAND INTERNATIONAL
Southern Cross Motors Ltd.
P.O. Box 218
Norfolk Island

Phone: 2022

Northern Ireland

KLOCKNER-HUMBOLDT-DEUTZ AG
Blackwood Hodge Ltd.
(Northern Ireland)
Dennison Industrial Estate
Ballyclare
Co. Antrim

Cables: Suntract Belfast
Phone: Ballyclare 3171

Norway

BOMBARDIER-ROTAX Gmbh
Colbjørnsen & Co. A/S
P.O. Box 80
N-1341 Bekkestua

Phone: 02-53 94 60

BRIGGS & STRATTON CORPORATION
Weswitco A/S
Olaf Schous Vei 4
Oslo 5

Telex: 18231
Cable: WESWI
Phone: (02) 351080

CATERPILLAR OVERSEAS SA
Pay & Brinck A/S
Brobekkeveen 62B
Postboks 65
Risløkka
Oslo

Telex: 11631
Cables: Pabrimas
Phone: (2) 15-92-50

DETROIT DIESEL ALLISION
Detroit Diesel Allison International – Europe
C30 General Motors Norge A/S
P.O. Box 205
2001 Lillestrom
Telex: 11682
Phone: 02 713860, 715860

DORMAN DIESELS LIMITED
Otto Thoresen & Moen A/S
Faberborggt 12,
P.O. Box 5080
Majorstua
Oslo 3
Telex: 19979 Thomo N
Cables: Ottot
Phone: (02) 46,69. 47.69. 52.80

FORD MOTOR CO. LIMITED
Ford Motor Co. (Norge) A/S
1410 Kolbotn
Telex: 11659
Phone: (02) 801560

L. GARDNER & SONS LIMITED
Norsk Gardner Diesel A/S
6715 Vaagsvaag

KLOCKNER-HUMBOLDT-DEUTZ AG
William Knudsen A/S
Fred Olsensgt. 1
P.B. 721 Sentrum
Oslo 1
Telex: 11309 Wilknud
Cables: Wilknud
Phone: 423910

LEYLAND INTERNATIONAL
British Leyland Norge A/S
Trondheimsveien 275
P.O. Box 55-Arvoll
Oslo 5
Telex: 18276 BLN N
Cables: Britleynor
Phone: 225500

NISSAN DIESEL MOTOR CO. LIMITED
A/S INOR
Postboks 1239
3001 Drammen

PERKINS ENGINES LIMITED
Universal Diesel A/S
Osterdalsgt 17
Oslo 6
Telex: 16855 Unidi
Cables: Unidiesel
Phone: 19 32 00

PETTERS LIMITED
Polytex A/S
Trondheimsveien 139
Oslo 5
Telex: 11613 Pindu
Cables: Polytex
Phone: Oslo 357780

RUGGERINI MOTORI SpA
Gentrade A/S
P.O. Box 297
Sandefjord
Telex: 16316 Agent N

VOLVO PENTA
Volvo Norge A/S
Lörenvangen 41
Postboks 31, Økern
Oslo 5
Telex: 16647
Phone: (02) 22 60 60

WHITE ENGINES INCORPORATED
B.M. Heede A/S
P.O. Box 4231
Sandakerveien 74
Oslo 4
Cables: Prosperity

Oman

CATERPILLAR OVERSEAS SA
Oasis Trading & Equipment Company
P.O. Box 1002
Muttrah
Muscat
Telex: 329 Alfaiha MB
Phone: 2160

KATOLIGHT
Oman Technical Centre
P.O. Box 1079
Muttrah

KIRLOSKAR OIL ENGINES LIMITED
Saeed Bin Nasser Al-Hashar
P.O. Box 331
Muscat
Telex: MB 245 Motorco
MB 294 Hashar
Cables: Al-Hashar

LEYLAND INTERNATIONAL
Darwish & Bros.
P.O. Box 75
Muscat
Telex: Lujaina MB 230 Muscat
Cables: Areej
Phone: 2703 2826

LUCAS SERVICE OVERSEAS LIMITED
W.J. Towell & Co.
P.O. Box 1061
Muttrah
Muscat
Telex: MN214
Cables: Towell
Phone: 772131/3/65
772261

NISSAN DIESEL MOTOR CO. LIMITED
Al-Hashar & Co.
P.O. Box 1028
Muttrah
Muscat

PERKINS ENGINES LIMITED
Waleed Associates
P.O. Box 437
Muscat
Telex: Waleed MB 270
Cables: Waleed
Phone: 722495 & 722592

PETBOW LIMITED
Zubair Enterprises (Electronics)
P.O. Box 127
Muscat

PETTERS LIMITED
Nasser Abdullatif & Co.
P.O. Box 3636
Ruwi
Muscat
Telex: MB 375
Cables: Nasser
Phone: Muttrah 2807

JOHN ROBSON (SHIPLEY) LIMITED
Nasser Abdullatif & Co.,
P.O. Box 3636
Ruwi
Muscat

Pakistan

ALLIS CHALMERS
Marghzar Industries Ltd.
Hab Chauki Road S.I.T.E.
P.O. Box 2706
Karachi
Cables: Crawler
Phone: 290111-19

BRIGGS & STRATTON
Peerless Industrial Engineering Co.,
501, 5th Floor Qamar House
M.A. Jinnah Road
P.O. Box 5394
Karachi 2
Phone: 230342

CATERPILLAR OVERSEAS SA
Republic Motors Ltd. (Tractor & Machinery Division)
D-2 S.I.T.E.
Manghopir Road
Karachi – 1
Cables: Repmotors
Phone: 29-01-02/3/4/5

DETROIT DIESEL ALLISON
Consolidated Equipment Co.,
805 Muhammadi House
1.1 Chundrigar Road
P.O. Box 4093
Karachi
Cables: Lesscost
Phone: 223529

DORMAN DIESELS LIMITED
Greaves Cotton & Co. (Pakistan) Ltd.
P.O. Box 4908
Volkswagen House
Beaumont Road
Karachi 4
Cables: Greaves
Phone: 512576/7/8/9

L. GARDNER & SONS LIMITED
Brady & Co. of Pakistan Ltd.,
Shernaz House
West Whart Road
Karachi 2

J.L.O.
Sibtain Brothers
Room No. 43 to 48
Ghafur Chambers
Abdullah Haroon Road
P.O. Box 4225
Karachi 3
Cables: Sibtain
Phone: 74177, 75177

KATOLIGHT
Platinum Traders
P.O. Box 4334
Karachi 2

KLOCKNER-HUMBOLDT-DEUTZ AG
Adamjee-Deutz (Pakistan) Ltd.
P.O. Box 4405
Adamjee House
4th Floor
J.J. Chundrigar Road
Karachi
Telex: 2831 Admji pw
Cables: Deutzmotor
Phone: 227731/33

KOHLER INTERNATIONAL LIMITED
Sai & Company
P.O. Box 3109
245-D/6 P.E.C.H.S.
Karachi 29
Pakistan

LEYLAND INTERNATIONAL
Republic Motors Limited
D-2 S.I.T.E.
Manghopir Road
Karachi-16
Cables: Repmotors
Phone: 291120, 290909

LUCAS SERVICE OVERSEAS LIMITED
Lucas Service (Pakistan) Ltd.
Al-Farid Centre
Moulvi Tamizuddin Khan Road
P.O. Box 506
Karachi 4
Cables: Lucasrep
Phone: 51 20 40
51 20 49

PERKINS ENGINES LIMITED
Shahnawaz Ltd.
P.O. Box 4766
19 West Wharf
Karachi-2
Cables: Autoshez
Phone: 223021

PETBOW LIMITED
Blackwood Hodge (Pakistan) Ltd.
19 West Wharf Road
Karachi

PETTERS LIMITED
Sibtain Brothers
P.O. Box 4225
43-48 Ghafur Chambers
Abdullah Haroon Road
Karachi 3
Cables: Sibtain Karachi
Phone: 74177, 75177

WHITE ENGINES INCORPORATED
International Steel &
General Agencies
125 Block "A"
S.M.C.H.S.
Karachi 6
Telex: Isga-KR 3606
Cables: Fuadsteel Karachi
Phone: 434641, 430300

Panama

ALLIS CHALMERS
25-45 Justo Arosemena Avenue
P.O. Box 2140
Zone 1
Panama
Telex: Eastern – 3598 Pn
Cables: Ficsa and Fivsa
Phone: 62-73-00

BRIGGS & STRATTON CORPORATION
Cardoze & Lindo S.A.
P.O. Box 7342
Panama 5
Telex: 715
Cables: Carlindo

CATERPILLAR OVERSEAS SA
Cardoze & Lindo S.A.
Calle 85 y Transistemica
Apartado 7342
Panama
Telex: TRT-PA2106
Comsa/Uwi 368715
Cables: Carlindo
Phone: 60-1155

DETROIT DIESEL ALLISON
Transporte Y Equipo S.A.
Boyd-Roosevelt Avenue
P.O. Box 2145
Panama 1
Telex: 3480089
Cables: Tesa
Phone: 60-2100

DORMAN DIESELS LIMITED
Maquinarias y Equipo Romero S.A.
Apartado U
Panama 4

KATOLIGHT
Motores Electicos
Apartado 10628
Panama 4

KLOCKNER-HUMBOLDT-DEUTZ AG
Servinaves Panama S.A.
Ave. Justo Arosemena
Calle 44
Bellevista
Apartado 031
Panama 5
Telex: 3689-75
Cables: Servinpan
Phone: 47 – 9663
9704
2621

KOHLER INTERNATIONAL LIMITED
F. Icaza & Co. Inc.
P.O. Box 2140
Avenida Justo Arosemena
No. 25-45
Panama City 1

LEYLAND INTERNATIONAL
Motores Britannicos S.A.
Apartado 1057
Panama 1
Cables: Heurtematte
Phone: 253900
250 644

LUCAS SERVICE OVERSEAS LIMITED
Motores Nacionales S.A.
Avenida Once No. 26, 93
P.O. Box L
Panama 4
Cables: Dieselco
Phone: 54510, 54514,
54592

NISSAN DIESEL MOTOR CO. LIMITED
Smoot y Paredes S.A.
Transistmica Ave.
Bolivar
Apartado "C"
Zone 4

PERKINS ENGINES LIMITED
Maquinarias y Equipo Romero S.A.
Apartado U
Panama 4
Telex: 368675
Cables: Romermaq
Phone: 25-5133/25-5134

PETTERS LIMITED
Distribuidora Cummins Diesel
de Panama S.A.
Apartado 6-1394
El-Dorado
Telex: 368690
Cables: Cumpan
Phone: 60.0666

VOLVO PENTA
Maquinarias y Equipo Romero S.A.
Apartado U
Panama 4
Telex: 368675
Phone: 25 51 30

WHITE ENGINES INCORPORATED
Agro S.A.
Calle Felipe Clemant
Apartado 3417
Panama 4
Telex: 3480146
Cables: Agrosa
Phone: 25-3245, 25-1446

Papua New Guinea

DETROIT DIESEL ALLISON
Detroit Diesel Allison International
c/o General Motors Holden's
Pty. Ltd.
Princess Highway
P.O. Box 163
Dandenong
Victoria 3175
Australia
Telex: 30168 or 30212
Cables: G.M.P. & I.P.
Phone: 792-0111

LEYLAND INTERNATIONAL
Burns Philip (New Guinea) Ltd.
Musgrave Street
Port Moresby
Papua New Guinea

LUCAS SERVICE OVERSEAS LIMITED
Ela Motors Ltd.
Scratchley Road
Badali
P.O. Box 75
Port Moresby
Telex: 116
Cables: Elamot
Phone: 54088

PERKINS ENGINES LIMITED
Tutt Bryant Pacific Limited
P.O. Box 326
Port Moresby
Cables: Tutby
Phone: 53522

PETTERS LIMITED
(New Guinea, British)
A.N.I. Australia Pty. Ltd.
A.N.I. Perkina Divisions
P.O. Box 117
16 Parramatta Road
Lidcombe
New South Wales

PETTERS LIMITED (Indonesia)
Technische Handelmaatchappij, N.V.
Akkerman & Company
Leekweg 17 (Binckshorst)
Gravenhage, Netherlands
Cables: Stampwals Den Haag
Phone: 390275-331277

Paraguay

BRIGGS & STRATTON CORPORATION
Villalonga Hermanos Comercial e Industrial, S.A.
Casilla de Correo 274
Asuncion
Cables: Villalonga
Phone: 49-1719

CATERPILLAR OVERSEAS SA
S.A.C.I. H. Petersen
Pte Eligio Ayala 175/1
Casilla de Correo 592
Asuncion
Cables: Pargtrade
Phone: 44-124/5/6

DETROIT DIESEL ALLISON
Automotores Comercial Industrial S.A.
P.O. Box 5
Asuncion
Telex: 99144 Metalpy
Cables: Acisa Asuncion
Phone: 41-838

DORMAN DIESELS LIMITED
Artaza Hermanos Comercial e Industrial S.A.
Casilla Postal 235
Asuncion
Cables: Artaza
Phone: 44-493

KLOCKNER-HUMBOLDT-DEUTZ AG
Juan Wiske
Palma 573
Casilla Correo 51
Asuncion
Cables: Juvico
Phone: 45-989

KOHLER INTERNATIONAL LIMITED
Agropecuaria F. Arza Encina
Casilla Correo 793
25 De Mayo No. 865
Asuncion

LEYLAND INTERNATIONAL
Gomal S.A.C.
Casilla de Correo 1201
Azara 172/6
Asuncion
Cables: Gomal

LUCAS SERVICE OVERSEAS LIMITED
Ipa Sociedad Anonima de Comercio e Industrias
Calle Chile 911/23
Casilla de Correo 211
Asuncion
Cables: Perersenco
Phone: 4-1272
4-6803
4-3351
4-8391
4-8765

NISSAN DIESEL MOTOR CO. LIMITED
15 de Agosto Esq,
Presidente Franco
Asuncion

PERKINS ENGINES
H. Petersen S.A.C.I.
Pt Elicio Ayala 481/483
Asuncion
Cables: Pargtrade
Phone: 4-4124/5/6

PETTERS LIMITED
Villalonga Hnos C.E.I.S.A.
Estrella 702
Casilla de Correo 274
Asuncion
Cables: Villalonga
Phone: 4408/09 – 8889-7298

Peru

ALLIS CHALMERS
Empeco S.A.
Parque Internacional D.E.
Industria Y Comercio
Calle 11 No 203 CDRA
8 Ave Oscar R.
Benavides (Colonial)
Telex: 3540081
Cables: Empeco
Phone: 510920

BRIGGS & STRATTON CORPORATION
Cia Importadora Derteano & Stucker, S.A.
Apartado 1663
Lima
Cables: Dest
Phone: 31-4766

CATERPILLAR OVERSEAS SA
Enrique Ferreyros & Cia S.A.
Ave. Industrial 675
Apartado 150
Lima
Telex: 35447, 20102
Cables: Ferreyco
Phone: 52- 0-70

DETROIT DIESEL ALLISON
CIPSA – Commercial Peruana S.A.
Av. Wicolas Ayllon No. 1928
Carretera Central KM 1.5
P.O. Box 2 94
Lima 3
Telex: 394-25333 Pu Milne
Cables: Aspic
Phone: 313070

DORMAN DIESELS LIMITED
Somerin S.A.
La Colmena
Oficina Colmena
Lima
Telex: Somerin PX5529
Cables: Somerin
Phone: 287715

FORD MOTOR COMPANY LIMITED
Ford Motor Co. del Peru S.A.
Apartado 4690
Lima

KATOLIGHT
Technocom
P.O. Box 11120
Lima 14 (Santa Beatriz)

KLOCKNER-HUMBOLDT-DEUTZ AG
Diesel Centro S.A.
Avenida Venezuela 2720
Casilla No. 789
Lima 1
Telex: 25782
Cables: Dicesa
Phone: 238712, 320820

KOHLER INTERNATIONAL LIMITED
Empeco S.A.
P.O. Box 5692
Omega 203-Cuadra 38
Oscar Benevides
Lima

LEYLAND INTERNATIONAL
British Leyland Del Peru
1647 Ave. Republica de Panama
Balconcillo
Lima 13
Cables: Leymotors
Phone: 31-3313 31-3955

LUCAS SERVICE OVERSEAS LIMITED
Compania Importadora
Derteano & Stucker S.A.
Casilla 1663
Avenida Mexico No. 363-373
Lima
Cables: Dest
Phone: 314766

PERKINS ENGINES LIMITED
Motores Diesel Andinos S.A.
Carolina Vargas 275
San Isidro
Lima
Telex: 21076 Pu Donalim
Phone: 406882

PETTERS LIMITED
Industrial Lima S.A.I.L.S.A.
P.O. Box 4946
Calle Omega 255
Parque Internacional Industrial
Lima
Telex: Berco PX5350
Cables: Ilsamac
Phone: 513990 – 514972

JOHN ROBSON (SHIPLEY) LTD
Comercial Gutierrez S.A.
Panamericana Norte s/n
Apartado 77
Pacasmayo
Peru

VOLVO PENTA
Volvo del Peru S.A.
Volvo Penta Division
Depto de Servicio
P.O. Box 815
Lima
Telex: PX-25395
Cables: Pervolvo
Phone: 32 15 70
32 17 83

WHITE ENGINES INCORPORATED
Luis Schydlowsky, S.A.
Casilla 2146
Avda. Benavides 195
Of. 101, Miraflores
Lima
Telex: 21098PE – Schydlo
Cables: Schydlowsky
Phone: 270692, 270052

Philippines

ALLIS CHALMERS
Chindisco (Phil) Inc.
2288 Pasong Tamo Ext
P.O. Box 1066 Commercial Center
Makati Rizal
Telex: ITT-7425193 RCA 7222360
Eastern PN 3598
Cables: Clindisco
Phone: 89-15-26

BRIGGS & STRATTON CORPORATION
Asia/Pacific Regional Office
P.O. Box 7417
Air Mail Exchange Office
M.I.A. 3120
Telex: 722 2023
Cables: Bands

CATERPILLAR OVERSEAS S.A.
Usiphil Inc
P.O. Box 55cc
Makati
Rizal
Cables: Usiphil

DETROIT DIESEL ALLISON
General Diesel Power Corp.
6305 South Xway
P.O. Box 2311 MCC
Makati
Rizal
Telex: 7420444 Gensel PM
Cables: Gendiesel, Manila
Phone: 88-15-12 or
88-15-33

DORMAN DIESELS LIMITED
Jardine Davies Incorporated
Jardine Davies Building
222 Buendia Avenue
P.O. Box 561
Makati Commercial Centre
Rizal
Telex: 7425170 via ITT
7222192 via RCA
Cables: Dracoman
Phone: 89-30-61

FORD MOTOR COMPANY LIMITED
Ford Philippines Inc.
P.O. Box 415
Makati Commercial Centre
Makati
Rizal

INDIAN NATIONAL DIESEL ENGINE CO. LTD.
Warner Burnes & Co Ltd
Makati
Rizal

J.L.O.
American Factors (Phil) Inc.
2310 Pasong Tamo Extension
P.O. Box 660
Makati
Rizal
D708
Telex: through Williams Equipment Co. Ltd – WECO 742530
Cables: Amfacphil
Phone: 88-89-31

KATOLIGHT
VS Energetics, Inc.
P.O. Box 1195
Makati
Rizal

KLOCKNER-HUMBOLDT-DEUTZ AG
Makati Machinery & Equipment Company Inc. (MMECI)
1120 Pres. Quirino Avenue
P.O. Box 1526
Makati CC
Paco
Manila
Telex: Resultu 7420318 Manila
Cables: Resolute
Phone: 587630, 587622

KOHLER INTERNATIONAL LIMITED
Atkins Kroll & Co. Inc.
P.O. Box 308
7232 Malugay Street
Makati
Rizal

LEYLAND INTERNATIONAL
Amalgamated Motors (Philippines) Inc.
P.O. Box PA 413
Port Area South 2803
Manila
Telex: Leyampi 3439
Cables: Leyampi
Phone: 407041/51

LUCAS SERVICE OVERSEAS LTD
Electro Diesel Sales & Service Inc.
Lucas House
1108 Pres. Quirino Ave
Malate
Manila P.O. Box 311
Telex: 2350, 3422, 5138
2300
Phone: 50.36.67

NISSAN DIESEL MOTOR CO. LIMITED
Columbian Motors Corp.
Subangdaku
Mandaue City

PERKINS ENGINES LTD
Clindisco (Phi.) Inc.
P.O. Box 1006 M.C.C.
2288 Pasong Tamo Ext.
Makati
Rizal
Telex: Eastern Ext. – PN3598
ITT7425193
RCA7222360
Phone: 89-15-26, 89-55-96

PETTERS LIMITED
Muller & Phipps Inc. Corp.
Pioneer Street
Mandaluyong
Rizal
Cables: Mulphico
Telex: RCA 8228657
Phone: 70-06-11, 78-19-12

JOHN ROBSON (SHIPLEY) LTD.
Schmid & Oberly Inc.
P.O. Box 1916
Manila

RUGGERINI MOTORI SpA
Newmark Commericla &
Industrial Corp.
855A & B de Los Santos Ave.
Cor. Panay Ave.
Quezon City

VOLVO PENTA The Edward J. Nell Company Makati Commercial Center Makati Rizal Manila	Telex: Cables: Phone:	7420234, 7425234 Neled 89 12 14/31/55/66/ 85 89 22 32/75
WHITE ENGINES INCORPORATED Warner Barnes & Co Inc P.O. Box 1191 Makati Commercial Center Makati	Telex: Cables: Phone:	Wareng PN 3536 Wareng Makati 89-4-61

Poland

CATERPILLAR OVERSEAS SA Bowmaker (Plant) Ltd. Polish Operations Division Watling St. Cannock Staffs. U.K.	Telex: Cables: Phone:	337548 Bowplant 2551
CATERPILLAR OVERSEAS SA Osrodek Informacjii Techniczenej (Technical Information Center) Bowmaker (Plant) Ltd. U1. Stawki 2 Warsaw	Telex: Phone:	814899 39-64-02 39-71-62 39-56-38
KLOCKNER-JUMBOLDT- DEUTZ AG Unitex S.A. Import-Export Stawki 2 00-950 Warszawa	Telex: Cables: Phone:	813751 Utx pl Unitex Warszawa 398832

Portugal

ALLIS CHALMERS Casa Americana Comercial S.A.R.L. Avenida de Republica 27 Caixa Postale 1219 Lisbon	Cables:	Renkers
BOMBARDIER-ROTAX GMBH Antonio Marques Boavida Almofala de Baixo Avelar	Phone:	32161 Indicativo 036
BRIGGS & STRATTON CORPORATION Mendes de Almeida Comercio de Industria de Maquinas e Equipamentos S.A.R.L. Av. 24 de Julho 52, A-G Lisbon 2	Telex: Cables: Phone:	12185 Dalmeida-Lisboa 667794
CATERPILLAR OVERSEAS SA Sociedade Tecnica de Equipamentos e Tractores S.A.R.L. Apartado 1351 Lisbon	Telex: Cables: Phone:	12778 Stetra Sacavem 251-1011
DETROIT DIESEL ALLISON Detroit Diesel Allison International Europe c/o General Motors de Portugal Ltda Av. Marechal Gomes Da Costa 33 Caira Postal 2484 Lisbon 6	Telex: Cables: Phone:	12599 Lisbon Portautoex, Lisbon 384201 thru 384209
DORMAN DIESELS LIMITED Turbodiesel Sociedade Importadora de Maquinas Ltda Largo do Mastro 29-2° Lisbon 2	Telex: Cables: Phone:	12622 Rotciv P – attention Turbo- diesel Turbodiesel Lisboa 53 96 63, 5572 67
FORD MOTOR CO. LIMITED Ford Lusitania S.A.R.L. Apartado 2248 Rua Rose Araujo 2 Lisbon 2	Telex: Phone:	12218 53914
L. GARDNER & SON LIMITED Conde Barao, Industrias Metalicas e Comercio de Representacoes S.A.R.L. Avenida 24 de Julho 62-64 Lisbon 2 Apartado 2734	Phone:	671011/8
KLOCKNER-HUMBOLDT- DEUTZ AG Motope Motores Oleos Pesados S.A.R.L. Rua da Vitoria 88-3° Apartado 2565 P – Lisbon 2	Telex: Cables: Phone:	Motope P 16582 Deutzmotor 327195/6/7
LEYLAND INTERNATIONAL British Leyland de Portugal Automoveis Ltda Avenido do Brasil 45 Apartado 5090 Lisbon	Telex: Cables: Phone:	12185 Sancin P Brileypor Lisbon 5 251 0171/2 251 0508 251 0984 251 0622
NISSAN DIESEL MOTOR CO. LIMITED Entreposto Comercial de Automoveis S.A.R.L. Praca de Moscavide 1-Olivais Sul Lisboa 6		
PERKINS ENGINES LIMITED Auto-Industrial S.A.R.L. Estrada da Circunvalacao Portela da Ajuda Lisbon 2	Telex: Cables: Phone:	12444 Auto LP Automoveis Lisbon 210021/2101617
PETTERS LIMITED E. Pinto Basto e Cia Ltd., Apartado 2200 1 Avenida 24 de Julho Lisbon 2	Telex: Cables: Phone:	Basto Lisbon 12201 Pinto Lisbon 361536/261581
RUGGERINI MOTORI S.p.A. Centro De Comercio Agricola de Santarem Apartado 72 Rua Serpa Pinto 84 Santarem		
VOLVO PENTA Agencia Maritima Transatlantica Lda Rua do Alecrim 20-F-1° 9-1 P.O. Box 2-536 Lisboa 2	Telex: Cables: Phone:	Amatra 18325 Transmara 324351/2

Principe Island

CATERPILLAR OVERSEAS SA Sorel S.A.R.L. P.O. Box 40 Luanda	Telex: Cables: Phone:	3229 Sorel 7-22-81/2/3/4/5

Puerto Rico

ALLIS CHALMERS Gonzalez Trading Inc. San Patricio & De Diego Avenues URB Industrial La Reviera P.O. Box 13067 Rio Piedras Puerto Rico 00921	Telex: Cables: Phone:	Gontra Gontra San Juan P.R. (809) 783-9380

BRIGGS & STRATTON CORPORATION
Badrena & Perez Inc.
P.O. Box 1839
Carpenter Road 225
Hato Rey 00919
Telex: 9047
Cable: Ferbad
Phone: 767-2455

CATERPILLAR OVERSEAS SA
USI Puerto Rico Inc.
John F. Kennedy Eve.
G.P.O. Box 2529
San Juan 00936
Telex: 325-330 (RCA)
Cables: Usipri Sanjuan Via RCA
Phone: (809) 782-4100

DETROIT DIESEL ALLISON
Diesel de Puerto Rico Div.
Truck Fleets Inc.
G.P.O. Box 4008
San Juan 00936
Telex: 3252367
Cables: Diesel San Juan
Phone: 809-764-5330

KATOLIGHT
Western Gas Products Corp.
P.O. Box 1270
Mayaguez 00708

KOHLER INTERNATIONAL LIMITED
Garcia Machinery Inc.
G.P.O. Box 3368
Road 2 KM 4-8
(Pueblo Viejo)
San Juan
Puerto Rico 00936

LEYLAND INTERNATIONAL
British Car Distributors Inc.
KM8.8 Carrera
Militar Numero 2
Bayamon
Puerto Rico 00620

PERKINS ENGINES LIMITED
Halco Sales Inc.
Ave. San Marcos
El Comandante Industrial Park
Carolina Heights 00122
Phone: 809/769-7575

Qatar

CATERPILLAR OVERSEAS SA
Mohamed Abdulrahman Al-Bahar
P.O. Box 2171
Doha
Telex: 4225 DH
Cables: Bahar
Phone: 21026

DETROIT DIESEL ALLISON
Mannai Trading Company
P.O. Box 76
Doha
Telex: 957-4208
Cables: Mannai
Phone: 26251

KOHLER INTERNATIONAL LIMITED
Al Khalaf Trading & Construction Co.
P.O. Box 226
Doha

LEYLAND INTERNATIONAL
Darwish Automobiles
P.O. Box 40
Doha
Telex: DH211 Aldarwish
Cables: Autodar
Phone: 23133

LUCAS SERVICE OVERSEAS LIMITED
Al Nasr Machinery Division
P.O. Box 28
Doha
Telex: 4242 Al. Nasr
Cables: Alnasr
Phone: 22284, 24451/2/3

NISSAN DIESEL MOTOR CO. LTD.
Jaidah Motors & Trading Co.
P.O. Box 150

PERKINS ENGINES LIMITED
Alnasr Machinery Division
P.O. Box 28
Doha
Telex: 4401 Macdir D.H.
Cables: Alnasr
Phone: 24451, 22284

PETTERS LIMITED
Oriental Trading Co.,
P.O. Box 96
Doha
Telex: D.H. 286
Cables: Oriental

JOHN ROBSON (SHIPLEY) LIMITED
Abdulrehman Bin Abdulla Abidan
Fakhroo & Sons
P.O. Box 254
Doha

Reunion

CATERPILLAR OVERSEAS SA
Ets Camille Macé & Cie
46 rue de la Bourdonnais
B.P. 57
St. Denis 97462
Telex: 38RE
Cables: Cemace
Phone: 216-700

KLOCKNER-HUMBOLDT-DEUTZ AG
Valcke Frères
Boussereau
B.P. 511
Z.1. Le Chaudron
Sainte Clotilde
Telex: REB St. Denis
Phone: 213588

LEYLAND INTERNATIONAL
Landis Madagascar
5 Rue Robert Ducrocq
Behorirka
P.O. Box 633
Tananarive
Malagasy
Cables: Lanbrocomp Tananarive

LUCAS SERVICE OVERSEAS LIMITED
Electro-Diesel Reunion S.A.
Route Nationale 2
St. Clotilde 97-4
B.P. 600
97473 St. Denis
Cables: Eldire

PERKINS ENGINES LIMITED
Electro Diesel Reunion S.A.
B.P. 600
97473 Saint Denis
La Reunion
Telex: 50 Re Ste-clotilde
Cables: Eldire
Phone: 21.07.78

PETTERS LIMITED
Sorequip
51 Rue Labourdonnais
B.P. 714
Saint Denis
Ile de la Reunion

Rio Muni

CATERPILLAR OVERSEAS SA
Finanzauto S.A.
Plaza de las Cortes 6
Madrid 14
Spain
Telex: 27752
Cables: Finanzauto Madrid
Phone: (91) 448-2700

Rwanda

CATERPILLAR OVERSEAS SA
Chanic
B.P. 930
Bujumbura
Burundi
Cables: Chanusa Bujumbura
Phone: 3284

DORMAN DIESELS LIMITED
S.A.R.L. Compagnie Industrielle
et Commerciale Du Rwanda "Circa"
B.P. 234
Kigali

KLOCKNER-HUMBOLDT-DEUTZ AG
Somirwa
B.P. 266
Kigali
Telex: 27
Phone: 5421

LEYLAND INTERNATIONAL
Compagnie De l'Afrique
Oricutale
'Old East' S.A.R.L.
Rue de la Revolution
P.O. Box 711
Kigali

Cables: Broche
Phone: 5240/5290

PETTERS LIMITED
N.A.H.V.
B.P. 626
Kigali

Sabah

LEYLAND INTERNATIONAL
Champion Motors (SABAH)
SDN. BHD
16 Prince Phillip Drive
Kota Kinabalu

PETTERS LIMITED
William Jacks & Co.
(Borneo) Ltd.
P.O. Box 278
8 Praya Road
Kota Kinablu

Sao Tome

CATERPILLAR OVERSEAS SA
Sorel S.A.R.L.
P.O. Box 408
Luanda
Angola

Telex: 3229
Cables: Sorel
Phone: 7-22 81/2/3/4/5

LEYLAND INTERNATIONAL
Champion Motors (Sarawak)
SDN. BHD.
Jalan Tunku Abdul Rahman
Padungan
Kuching

Saudi Arabia

ALLIS CHALMERS
General Machinery Agencies
P.O. Box 139
Jeddah

Telex: 40178 Saladin SJ
Cable: Trust
Phone: 29501/2

BRIGGS & STRATTON
CORPORATION
Sami Kutbi & Co.
Medina Street
P.O. Box 247
Jeddah

Cables: Sami
Phone: 29057 C.R. 470

DETROIT DIESEL ALLISON
General Machinery Agencies
P.O. Box 139
Jeddah

Telex: 92-40178
Cable: Trust
Phone: 29501/2

DORMAN DIESELS LIMITED
Yusuf Bin Ahmed Kanoo
Airport Road
P.O. Box 753
Riyadh

Cables: Kanoo
Phone: 27132, 28942

Hussein M Fayez & Sons
P.O. Box 11
Jeddah

Cables: Fayez
Phone: 2575

L. GARDNER & SON LIMITED
General Engineering Trading Co.
P.O. Box 4586
Jeddah

KATOLIGHT
Mohamed Ali Maghrabi & Sons
P.O. Box 61
Jeddah

KIRLOSKAR OIL ENGINES
LIMITED
Abdullah & A.M. Bahamdein
P.O. Box 13

Cable: Albahamdein
Phone: 26784

KLOCKNER-HUMBOLDT-
DEUTZ AG
A.S. Bugshan & Bros
P.O. Box 378
Jeddah

Telex: 40179 Bugshan SJ
Cables: Bugshan
Phone: 25250

KOHLER INTERNATIONAL
LIMITED
Sami Kutbi & Co.
P.O. Box 247
Jeddah

LEYLAND INTERNATIONAL
Darwish Ben Abdullah Darwish
Darwish Building
P.O. Box 153
Dammam

Cables: Aldarwish

LUCAS SERVICE OVERSEAS
LIMITED
Alissa
Alissa Building
Street No. 7
P.O. Box 192
Alkhobar

Telex: 67022 Alissa SJ
Cables: Lateef
Phone: 41510

NISSAN DIESEL MOTOR CO. LTD.
Rolaco Trading & Contracting
P.O. Box 222
Jeddah

PERKINS ENGINES LIMITED
Abdulaziz & M.A. Alhomaih
Aljomaih Building
P.O. Box 132
Riyadh

Telex: 20023 Jomaih S.J.
Cables: Aljomaih

PETBOW LIMITED
Abdul Rahman Dawood Al-Gilani
P.O. Box 1159
Dammam

PETTERS LIMITED
Ali Baobeid
Ben Qubos Establishment
for Trading & Imports
P.O. Box 5440
Jeddah

Cables: Luckytravel
Phone: 24033

JOHN ROBSON (SHIPLEY)
LIMITED
Baghanem Corporation for
Agric. & Commerce
P.O. Box 14
Jeddah

RUGGERINI MOTORI S.p.A.
Ghaleb Abdullah Obeid
P.O. Box 484
Riyadh

Telex: 20112
Tricorp SJ

WHITE ENGINES
INCORPORATED
Mohamed Ali Maghrabi & Son,
P.O. Box 61
Jeddah

Telex: Mamtex 400766
Phone: 22636, 22870

Senegal

KLOCKNER-HUMBOLDT-
DEUTZ AG
Valcke Freres
Sté Sofica
B.P. 1783
KM 4 — Route de Rufisque
Dakar

Telex: 663 Sofica SG
Phone: 50063
50537

KOHLER INTERNATIONAL LIMITED
Senegal Agricole Material (S.A.M.)
Km 7 Route de Rufisque
B.P. 229
Dakar

LEYLAND INTERNATIONAL
Societe Commerciale De L'Ouest Africain
P.O. Box 50
Dakar
Cables: Torfyr

LUCAS SERVICE OVERSEAS LIMITED
Nouvelle Societe Commerciale Africaina (NOSOCO)
2 Rue Faidherbe
B.P. 791
Dakar
Telex: 503
Cables: Unamotors
Phone: 23350

PERKINS ENGINES LIMITED
Matforce Nosoco
B.P. 791
Dakar
Telex: 503
Phone: 369-99 and 233-50

PETTERS LIMITED
Hamelle-Afrique S.A.
KM 3 Route de Rufisque
B.P. 162
Dakar
Telex: Torfyr 544
Cables: Hamelaf
Phone: 367.46.92

Seychelles

LEYLAND INTERNATIONAL
British Motors (Seychelles) Pty. Ltd.
P.O. Box 83
Victoria
Mahe
Cables: Brimotors Seychelles

PERKINS ENGINES LIMITED
Engineering Supplies Ltd
P.O. Box 340
New Port
Victoria
Mahe
Phone: 22183

Sierra Leone

DETROIT DIESEL ALLISON
Blackwood Hodge (Sierra Leone) Ltd
P.O. Box 1456
Freetown
Telex: Freetown 989-3272
Cable: Suntract
Phone: 4593, 4385

DORMAN DIESELS LIMITED
Holman Brothers Ltd.
P.O. Box 465
Freetown
Cables: Airdrill
Phone: 4029

LEYLAND INTERNATIONAL
Compagnie Francaise de l'Afrique Occidentale
Motors Dept.
P.O. Box 823
Freetown
Cables: Senafrica
Phone: 3088

LUCAS SERVICE OVERSEAS LIMITED
The United Africa Co. of Sierra Leone Ltd.
Lucas House
8 Wilberforce Street
P.O. Box 418
Freetown
Cables: Unamotors
Phone: 2437

PERKINS ENGINES LIMITED
African Commercial and Agricultural Enterprises Ltd.
P.O. Box 1058
18 Light-Foot Boston Street
Freetown
Telex: 237
Cables: Afrocom
Phone: 6109-6244

PETTERS LIMITED
Cie Francaise de l'Afrique Occidentale
Technical Dept.
P.O. Box 823
Freetown
Cables: Senafrica
Phone: 3088

Singapore

BRIGGS & STRATTON CORPORATION
United Motor Works (S) Pte Ltd.
12 Jurong Town
Singapore 22
Telex: RS 21547
Cables: Unity
Phone: 653155

CATERPILLAR OVERSEAS SA
Tractors Malaysia Berhad
10km Bukit Timah Road
G.P.O. Box 2977
Singapore 21

DETROIT DIESEL ALLISON
General Diesel Supplies (S) Pte. Ltd.
32 Jurong Pier Road
Singapore 22
Telex: 786-21962
Cables: Gedes
Phone: 655222

DORMAN DIESELS LIMITED
The GEC Company of Singapore Pte. Ltd.
Diesel Division
Magent House
P.O. Box 4046
Bukit Timah
Singapore 21
Telex: RS 21508 A/B Gecsing
Cables: Enelectico or Gecsingapore
Phone: 663011

FORD MOTOR CO. LIMITED
Ford Motor Co. Private Ltd.
P.O. Box 4047
8½ Mile Bukit Timah Road
Singapore
Telex: R.S. 21616 (Sinford)
Phone: 666111

L. GARDNER & SONS LIMITED
Malayan Motors (Pte) Ltd.
14-20 Orchard Road
P.O. Box 394
Singapore 9
Phone: 332085 (Automotive & Rail Traction)

INDIAN NATIONAL DIESEL ENGINE CO. LIMITED
K.S.M. International
15-B/4 Amber Mansion
Orchard Road
Singapore 9

KATOLIGHT
Industrial Machinery & Electric Ltd.
65 Namly Road
Singapore 10

KLOCKNER-HUMBOLDT-DEUTZ AG
Deutz Far East (Pte) Ltd.
Jurong Town Hall (1st Fl.)
Jorong Town Hall Road
P.O. Box 98
Jurong
Singapore 22
Telex: RS 21961 DFE
Cables: Deutzmotor
Phone: 654044

KOHLER INTERNATIONAL LTD.
Tanglin
P.O. Box 12
Singapore 10
Telex: 23946
Cable: Kohlering Singapore

LEYLAND INTERNATIONAL
Malayan Motors (Pte) Ltd.
45 Orchard Road
P.O. Box 441
Singapore
Telex: RS 21772
Cables: Malaymotor
Phone: 332085

LUCAS SERVICE OVERSEAS SA
Far East Motors (Pt.) Ltd.
54 Orchard Road
P.O. Box 2697
Singapore 9
Telex: 21772 Wearnes
Cables: Ignition
Phone: 328341 to 8

NISSAN DIESEL MOTOR CO. LTD.
SM Mechanical (S) Pte Ltd.
9 & 11 Sin Ming Road
Bik 23
Singapore 20

PERKINS ENGINES LIMITED — Telex: Perkoil R23360; Phone: 594471/2/3
Perkins Engines Eastern Ltd.
549 Upper Thomson Road
Singapore 20

PETTERS LIMITED — Telex: 8721324; Cables: Expanded; Phone: 660011
William Jacks & Co.
(Singapore) Ltd
11km Milestone
Bukit Timah Road
P.O. Box 4049
Singapore 21

JOHN ROBSON (SHIPLEY) LTD
Inter-Golden Wood Pte. Ltd.
5616, 5th Floor
Woh Hup Complex
Beach Road
Singapore 7

RUGGERINI MOTORI S.p.A. — Telex: 21390 Oldarno
Bm Benh Meyer & Co.,
P.O. Box 2000
100 Pasir Panjang Road
Singapore 5

VOLVO PENTA — Telex: 21772 (Wealco Singapore); Cables: Wealco; Phone: 65 57 44, 65 50 87, 65 54 32
Wearne's Equipment and Land
Company (P) Ltd
No. 3 Gul Circle
Jurong
Singapore 22

Solomon Islands

LEYLAND INTERNATIONAL — Cables: Trade
British Solomons Trading Co. Ltd.
P.O. Box 114
Honiara

PERKINS ENGINES LIMITED — Cables: Solmot; Phone: 313.374
Solomon Motors Limited
P.O. Box A61
Honiara

PETTERS LIMITED
E.V Lawson Ltd.
P.O. Box A144
Honiara

Somalia

DORMAN DIESELS LIMITED — Cables: Busharo; Phone: 3919
Islah Commercial Agency
P.O. Box 1035
Mogadishu

KLOCKNER-HUMBOLDT-DEUTZ AG — Telex: 37606 Imellom; Cables: Imelombarde; Phone: 587291
Zustandig
IML
Via G. Lorca 25
22050 Lomagna (Como)
Italy

LEYLAND INTERNATIONAL — Telex: 607 Fiat; Cables: Waddada Muqdisho; Phone: 3666-3135-2785
Government Trading Agency
for Vehicles and Accessories
P.O. Box 390
Mogadishu

LUCAS SERVICE OVERSEAS SA — Cables: Fima; Phone: 3803, 2089
Fima S.P.A.
Casella Postale 344
Mogadishu

PETTERS LIMITED — Cables: Loche; Phone: 142
Pratelli Loche
P.O. Box 359
Mogadishu

RUGGERINI MOTORI S.p.A.
Gibril Ali Abdulle
C.P. 303
Afgoi
Mogadishu

South Africa

BRIGGS & STRATTON CORPORATION — Telex: 43-7516; Cables: Lecauto; Phone: 838-3111
Autolec Ltd.
Corner La Rochelle &
Crystal Roads
Springfield (Eloff St.
Ext. South)
P.O. Box 2964 (Johannesburg
2000)
Johannesburg 2001

CATERPILLAR OVERSEAS SA — Telex: 43-7878/9; Cables: Isacat; Phone: 36-10-11
Caterpillar (Africa) (Pty) Ltd.
Anvil Road
P.O. Box 11481
Johannesburg

CATERPILLAR OVERSEAS SA — Telex: 2-632; Cables: Shipments; Phone: 82721
Barlow's (O.F.S.) Ltd.
Nuffield Street
Hamilton
Bloemfontein 9301

DETROIT DIESEL ALLISON — Telex: 74-7684; Cables: Portautoex, Port Elizabeth; Phone: 6-2384, 6-1131 Port Elizabeth
Detroit Diesel Allison International
c/o General Motors South Africa
(Pty) Ltd.
Aloes Plant (near Port
Elizabeth)
P.O. Box 1137
Port Elizabeth 6000

DORMAN DIESELS LIMITED — Telex: 8-0267 S.A.; Cables: Kopelrail,
Ruston Dorman Diesels
(Pty) Ltd
P.O. Box 6001
Dunswart
Transvaal 1508

FORD MOTOR CO. LTD. — Telex: 74 7411; Phone: 2 7011
Ford Motor Co. of South
Africa (Pty) Ltd.
P.O. Box 788
Ford House
Port Elizabeth

L. GARDNER & SON LIMITED — Telex: 87353; Phone: 616.2130
Gardner Engines South Africa
(Pty) Ltd.
350 Main Reef Road
P.O. Box 25619
Denver
Transvaal

J.L.O. — Telex: 43-0572sa; Cables: Mosems; Phone: 36-5235
Diesel Installations (Pty) Ltd.
Brewery Street & Anvil Road
P.O. Box 104
Isando
Transvaal

KOHLER INTERNATIONAL LIMITED
Hessen Engineering Co.
(Pty) Ltd.
P.O. Box 39066
Bramley
Johannesburg 2018

LEYLAND INTERNATIONAL
Leyland South Africa Ltd.
(Car and Light Commercial
Vehicle Division)
P.O. Box 190
Goodwood 7460
Cape Town

Telex: 57-7587
Cables: Brimocor
Phone: Cape Town 938 3661

LUCAS SERVICE OVERSEAS LIMITED
Lucas Industries South Africa (Pty) Ltd.
Head Office
8/10 Simmonds Street South
P.O. Box 360
Johannesburg

Telex: 8-7795
Cables: Lucassa
Phone: 836.3044

PERKINS ENGINES LIMITED
Perkins Engines (Pty) Ltd.
P.O. Box 31285
6th Floor
Noswal Hall
Braamfontein
2017 Johannesburg

Telex: 8-0037SA
Cables: Perkoil
Phone: 725-5715

PETTERS LIMITED
Northfield Engineering (Pty) Ltd.
P.O. Box 3119
13/15 Dobson Street
Port Elizabeth

Cables: 4-13805
Phone: 4-13644/5/6

PETTERS LIMITED
Vincent & Pullar (Pty) Ltd.
P.O. Box 1146
5 Eaton Road
Durban
Natal

Telex: 6-7444DN
Cables: Electrical
Phone: 357166

RUGGERINI MOTORI S.p.A.
Industrial Units Ltd.
P.O. Box 5945-2000
7 Heidelberg Road
Village Main
Johannesburg

Telex: 43-7772 Induna

VOLVO PENTA
Lawson's Marine & Industrial (Pty) Ltd.
59 Beaumont St.
Booysens
P.O. Box 10599
Johannesburg

Telex: 95-437638
Cable: Lavolvo
Phone: 32-6201/2/3/4

VOLVO PENTA
Lawson's Autodiesels
P.O. Box 93
Sanlamhof
Cape Province

Telex: 570258
Cables: Lavolvo Cape Town
Phone: 970032

VOLVO PENTA
Lawson's Autodiesels
P.O. Box 3460
Durban
Natal

Telex: 67474
Cables: Lavolvo Durban
Phone: 333771

WHITE ENGINES INCORPORATED
Mosethals Engineering Co. (Pty) Ltd.
P.O. Box 104
Corner Brewery Street & Anvil Road
Isando

Cables: Mathner
Phone: 36-5235

South West Africa

LUCAS SERVICE OVERSEAS LIMITED
Lucas Service SWA (Pty) Ltd.
28 Bell Street
P.O. Box 5015
Windhoek

Cables: Lucassa
Phone: 5973

Spain

BRIGGS & STRATTON CORPORATION
Industrias Fita S.A.
Calle San Lazaro, 54
Figueras
Gerona

Telex: 57177
Cables: Fita
Phone: 241300

CATERPILLAR OVERSEAS SA
Finanzauto S.A.
Dos Hermanas
Sevilla

Telex: 72230
Phone: (954) 721-350

Finanzauto S.A.
Plaza de las Cortes 6
Madrid 14

Telex: 27752
Cables: Finanzauto
Phone: (91) 448-2700

DETROIT DIESEL ALLISON
Detroit Diesel Allison International — Europe
c/o General Motors France
56a 58 Ave. Louis Roche
92231 Gennevilliers
France

Telex: Genlmot 62050
Cables: Parautexap, Gennevilliers
Phone: 790-70-00

DORMAN DIESELS LIMITED
Guascor S.A.
Guascor Building
P.O. Box 30
Zumaya

Telex: 36310 Guazo E
Cables: Guascor Sa Zumaya
Phone: (43) 86 12 48
(43) 86 12 49

KATOLIGHT
Atlantic Industries S.L.
Buenos Aires 12 — 1° Local Num. 5
Bilbao

KLOCKNER-HUMBOLDT-DEUTZ AG
Cia. Española de Motores Deutz
Otto Legitimo S.A.
Av. Pio XII., No. 100
Apartado 50.938
Madrid 16

Telex: 43739
Cables: Deutzmotor
Phone: 2022240, 2023240

KOHLER INTERNATIONAL LIMITED
Coprima Ltd.,
Zurbano 56-2
Madrid 10

LEYLAND INTERNATIONAL
Leyland España S.A.
Apartado 14.845
Madrid

Telex: 22840-E
Cables: Leyesp
Phone: 675 01 50
675 29 50
675 16 50

PERKINS ENGINES LIMITED
Motor Iberica S.A.
(Division Zona 2)
Carretera del Aero-Club
Carabanchel Alto
Madrid

Telex: 27324
Cables: Perki-e
Phone: 208 52-40
208-96-40
208-98-40

PETTERS LIMITED
Alton S.A.
Avenida Generalissimo 84
Madrid 16

Cables: Caralton Madrid
Phone: 2592557, 2500046

RUGGERINI MOTORI S.p.A.
L. Zabala S.A.
Astola 6
Abadiano (Vizcaya)

VOLVO PENTA
Volvo Concesionarios S.A.
Avda Generalisimo 20
Madrid 16

Telex: 23296 volco e
Cables: Volvosa
Phone: 262 22 07

Sri Lanka

DETROIT DIESEL ALLISON
C.I.A.S. Ltd.
150 Ward P1.
P.O. Box 610
Colombo

Cables: Ciastri
Phone: 94191

DORMAN DIESELS LIMITED
Brown & Co. Ltd.
P.O. Box 200
481 Darley Road
Colombo 10
Telex: 1111
Cables: Metal
Phone: 91171

L. GARDNER & SONS LIMITED
Taos Ltd.
Kew Road
Colombo 2
Phone: 79629

J.L.O.
Mackwoods Limited
38 D.R. Wijewardena Mawatha
P.O. Box 91
Colombo
Telex: 1153
Phone: 32941-5

KIRLOSKAR OIL ENGINES LTD.
Samuel Sons & Co. Ltd
P.O. Box 46
Colombo 12
Cables: Echiron
Phone: 32341

LEYLAND INTERNATIONAL
Walker Sons & Co. Ltd.
P.O. Box 166
Colombo
Cables: Walkers
Phone: 28441

LUCAS SERVICE OVERSEAS LIMITED
Walker Sons & Co. Ltd.
P.O. Box 166
Colombo
Telex: Colombo 1118
Cables: Walkers
Phone: Fort No. 28441/9

NISSAN DIESEL MOTOR CO. LIMITED
Associated Motorways Ltd.
185 Union Place
Colombo 2

PETBOW LIMITED
Blackwood Hodge (Ceylon) Ltd
Nadaraja Building
Galle Road
P.O. Box 688
Kollupitiya
Colombo 3

PERKINS ENGINES LIMITED
The Ceylon Motor Transit Co. (1955) Ltd.
P.O. Box 299
101 D.S. Senanayake Mawatha
Colombo 8
Cables: Bonanza
Phone: 96101/5, 93401

PETTERS LIMITED
Sri Lanka State Trading (General) Corporation
P.O. Box No. 1686
No. 119 Wekande Road
Colombo 2
Cables: Rajawasa
Phone: 36233/4/5

St. Helena

LEYLAND INTERNATIONAL
Soloman & Co. (St. Helena) Ltd.
Jamestown
Island of St. Helena
Cables: Solomon St. Helena

St. Kitts

DETROIT DIESEL ALLISON
S.L. Horsford & Co. Ltd.
Agent of Tugs & Lighters Ltd.
Trinidad
P.O. Box 45
Basseterre
Telex: 387284 Trinidad
Cables: Tuglighter, Port-of-Spain
Phone: 2085 St. Kitts

LEYLAND INTERNATIONAL
S.L. Horsford & Co. Ltd.
P.O. Box 45
Basseterre
Cables: Horsford

PERKINS ENGINES LIMITED
TDC
P.O. Box 142
Basseterre
Cables: Tadco
Phone: 2511

PETTERS LIMITED
S.L. Horsford & Co. Ltd.
P.O. Box 45
Marshal House
Basseterre
Cables: Horsford
Phone: 2085

St. Lucia

DETROIT DIESEL ALLISON
St. Lucia Yacht Services
Agent of Tugs & Lighters Ltd.
Trinidad
Castries
Telex: 387284 Trinidad
Cables: Tuglighter, Port-of-Spain, Trinidad
Phone: 2879 St. Lucia

LEYLAND INTERNATIONAL
J.Q. Charles Ltd.
P.O. Box 279
Castries
Cables: Jayqu St. Lucia
Phone: 2721-4

LUCAS SERVICE OVERSEAS LIMITED
Peter & Co. Ltd.
P.O. Box 84
Castries
Cables: Peters

PERKINS ENGINES LIMITED
Universal Services Limited
Castries
P.O. Box 463
Phone: 4109-2797 St. Lucia

PETTERS LIMITED
Morne Dudon Industries Ltd.
P.O. Box 397
Castries
Telex: LC 317
Cables: Witeco
Phone: 8657

St. Martin

PERKINS ENGINES LIMITED
Caribbean Supply Co. N.V.
Pointe Blanche
P.O. Box 18
Phillipsburg
St. Maarten
Telex: 8000/1
Cables: Carsucon

St. Vincent

DETROIT DIESEL ALLISON
St. Vincent Motors
Agent of Tugs & Lights Ltd.
Trinidad
Arnos Vale
Telex: 387284 Trinidad
Cables: Tuglighter Port-of-Spain, Trinidad
Phone: 61557 St. Vincent

LEYLAND INTERNATIONAL
Hazells Ltd.
P.O. Box 108
Kingstown

LUCAS SERVICE OVERSEAS LIMITED
St. Vincent Sales & Services Ltd.
P.O. Box 599
Kingstown
Cables: Saleserve
Phone: 71820

PERKINS ENGINES LIMITED
St. Vincent Sales & Service
P.O. Box 599
Kingstown
Phone: 71820
Cables: Saleserve St.

PETTERS LIMITED
Hazells Ltd.
Kingstown
Telex: BOAC WB512
Cables: Hazel

Sudan

CATERPILLAR OVERSEAS SA
Sudanese Tractor Company Ltd.
74 Barlaman Avenue
P.O. Box 1840
Khartoum
Cables: Tractors
Phone: 72828

CATERPILLAR OVERSEAS SA
Sudanese Tractor Company Ltd
P.O. Box 301
Wad Medani
Cables: Tractors
Phone: 639

DETROIT DIESEL ALLISON
Taha El Sayed El Roubi & Co.
P.O. Box 467
Khartoum
Telex: 970-293
Cables: Telroubi
Phone: 71308, 76449, 33145 or 31923

DORMAN DIESELS LIMITED
Sharaf Modern Engineering Agencies
P.O. Box 1701
Khartoum
Telex: 280 Sharaf KM
Cables: Engincom
Phone: 75192 & 81773

L. GARDNER & SONS LIMITED
El Laboudi Commercial House
P.O. Box 1338
Khartoum
Phone: 41204

KATOLIGHT
Asim Trading Enterprises
P.O. Box 1354
Khartoum

KIRLOSKAR OIL ENGINES LIMITED
Jamnadas Odhavji Sheth
P.O. Box 648
Omdurman
Cables: Jaytex
Phone: 52308/56951

KLOCKNER-HUMBOLDT-DEUTZ AG
Dalia Commercial Enterprises
El Lew El Abiad Building
Flat 29
P.O. Box 2477
Khartoum
Telex: 401 Dalia
Cables: Yawadalla
Phone: 72965, 80042, 72557

LEYLAND INTERNATIONAL
El Gezira Automobile Co.,
P.O. Box 232
Khartoum
Telex: 455 Gezcar
Cables: Gezcar
Phone: Khartoum 78555-41867

LUCAS SERVICE OVERSEAS LIMITED
Angelo & Co. Spare Parts
P.O. Box 582
Khartoum
Phone: 43026

NISSAN DIESEL MOTOR CO. LIMITED
Hassouna Auto Work Shops & Stores
P.O. Box 380
Khartoum

PERKINS ENGINES LIMITED
Aboulela Engineering Co. Ltd.
P.O. Box 1341
Aboulela Building
U.N. Square
Khartoum
Telex: 421 Handasiya
Cables: Handasiya
Phone: 77257/9

PETTERS LIMITED
Bittar Engineering Co. Ltd.
P.O. Box 1011
407 Gamhuria Avenue
Khartoum
Telex: 285
Cables: Engbox
Phone: 70952, 71245, 71045

JOHN ROBSON (SHIPLEY) LIMITED
J.N. Valvis Ltd.,
P.O. Box 247
Khartoum

Surinam

CATERPILLAR OVERSEAS SA
Surinaamse Machinehandel N.V.
P.O. Box 1808
Paramaribo
Telex: PBO 42 CKC SME 182 Surmac
Cables: Surmac
Phone: 82222

DETROIT DIESEL ALLISON
N.V. Ingenieursbureau
H.N. Van Dijk
Industrieweg 300N
P.O. Box 447
Paramaribo
Telex: 397-160 SME VAN DYK
Cables: Vandijk
Phone: 89163

KATOLIGHT
George H. Knoppel
P.O. B. 222
Paramaribo

KLOCKNER-HUMBOLDT-DEUTZ AG
N.V.C.E. Vervuurts
Handelssonderneming
Beekhuizenweg No. 211-212
Livorno
Paramaribo
Telex: 151
Cables: Fructomel
Phone: Paramaribo 81515

LEYLAND INTERNATIONAL
NV Handelmaatschappij H.J. de Vries
P.O. Box 1849/1850
Waterkant 90/94
Paramaribo
Cables: Petzoldt
Phone: 4242

LUCAS SERVICE OVERSEAS LIMITED
Handelmaatschappij I Fernandes & Son N.V.
Corner Keizer en Klipsteenstraat 2-3-4-6-8-10
P.O. Box 1834
Paramaribo
Cables: Ferson
Phone: 76844, 71313, 74844

NISSAN DIESEL MOTOR CO. LIMITED
N.V. Reli
Steenbakkerijstract 56-60
P.O. Box 632
Paramaribo

PERKINS ENGINES LIMITED
N.V. Handelmaatschappij
H.J. de Vries
P.O. Box 1848-1850
Saramaccastraat
Paramaribo
Telex: 128
Cables: Petzoid
Phone: 71222-

PETTERS LIMITED
F.E. Van Der Jagt Agencies
Saramaccastraat 92
Paramaribo
Cables: Najagt
Phone: 71734-71117

VOLVO PENTA
H. Bromet
Domineestraat 34
P.O. Box 2924
Paramaribo
Telex: 141
Cables: Bromet
Phone: 73512/3

Swaziland

CATERPILLAR OVERSEAS SA
Barlow's (Swaziland) (Pty) Ltd.
P.O. Box 120
Manzini
Telex: SMX63
Cables: Shipments
Phone: 2363

Sweden

BOMBARDIER-ROTAX Gmbh
AB E. Fleron
P.O. Box 186
S-201 21 Malmo
Telex: 32112
Phone: 040-160000

BRIGGS & STRATTON CORPORATION
Axelson & Engwall K/B
P.O. Box 42036
S12612 Stockholm 42
Telex: 10606
Cables: Greiffcomp
Phone: 08-188460

CATERPILLAR OVERSEAS SA
Engson
Engström & Nilson Maskin AB
Fack
17220 Sundbyberg
Telex: 1544 (General)
1506 (Par)
Cables: Engson Stockholm
Phone: (8) 28-25-60

DETROIT DIESEL ALLISON
Detroit Diesel Allison
International — Europe
c/o General Motors Nordiska A.B.
Motorvagen 1
Fack
S-104-60 Stockholm
Telex: 1569
Cables: Genmotnord
Phone: 08-44-01-08

DORMAN DIESELS LIMITED
Eivind K. Son Sylvan A.B.
St. Badhusgatan 20
41121 Goteborg C.
Sweden
Telex: 20731
Cables: Sylvano
Phone: Goteborg
(031) 17 12 30

FORD MOTOR CO. LIMITED
Ford Motor Company A.B.
Fack
S102-50 Stockholm 27
Sweden
Telex: 19451
Phone: (08) 679800

KLOCKNER-HUMBOLDT-DEUTZ AG
Svenska Deutz A.B.
P.O. Box 6037
S-17506 Jaerfaella 6
Telex: Stockholm 17284
Cables: Deutzmotor
Phone: 08/7600400

LEYLAND INTERNATIONAL
British Leyland Sweden A.B.
Gamia Tuvevaegen 15B
Fack
402 70 Gothenburg 8
Telex: 20941 Leyscan S
Cables: Leyscan
Phone: 51.38.00

PERKINS ENGINES LIMITED
Malte Manson A.B.
Fack
S-58101 Linkoping
Telex: 50049 Manson S
Cables: Motorverktyg
Phone: 013/150000

PETTERS LIMITED
Aktiebolaget Transfer
Reserve Power
Prastgardsgatan 9
Postfack
172-20 Sundbyberg 1
Telex: 19339
Cables: Transfer-Stockholm
Phone: 98-16-20

RUGGERINI MOTORI S.p.A.
AB Sterner Blomouist
N. Grangesbergsgatan 28
S-214 50 Malmo
Telex: 32127 Sterner

VOLVO PENTA
Volvex Motor AB
Rundbacksgatan 1-5
417 05 Goteborg
Phone: 031/51 39 80

Switzerland

ALLIS CHALMERS
Aksa-AG Ingenievr Bureau — 012787
CH-5430
Wettingen 3
Telex: 55307
Cables: Aksa
Phone: 056-6-36 58

BOMBARDIER-ROTAX GMBH
Ernst Messer AG
CH-4704
Niederbipp BE
Telex: 62806
Phone: 065-732363

BRIGGS & STRATTON CORPORATION
European Office
Steinenring 12
4051 Basel
Telex: 62616
Cables: Brigs
Phone: 061/39 11 29

CATERPILLAR OVERSEAS SA
118 Rue Du Rhone
1211 Geneva 3
Telex: 22706 & 22833
Cables: Catoversea
Phone: (022) 20-62-22

DETROIT DIESEL ALLISON
Detroit Diesel Allison
International — Europe
c/o General Motors Suisse S.A.
Salzhausstrasse 21
2501 Biel-Bienne
Telex: 34217 GMS-CH
Cables: GMS-CH
Phone: 032-215111

DORMAN DIESELS LIMITED
Ultrich Rohrer-Marti AG
3052 Zollikofen
Berne
Telex: 32262
Cables: Rohrermarti
Phone: 031 571157

FORD MOTOR CO. LIMITED
Ford Motor Co.
(Switzerland) S.A.
Kurvenstrasse 35
CH 8021 Zurich
Telex: 52230
Phone: 601110

KLOCKNER-HUMBOLDT-DEUTZ AG
Hans F. Wurgler
Ing.-Buro
Rautistr. 31
CH-8047 Zurich-Albisrieden
Telex: 53868
Cables: Maschine
Phone: 526655/6

LEYLAND INTERNATIONAL
British Leyland (Switzerland) AG
Herostrasse 7
Postfach
8048 Zurich
Telex: 56762 55114
Cables: Motofrey
Phone: 010 4151 62
9090

PERKINS ENGINES LIMITED
Promot A.G.
CH-5745 Safenwil
Telex: 68134 (pmot ch)
Cables: Promot
Phone: 062-679211

PETTERS LIMITED
Kofel A.G.
Baumaschinen
Aegertstrasse 11
CH-8305 Industria Sued
Dietlikon
Telex: 54242
Phone: 01-833 1077

RUGGERINI MOTORI S.p.A.
Agrigid S.A.
P.B. 166
Chemin du Devin 51
Ch-1012 Lausanne
Telex: 25398 Agrid

VOLVO PENTA
Elektro Mechanik
Hans Moichtry
Gibraltarstrasse 13
6000 Luzern
Telex: 78256
Phone: 041/221222/23

Syria

BOMBARDIER-ROTAX GMBH
Rinaldo Draghi
B.P. 521
Alep
Phone: 14070-19120-26144

CATERPILLAR OVERSEAS SA
M. Ezzat Jallad & Fils
Pour le Commerce et la Representation
(M. Dureid Jallad & Cie)
Aleppo Street
Al-Qaboun
P.O. Box 23
Damascus
Cables: Jallad
Phone: 550-321
555-012

DORMAN DIESELS LIMITED
Nagib Baki Trading Co. S.A.
Fardoss Street
Baki Building
P.O. Box 135
Damascus
Cables: Baki
Phone: 11 343-4

INDIAN NATIONAL DIESEL ENGINE LIMITED
Hijjar & Massamari
Damascus

KATOLIGHT
Mouhamed Al-Khouja
P.O. Box 4911
Damascus

KIRLOSKAR OIL ENGINES LIMITED
Mohamed Ali Mardini
P.O. Box 2601
Damascus
Cables: Molimar
Phone: 223442
448422 – R

KLOCKNER-HUMBOLDT-DEUTZ AG
Dajani u.Farra AG
P.O. Box 109
Fardoss Street
Damascus
Cables: Tasada
Phone: 115678

KOHLER INTERNATIONAL LIMITED
Siedco
P.O. Box 363
Damascus

LEYLAND INTERNATIONAL
Aftomachine
P.O. Box 3130
Fardous Street
Damascus
Telex: Aftoma 11036SY
Cables: Aftoma Machine

LUCAS SERVICE OVERSEAS LIMITED
Automotive and Industrial Equipment Co. Inc.
El Fourat Street
P.O. Box 789
Damascus
Telex: 11357
Cables: Indeco
Phone: 111044-221961

PERKINS ENGINES LIMITED
Roy David Homsi
P.O. Box 295
Damascus
Cables: Dagsi

PETTERS LIMITED
Paysir Khaldi
Fardoss Street
Balkis 43
P.O. Box 3371
Damascus
Cables: Taldico
Phone: 113722

JOHN ROBSON (SHIPLEY) LIMITED
Consulting Engineering & Trading Bureau
P.O. Box 3526
Fardoss Street
Damascus

RUGGERINI MOTORI S.p.A.
Mahmoud Chaaban
B.P. 813
Alep

Tahiti

ALLIS CHALMERS
Etablissements Philippe J. Lucas
Boite Postale 483
130 Avenue De Marechal Foch
Papeete
Cables: Phillucas

DETROIT DIESEL ALLISON
Ets Bredin Freres
P.O. Box 21
Papeete
Phone: 20258

KLOCKNER-HUMBOLDT-DEUTZ AG
René Solari
7 Rue du Général de Gaulle
B.P. No. 2
Papeete
Phone: 259

KOHLER INTERNATIONAL
Lassere Industrie S.A.R.L.
B.P. 487
Avenue Georges Clemeneau
Papette, Tahiti

LEYLAND INTERNATIONAL
Intercar S.A.R.L.
P.O. Box 12
Papeete
Cables: Intercar
Phone: 2.04.35

LUCAS SERVICE OVERSEAS LIMITED
Ets Rene Solari et Fils
7 Rue du General de Gaulle
B.P. 2
Papeete
Telex: Soldo 055 FP
Cables: Soldo
Phone: 2.53.59

PERKINS ENGINES LIMITED
Monsieur H. Lombard
P.O. Box 6
Papeete
Phone: 20484

RUGGERINI MOTORI SpA
Sablage Metallisation
Peinture du Pacifique
P.O. Box 596
Papeete

Taiwan

ALLIS CHALMERS
112 Hsin Sheng South Road
Section 1
P.O. Box 1-41
Taipei
Telex: 11731 Chiaoltd
Cables: Chiaoltd
Phone: 350220-2

BRIGGS & STRATTON CORPORATION
William Chang & Co. Ltd.
293-911 Sung Kiang Road
P.O. Box 577
Taipei
Telex: 11743 Goldgate Wmchang
Cables: Wmchang
Phone: 522330, 522336

CATERPILLAR OVERSEAS SA
TTC Equipment & Service Ltd.
79 Chung Shan Road North
Section 2
P.O. Box 68-245
Taipei
Cables: Taitradco

DETROIT DIESEL ALLISON
Summit Engineering Co. Ltd.
Room C 3rd Floor
Morrison Plaza
No. 25 Jen Ai Road
4th Section
P.O. Box 53-5
Taipei
Telex: 785-11156 Sumiteng
Cables: Sumiteng
Phone: 722111

DORMAN DIESELS LIMITED
Winston & Co. Ltd.
Room 702 Traders Building
65 Nanking E Road
Section 3
P.O. Box 59826
Taipei
Cables: Winstoncol
Phone: 321373

KATOLIGHT
A1-Reliable Industries Inc.
P.O. Box 9-086
Taipei

KLOCKNER-HUMBOLDT-DEUTZ AG
Cosa Liebermann Taiwan Ltd.
107 Section 2
Chungshan N. Road
2nd Floor
P.O. Box 1756
Taipei
Telex: 11312 Cosaco
Cables: Coasco
Phone: 51171218

KOHLER INTERNATIONAL LIMITED
Solomon Enterprises Co. Ltd.
P.O. Box 7-154
Room 703
129 Sung Chiang Road
Taipei

LEYLAND INTERNATIONAL
Yung Wei Tung Trading Co. Ltd.
No. 141 Tun Hua North Road
P.O. Box 80
Taipei
Telex: 22545 Weitung
Cables: Weitungmo
Phone: 755611, 730188
743511, 743998

NISSAN DIESEL MOTOR CO. LIMITED
The Chinese Automobiles Trading Co. Ltd.
169 Section 2
Nanking East Road
Taipei

PERKINS ENGINES LIMITED
Golden Gate Engineering & Development Co. Ltd.
123 Nanking East Road
3rd Section
P.O. Box 762
Taipei
Telex: 1174 Goldgate
Phone: 551 6733
571 9216
511 8397

RUGGERINI MOTORI S.p.A.
Fuh Yang Trading Co.
P.O. Box 9-086
93 Shao–Tsuang Street
Kaohsiung
Telex: 71174 Unitrade

WHITE ENGINES INCORPORATED
Solomon Enterprise Co. Ltd.
Room 703
P.O. Box 7-154
129 Sung-Chiang Road
Taipei
Telex: 22218 Padobebe
Cables: Solo
Phone: 560368/69

Tanzania

ALLIS CHALMERS
Holman Bros. (East Africa) Ltd.
P.O. Box 1938
Dar-Es-Salaam

CATERPILLAR OVERSEAS SA
Construction Equipment Division of the UAC of Tanzania Ltd.
P.O. Box 2568
Dar-Es-Salaam
Cables: Afritrak
Caressalaam
Phone: 63355

DETROIT DIESEL ALLISON
Blackwood Hodge Group Services Ltd.,
Hunsbury Hill Ave.,
Northampton NN4 9QT
U.K.
Telex: 31476
Cables: Suntract,
Phone: 0604-61111

DORMAN DIESELS LIMITED
Agricultural & Industrial Supplies Co. Ltd.
P.O. Box 4797
Dar-Es-Salaam
Cables: Agrind
Phone: 25201

KLOCKNER-HUMBOLDT-DEUTZ AG
Achelis (Tanganyika) Ltd.
Karim Ladha Mansion
P.O. Box 9003
Makunganya Str.
Dar-Es-Salaam
Telex: 41067
Cables: Achelissons
Phone: 21078, 24011

LEYLAND INTERNATIONAL
Leyland Albion (Tanzania) Ltd.
P.O. Box 2388
Dar-Es-Salaam
Cables: Leyalb
Phone: 63439, 63449

LUCAS SERVICE OVERSEAS LIMITED
Delta (Tanzania) Ltd.
Pugu Road
P.O. Box 44
Dar-Es-Salaam
Telex: 41011
Cables: Delta
Phone: 20129, 22336

PERKINS ENGINES LIMITED
Century Tractor & Implements (Tanzania) Ltd.
P.O. Box 21011
Dar-Es-Salaam
Telex: 41058
Phone: 27727/8

PETTERS LIMITED
Agricultural & Industrial Supplies Co. Ltd.
P.O. Box 4797
Upanga Road
Dar-Es-Salaam

Thailand

ALLIS CHALMERS
89 Khor
Super Highway 17½km
Bangkok 9
Telex: BK 2740 Italthai
Cables: Titrac
Phone: 791136-8

BOMBARDIER-ROTAX GMBH
Jones Company Ltd.
P.O. Box 686
Bangkok
Phone: 31267, 38628

BRIGGS & STRATTON CORPORATION
United Motor Works (Siam) Ltd.
156 Suriwongse Road
P.O. Box 370
Bangkok
Telex: 7882629
Cables: Yeesoh
Phone: 32987-88-89

CATERPILLAR OVERSEAS SA
The International Engineering Co. Ltd.
614 Sukhumvit Road
P.O. Box 39
Bangkok
Telex: Intenco Th2626
Cables: Gysom

DETROIT DIESEL ALLISON
Borneo Engineering Co. Ltd.
226 Nares Road
Bangkok
Telex: 788-2826
Cables: Deceja
Phone: 38981/38989

DORMAN DIESELS LIMITED
Term Engineering Co. Ltd.
P.O. Box 1513
No. 302/2 Therd Thai Road
Bangkok 6
Telex: BK 2588 Nuclear Bk
Cables: Nuclear
Phone: 667055, 665335, 665422-3

L. GARDNER & SONS LIMITED
Khin Nam Huat Limited Partnership
P.O. Box 255
361/363 Mahaputtarum Road
Bangkok
Telex: Khinnam Bk 2560
Phone: 33542 39211

INDIAN NATIONAL DIESEL ENGINE CO. LTD.
Siamlester & Co. Ltd.
Uchuliang Foundation Building
6th Floor,
968 Rama IV Road
Bangkok

J.L.O.
Hamlet & Co. Ltd.
1999 Petchburi Road Extension
P.O. Box 122
Bangkok 10
Cables: Hamlet
Phone: 927478-9, 92251

KATOLIGHT
Siri Sakol Co. Ltd.
132-134 Samyod New Road
Bangkok

KIRLOSKAR OIL ENGINES LIMITED
Mechanical Equipment Co.
295 Siphya Road
Mequip Building
P.O. Box 550
Bangkok
Telex: BK 2509
Cables: Mequip
Phone: 37961-5

KLOCKNER-HUMBOLDT-DEUTZ AG
B. Grimm & Co.
1643/4 Petchburi Road (Extension)
P.O. Box 66
Bangkok
Telex: BK 2614 Thaigrimm
Cables: Thaigrimm
Phone: 2524081

KOHLER INTERNATIONAL LIMITED
The East Asiatic Co.
Oriental Machinery Stores
60/1 Super Highway Route 31
Bangkok

LEYLAND INTERNATIONAL
Ley-Thai Motors Limited
144 Sukhumvit Road
Bangkok
Cables: Leythai
Phone: 54061-7

LUCAS SERVICE OVERSEAS LIMITED
Oriental Machinery Stores
Worachakr Dept.
539 Worachakr Road
Bangkok
Telex: Asiatic TH 2615
Cables: Machineco
Phone: 218161, 225102

NISSAN DIESEL MOTOR CO. LTD.
Siam Motors Co. Ltd.
865 Rama 1 Road
Bangkok

PERKINS ENGINES LIMITED
Mechanical Equipment Co. Ltd.
Perkins Engines Division
28 Soi Asoke Sukumvit Road
Bangkok
Telex: TH 2509
Cables: Mequip
Phone: 3912680, 3925571

PETBOW LIMITED
Loxley (Bangkok) Ltd.
G.P.O. Box 214
Loxley Building
304 Suapeh Road
Bangkok

PETTERS LIMITED
Loxley (Bangkok) Ltd.
G.P.O. Box 214
Loxley Building
304 Suapah Road
Bangkok
Telex: TH2808 Castrol & 2627 Castrol
Cables: Loxley
Phone: 26641

JOHN ROBSON (SHIPLEY) LIMITED
Lampangchai Co. Ltd.
46/7-14 Petchburi Road
Bangkok

RUGGERINI MOTORI S.p.A.
Saha Thai Machinery
133-135 Sam Yod. New Road
Bangkok

VOLVO PENTA
The East Asiatic Company Ltd.
Super Highway Branch
Oriental Machinery Stores
60/1 Highway Route 31
Din Daeng-Ladprao
Bangkok 4
Telex: BK 2615
Cables: Industreac
Phone: 770176, 770183, 770185, 770214, 771784, 773227

WHITE ENGINES INCORPORATED
Dodge & Seymour (Far East) Inc.
P.O. Box 192
942/16 Rama IV Road
4th Floor, S.C. Building
(Opposite Chulalongkon Hospital)
Bangkok
Cables: Eximco
Phone: 33557

Timor

PETTERS LIMITED
Sociedade Oriental de Transportes
E. Armazaens Lda.,
Avenida Marginal sa da Bandiera
Dili

Togo

CATERPILLAR OVERSEAS SA
Gastonegre S.A.
B.P. 134
Lome
Telex: 5231
Phone: 22-81/82, 32-70

LEYLAND INTERNATIONAL
Cie Francaise de L'Afrique Occidentale
Agence Centrale
Agence Centrale
B.P. 1246
Lome
Cables: Senafrica

LUCAS SERVICE OVERSEAS LIMITED
U.A.C. Togo
B.P. 345
Lome
Telex: 5219
Phone: 55.82, 58.15

PETTERS LIMITED
DTG (Departement Hamelle Afrique) S.A.
14 Avenue de la Liberation
B.P. 129
Lome
Telex: Detege 5241 Departement 9
Cables: Hamalaf
Phone: 62.45

PERKINS ENGINES LIMITED
Ets Gastonegre S.A.
14 bis
rue de Commerce
B.P. 134
Lome
Telex: 231 Gastonegre
Cables: Gastonegre
Phone: 22.81

Tonga

LEYLAND INTERNATIONAL
E.M. Jones Ltd.
P.O. Box 4
Nuku'alofa
Cables: Jones Nuku'alofa

Suva Motors Ltd.
P.O. Box 4
Suva
Fiji

Tuvalu Islands

LEYLAND INTERNATIONAL
Millers Motors
P.M.B.
Suva
Fiji
Telex: 2195
Cables: Millmot
Phone: 23031

Trinidad

BRIGGS & STRATTON CORPORATION
West Bend Sales Ltd.
P.O. Box 978
Port of Spain
Cables: Wends
Phone: 62-24057/62-23245

CATERPILLAR OVERSEAS SA
Tractors and Machinery (Trinidad) Ltd.
P.O. Box 945
Port of Spain
Telex: 308
Cables: Tracmac Port-of-Spain Trinidad
Phone: 62-1545, 62-31431

DETROIT DIESEL ALLISON
Tugs & Lighters Ltd. (Main Office)
69 Independence Square
P.O. Box 600
Port of Spain
Telex: 387284 Alston WG
Cables: Tuglighter Port-of-Spain
Phone: 62-24127

DORMAN DIESELS LIMITED
Alstons Marketing Co. Ltd.,
(Wilsons & Johnstone Division)
P.O. Box 431
Port of Spain

KATOLIGHT
Tatec
P.O. Box 355
Port of Spain

KOHLER INTERNATIONAL
Kirkconnell Brothers Ltd.
P.O. Box 72
School House Road
Kirk Plaza
Grand Cayman
West Indies

Thomas Peake & Co. Ltd.
P.O. Box 301
76 Henry & 2a Charlotte St.
Port of Spain

LEYLAND INTERNATIONAL
Trinity Motors (Robinsons) Ltd.
32 South Quay
P.O. Box 241
Port of Spain
Cables: Trinmot Port-of-Spain

LUCAS SERVICE OVERSEAS LIMITED
Laughlin & De Gannes Ltd.
37 Dundonald Street
P.O. Box 1254
Port of Spain
Cables: Bengaston Trinidad
Phone: PBX 51712/17

NISSAN DIESEL MOTOR CO. LIMITED
Neal & Massy Ltd.
P.O. Box 1298
Port of Spain

PERKINS ENGINES LIMITED
Tractors and Machinery
(Trinidad) Ltd.
Eastern Main Road
Laventille
Trinidad
Telex: Tracmac 308
Cables: Tracmac
Phone: Laventille 62-31431
San fernando
65-78840

PETTERS LIMITED
Charles McEnearney & Co. Ltd.
25 Richmond St.
P.O. Box 178
Port of Spain
Telex: CMCE 256
Cables: McEnearney
Phone: 32731-9

Tunisia

CATERPILLAR OVERSEAS SA
Ets. P. Parrenin S.A.
91 Avenue de Carthage
Tunis
Telex: 12422
Cables: Parnin
Phone: 256577

DETROIT DIESEL ALLISON
Detroit Diesel Allison International – Europe
c/o General Motors France
56 à 58 Ave. Louis Roche
92231 Gennevilliers,
France
Telex: Genlmot 62050
Cables: Parautexap
Phone: 793-3450
Gennevilliers

KLOCKNER-HUMBOLDT-DEUTZ AG
Societe Tunisienne
Sotradies
16 Rue Ahmed Tlili
Tunis
Telex: 12255 Tunis
Cables: Sotradies
Phone: 247033
243340

LEYLAND INTERNATIONAL
Societe de L'automobile et
du Materiel
10 bis Avenue de Ghana
Tunis
Telex: Sami 2252 TN
Cables: Sama
Phone: 242 155

LUCAS SERVICE OVERSEAS LIMITED
Soutdiem
35 Rue de Marseille
Boite Postale 773
Tunis
Telex: 12180
Cables: Sotudiem
Phone: 241090 245113

PERKINS ENGINES
Ets Louis Montenay
49 Avenue de Carthage
Tunis
Telex: 743
Cables: Secfa
Phone: 240-312/240-324

PETTERS LIMITED
Agrimotor S.A.
30 Rue Ali Dargouth
B.P. 415
Tunis
Telex: Agrimo 12014
Cables: Etaboulaire
Phone: 247.060

RUGGERINI MOTORI S.p.A.
Societe de L'Automobile et
du Materiel
10 Bis
Avenue de Ghana
Tunis
Telex: 12252 SAM TN

WHITE ENGINES INCORPORATED
Ets Louis Montenay
49, Avenue de Carthage
Boite Postale 862
Tunis
Telex: 743
Cables: Secfa
Phone: 240.324, 240.312

Turkey

ALLIS CHALMERS
Kutsi Begdes, Mumessillik
Ithalat-Ihracat
Barbaros Bulvari
Ikinci Yol-Besiktas
P.O. Box 125 Besiktas
Istanbul
Telex: 22391 Kube Tr
Cables: Kutsi-Besiktas
Phone: 47 89 50

BOMBARDIER-ROTAX Gmbh
Sefik Soyuyuce
Sitam Ticaret
Karanfil Sokak 30/5-6
Yenisehir
Ankara
Phone: 17 49 09

BRIGGS & STRATTON CORPORATION
Alagoz ve Ort. Alem Kol. Sti.
Persembepazari-Omeraga
Sokak No. 4/3
Karakoy
Istanbul
Telex: 23472
Cables: Alempomp
Phone: 449433

CATERPILLAR OVERSEAS SA
Cukurova Ithalat ve Ihracat T.A.S.
Buyukdere Caddesi
P.O. Box 124
Sisli
Telex: 22693
Cables: Cukurtas
Phone: 47-48-30

DETROIT DIESEL ALLISON
Kurt Basakinci Co.
Vali Dr. Resit Cad, 52
Cankaya
P.O. Box 55
Bakanliklar
Ankara
Telex: 821-42289, Kurt, TR

DORMAN DIESELS LIMITED
Ahmet Buldanlioglu & Co.
Gazi Bulvari 26
P.O. Box 1
Izmir
Telex: Istanbul 22603
Nakltr
Cables: Ehem
Phone: 23991, 36392

KATOLIGHT
Kopsan
Makine Sanayi ve Ticaret Ltd. Sti.
Buyudere Caddesi No. 66
Daire 6 Mecidiyekoy
Istanbul

KLOCKNER-HUMBOLDT-DEUTZ AG
Deutz Humboldt
Makina Ticaret Ltd., Sti
P.K. 226 Beyoglu
Siraselviler Caddesi Soganci Sokak
No. 19/2 Ittihad Sigorta Apt.
Cihangir
Istanbul
Telex: 22239 dhit tr.
Cables: Deutzhumboldt
Phone: 459166

LEYLAND INTERNATIONAL
B.M.C. Sanayi Ve Ticaret AS
Gazi Bulvari 47/49
P.K. 260
Izmir
Telex: 52320 BMCE TR
Cables: Bemece
Phone: 39780

LUCAS SERVICE OVERSEAS LIMITED
Elektro-Dizel Motor Sanayi ve Ticaret A.S.
Elmadag Caddesi No. 54-56
Harbiye
Istanbul
Telex: 23435
Cables: Simpar

PERKINS ENGINES LIMITED
Dizel Makina Ltd., Sirketi
Cumhuriyet Caddesi No. 10/c
Sisli
Istanbul
Telex: Eligold 22357 Istanbul
Cables: Dizlimited
Phone: 473675, 460941

PETBOW LIMITED
Tarmo Tarim Motor
Sanayi ve Ticaret A.S.,
Necatibey Cad
54 Karakoy
Istanbul

PETTERS LIMITED
Aziz Isvan Ticaret Ltd.
P.O. Box 82
Sishane Mesrutiyet Caddesi 184
Beyoglu
Istanbul
Cables: Arlim
Phone: 4.419.77

RUGGERINI MOTORI S.p.A.
Alveris
Kurekciler Caddesi No. 50
Karakoy
Istanbul

VOLVO PENTA
Erk Mühendislik Sanayi ve Ticaret Ltd.
Yildiz Caddesi Akdogan Sokak No. 27
P.K. 535 Karaköy
Istanbul
Telex: 22711 Orya TR
Cables: Erklimitsan
Phone: 40 31 36

WHITE ENGINES INCORPORATED
Metya Limited Sirketi
Methatpasa Cadessi No. 39/7
Ankara
Cables: Metyaticaret
Phone: 17.87.04

Uganda

ALLIS CHALMERS
Holman Bros. (East Africa) Ltd.
P.O. Box 2790
Kampala

CATERPILLAR OVERSEAS S.A.
Construction Equipment Division
of Gailey & Roberts (Uganda) Ltd.
P.O. Box 7123
Kampala
Cables: Afritak
Phone: 59441

DETROIT DIESEL ALLISON
Wiggleworth & Co. Uganda Ltd.
P.O. Box 7127
Kampala
Cables: Pettinato, Mombasa

KLOCKNER-HUMBOLDT-DEUTZ AG
Achelis (Uganda) Ltd
24 Seventh Street (Industrial Area)
P.O. Box 7198
Kampala
Cables: Achelissons
Phone: 53751

LEYLAND INTERNATIONAL
Leyland Albion (Uganda) Ltd.
P.O. Box 3778
Kampala
Cables: Leyalb

LUCAS SERVICE OVERSEAS LIMITED
Delta Ltd.
P.O. Box 28
Kampala
Cables: Delta
Phone: 42421/2

PERKINS ENGINES LIMITED
Farm Machinery Distributors (U) Ltd.
P.O. Box 5763
Kampala
Cables: Farmec
Phone: 41237

PETTERS LIMITED
Mackenzie Technical Services Ltd.
43 Roseberry Road
P.O. Box 7011
Kampala
Telex: 61080
Phone: 43326/7

United Kingdom

ALLIS CHALMERS
Deekay House
North Street
P.O. Box 14
Havant PO9 1QH
Hampshire
England
Telex: 86491 Deekay Havant
Cables: Autogen
Phone: Havant 74122 0712

BRIGGS & STRATTON CORPORATION
Autocar Electrical Equipment Co. Ltd.
16 Rippleside Commercial Estate
Ripple Road
Barking Essex
England
Telex: 885060
Cables: Autodulbee
Phone: 01-592 2112/3/4

DETROIT DIESEL ALLISON
Detroit Diesel Allison International – Europe
Div. of General Motors Ltd.
London Road
P.O. Box 6
Wellingborough
Northamptonshire NN8 2DL
England
Telex: 851-31329 Wellingborough
Cables: Genmopower,
Phone: 71122 Wellingborough

KATOLIGHT LIMITED
Tatton Electric Co. Ltd.
501 Staines Road West
Ashford
Middlesex TW15 2AB
England

KATOLIGHT LIMITED
W.S.M. Electric and Electronic Services
12 Campbell Street
Hamilton ML3 6AS
Scotland

KIRLOSKAR OIL ENGINES LIMITED
P.M. Engine Ltd.
35 Piccadilly
London W1V 0JD
England
Telex: 266250
Cables: Dieselpicc
Phone: 734-1066/7/8

KLOCKNER-HUMBOLDT-DEUTZ AG
Deutz Engines Limited
Riverside Road
Wandsworth
London SW17
England

Telex: 928373
Cables: Deutz
Phone: 9469161

KOHLER INTERNATIONAL LIMITED
J.H. Hancox Ltd.
Wood Lane
Earlswood
Solihull
Warwickshire B94 5JW
England

KOHLER INTERNATIONAL LIMITED
Hydraulic Cranes (Scotland) Ltd
57-61 Vere Road
Blackwood JL 11 9 RR
Lanarkshire
Scotland

LEYLAND INTERNATIONAL
Leyland House
174 Marylebone Road
London NW1 5AA
England

PERKINS ENGINES LIMITED
Peterborough PE1 5NA
England

Telex: 32501
Cables: Perkoil
Phone: 67474

PETBOW LIMITED
Marketing Advisory Services
329 Grand Building
Trafalgar Square
London W.C.2
England

PETTERS LIMITED
Hamble
Southampton SO3 5NJ
England

Telex: 47626
Phone: Hamble 2061

RUGGERINI MOTORI S.P.A.
Glanfield Lawrence Concessions
Victoria Road
Portslade
Brighton
Sussex BN4 1XY
England

VOLVO PENTA
Bolinders Co. Ltd.
150/8 Kings Cross Road
London WC1X 9DN
England

Telex: 24768
Phone: 01-278 2711

VOLVO PENTA
Robert Craig & Sons Ltd.
Great George's St.
Belfast
Northern Ireland BT15 1BW

Phone: Belfast (0232) 32971

WHITE ENGINES INCORPORATED
Diamond T. Motors
411 London Road
Isleworth
Middlesex U.K.
England

Telex: 935576
Cables: Diamond Isleworth Hounslow
Phone: Isleworth 71534

Upper Volta

CATERPILLAR OVERSEAS SA
Manutention Africaine
B.P. 636
Ouagadougou

Telex: 5230
Phone: 2155

LEYLAND INTERNATIONAL
Dima — HV
B.P. 25
Ougadougou

LUCAS SERVICE OVERSEAS LIMITED
Scoa
Service Auto
B.P. 32
Ougadougou

Uruguay

ALLIS CHALMERS
Cassarino Hermandos S.A.
Galica 1069
Casillad de Correo 153
Montevideo

Telex: Public Booth No. 702
Cables: Casarihnos Montivideo
Phone: 9122-22

BRIGGS & STRATTON CORPORATION
Manuel Guelfi y Cia
Av. Agraciada 1777
Montevideo

Telex: 398220
Cables: Maguel

CATERPILLAR OVERSEAS SA
General Machinery Co. S.A.
Avenida Joaquin Suarez 2856
Montevideo

Telex: 398.730 Gemcosa UY730
Cables: Gemcosa Montevideo
Phones: 20-99-21/2/3/4/5

DETROIT DIESEL ALLISON
Mapell Ltda
Cerrito 352
P.O. Box 441
Montevideo

Telex: 398-330
Cables: Mapell
Phone: 840223

DORMAN DIESELS LIMITED
Maquinas & Materiales SA
18 de Julio 2302
Montevideo

Cables: Maquinas Montevideo
Phone: 46171, 414336

FORD MOTOR CO. LIMITED
Ford (Uruguay) S.A.
Casilla de Correo 296
Montevideo

Telex: 298 096
Phone: 207521

KLOCKNER-HUMBOLDT-DEUTZ AG
Arfil S.A.
Casilla Correo 21
Tala 2270
U-Montevideo

Telex: Arfilsa UY 387
Cables: Micarta
Phone: 293814

KOHLER INTERNATIONAL LIMITED
Hidrotecnica Sociedad Anonima
General San Martin 2505/07
Montevideo

LEYLAND INTERNATIONAL
Frank Surgey S.A.
Casilla de Correo 704
Montevideo

Cables: Fransur

LUCAS SERVICE OVERSEAS LIMITED
Automotive Accessories S.A.
Cuareim 2179
Montevideo

Cables: Socra Montevideo
Phone: 27849, 20.62-10

PERKINS ENGINES LIMITED
Perkins Rio de la Plata S.A.
Av. Juan Carlos Gomez 1492
Montevideo

Telex: 278
Phone: 916254/88844

PETTERS LIMITED
Horacio Torrendell S.A.
Cuareim 2050 al 2082
Montevideo

Cables: Horatorsa Montevideo
Phone: 20.13.01

U.S.A.

ALLIS CHALMERS
P.O. Box 563
Harvey
Illinois 60426
Telex: 910-2572135
Hvyilep
Phone: (312) 339-3300

BRIGGS & STRATTON CORPORATION
Milwaukee
P.O. Box 702
Wisconsin 53201
Telex: 02-6776
Cables: Basco
Phone: (414) 461-1212

CATERPILLAR AMERICAS CO.
Peoria
Illinois 61629

DORMAN DIESELS LIMITED
Hemisphere Engine Sales Co. Inc.
P.O. Box 12385
Fort Worth
Texas 76116

DORMAN DIESELS LIMITED
Bay Engine & Parts Co.
1640 Evans Avenue
San Francisco
California 94124
Phone: (415) 826-7400

FORD MOTOR CO. LIMITED
Parts and Service Division
Industrial Engine Operations
P.O. Box 3080
Livonia
Michigan 48151
Phone: 32 32808

L. GARDNER & SONS LIMITED
L. Klindt
186 Venus Lane
Tamarac Park
Key West
Florida 33040
Phone: (Area code)
305 294 9434

KLOCKNER-HUMBOLDT-DEUTZ AG
Deutz Corporation
7585 Ponce de Leon Circle
Atlanta
Georgia 30340
Telex: 0707478
Cables: Oilengine
Phone: (404) 4496140

KOHLER INTERNATIONAL LIMITED
Kohler
Wisconsin 53044
Telex: 26888
Cables: Kohlerint
Phone: (414) 457-4441

LEYLAND INTERNATIONAL
British Leyland Motors Inc.
600 Willow Tree Road
Leonia
New Jersey 07605

LUCAS SERVICE OVERSEAS LIMITED
Lucas Industries America Inc.
Two Northfield Plaza
Troy
Michigan 48084
Telex: 8102321644
Phone: 3138791920

NISSAN DIESEL MOTOR CO. LIMITED
Chrysler Corporation
Marine and Industrial Products Div.
P.O. Box 1
Marysville
Michigan 48040

PERKINS ENGINES LIMITED
P.O. Box 283/24175 Research Drive
Farmington
Michigan 48024
Telex: 023.5300
Cables: Perkoil Farmington
Phone: 313-477-3900

PERKINS ENGINES LATIN AMERICA INC.
2600 Douglas Road
Suite 600
Coral Gables
Florida 33134
Telex: 515880
Cables: Perklac
Phone: (305) 4484884

PETTERS LIMITED
Onan Division
Onan Corporation
1400 73rd Avenue N.E.
Minneapolis
Minnesota 55432
Telex: 29-0476
Phone: (612) 574-50000

VOLVO PENTA
Volvo Penta of America Inc.
P.O. Box 12758
Koger Executive Center
Building No. 18
Suite 106
Norfolk
Virginia 23502
Telex: 823661
Phone: (804) 461-1552

WHITE ENGINES INCORPORATED
XM World Trade Inc.
One World Trade Center
Suite 4501
New York
New York 10048
Telex: 232886
Cables: Eximco
Phone: (212) 432-5200

U.S.S.R.

CATERPILLAR OVERSEAS SA
Pokrovsky Boulevard 4/17
Apartment 13
Moscow 101000
Telex: 7802 SU
Phone: 207-5658
207-1007
207-2625
207-2982

Venezuela

ALLIS CHALMERS
Sanchez & Cia S.A.
Avda Rossevelt — Prado de Marie
Apartado 1006
Caracas 101
Telex: 21235
Phone: 62.99.11 A1 19

BRIGGS & STRATTON CORPORATION
Sanchez y Cia S.A.
Apartado Postal 1006
Caracas
Telex: 21235
Cables: Sanchez
Phone: 62.99.11
62.45.01

CATERPILLAR OVERSEAS S.A.
General Electric de Venezuela S.A.
Calle Real — Sabana Grande
Apartado 1666
Caracas 101
Telex: 22724
Cables: Ingenetric
Phone: 71-98-11

DETROIT DIESEL ALLISON
Stewart & Stevenson De Venezuela S.A.
Apartado 1809
Av. 2 (El Milagro) No. 86A-27
Maracaibo
Telex: 395-61140
Savenmex
Phone: 223966 or 223251

DORMAN DIESELS LIMITED
Edificio Halven Esq Monroy
Avenida Universidad
Aparto 808
Caracas
Telex: 22567 A/B Halven
Cables: Halven
Phone: 54 312 1/25

FORD MOTOR CO (VENEZUELA) S.A.
Avenida Henry Ford
Zona Industrial Sur
Apartado 354
Valencia
Telex: Caracas 765377

KATOLIGHT
Electronica Osorio S.R.L.
Apartado 75776
Caracas 107

KOHLER INTERNATIONAL LIMITED
Lorenzo M. Bustillos &
Cia. Sucs. C.A.
Apartado No. 234
Av. Romulo Gallegos – Boleita
Caracas 10

KLOCKNER-HUMBOLDT-DEUTZ AG
Navimeca
Calle Comercio 103
YV – Punto Fijo – Edo. Falcón
Telex: 54164
Cables: Navimeca
Phone: 2343

LEYLAND INTERNATIONAL
Compania Anonima De Automoviles Y Repuestos
Apartado 6561
Pilita A. Mamey 95-1
Caracas
Telex: 21337 Tedesur
Cables: Cadayr
Phone: 429371 & 73

LUCAS SERVICE OVERSEAS LIMITED
Core Compania Anonima
Apartado 3969
Caracas
Telex: 21322
Cables: Core
Phone: 62.46.62/3/4/5

PERKINS ENGINES LIMITED
Corporacion Venezolana Del Motor
Division de Maquinarias
Apartado Postal No.74
Avenida Principal de los Ruices
Caracas
Telex: 22741 Sacco
Cables: Luap
Phone: 34.75.71

PETTERS LIMITED
Corporacion Mercantil
Venezolana S.A.
(Comersa)
Apartado 20
Avenida Diaz Moreno
c/c Cantaura
Con Antiguas Telares Karam
Valencia
Cables: Comersa
Phone: 89091/4

JOHN ROBSON (SHIPLEY) LIMITED
Corporacion Mercantil
Venezolana S.A.
Av. 100 (Constitucion) No. 92-20
Apartado No. 20
Valencia

RUGGERINI MOTORI S.p.A.
Befransa
Apartado de Correos
Los Ruices 70029
Final Segunda Transversal de
Los Ruices
Caracas
Telex: 25208 Befransa

VOLVO PENTA
Nautiven
Nautica Venzolana CA
Av. Los Samanes
Apartado 50925
Caracas
Telex: 21252
Cables: Nautiven
Phone: 74 05 15
74 17 91

WHITE ENGINES INCORPORATED
Servicios Tecnomat
Calle Branger No. 14
Los Rosales
Apartado (P.O. Box) 10.294
Caracas
Cables: Temaser
Phone: 61-12-24 or
61-12-25

Vietnam

LEYLAND INTERNATIONAL
Societe Des Garages Charner
131-133 Boulevard Nguyen Hue
Saigon
Cables: Charnerco

Societe D'Exploitation
Industrielle Commerciale
Saigon-Garaga
P.O. Box 47
Saigon
Cables: Losseic
Phone: 20.603

Vietnam KY-XA Cong-TY
155 Bd. Hue
Saigon

Societe Indochinoise de
Transports SA
4 Rue Nguyen
Trung Truc
B.P. 422
Saigon
Cables: S.I.T.

Virgin Islands

BRIGGS & STRATTON CORPORATION
Midwest Corp.
4 Strand Street
Box 1107
Frederiksted
St. Croix 00840

KOHLER INTERNATIONAL LIMITED
Spesco
P.O. Box 3127
St. Thomas
Virgin Islands

LEYLAND INTERNATIONAL
Hayes & Co.
P.O. Box 430
Christiansted
St. Croix
Virgin Islands 00820

LEYLAND INTERNATIONAL
Motor Holdings Ltd.
Quay Motor Co. Ltd.
P.O. Box 117
Tortola
Telex: Anebank VB 918
Cables: Quaymo
Phone: 2470

PETTERS LIMITED
J.R. O'Neal
P.O. Box 127
Road Town
Tortola
Cables: O'Neal

Western Sahara

LEYLAND INTERNATIONAL
Metalurgica de Santa Ana S.A.
Delegacion En El Aaium
Apartado 8
El Aaiun

Western Samoa

L. GARDNER & SONS LIMITED
I.H. Carruthers Ltd.
P.O. Box 578
Apia
Phone: Apia 88

LEYLAND INTERNATIONAL
Motor Distributors (Samoa) Ltd.
Main Beach Road
Apia
Cables: Modist

Yemen Arab Republic

CATERPILLAR OVERSEAS SA
The Tehama Trading Co. Ltd.
P.O. Box 337
Hodeidah
Telex: For the Tehama Trading Co. Ltd. c/o Cwbooth 501/502 Hodeidah
Cables: Tradco

DORMAN DIESELS LIMITED Salem Saleh Bahaj P.O. Box 895 Hodeidah	Cables: Bahaj Phone: 2224
KIRLOSKAR OIL ENGINES LIMITED Adhban Trading Corporation Republican Palace Street P.O. Box 1105 Sanaa	Telex: 260 Adhban YE Cables: Adhban Phone: 2918
KLOCKNER-HUMBOLDT-DEUTZ AG Tihama Tractors and Engineering Co. Ltd. P.O. Box 49 Sanaa	Telex: 217 ye Cables: Tiham Phone: 2462
LEYLAND INTERNATIONAL Amin Kassem M. Sultan & Co. P.O. Box 888 Taiz	Cables: Alamin Hodeida
NISSAN DIESEL MOTOR CO. LIMITED Adhban Trading Corporation P.O. Box 1105 Sanaa	
PERKINS ENGINES The Tehama Trading Co Ltd. P.O. Box 3337 Hodeidah	Telex: 218 Hoship ye (Sanaa) Cables: Tradco Phone: 2406
PETTERS LIMITED Awadh Salem Baobeid & Sons P.O. Box 3786 Hodeidah	Telex: 513 Ye Baobeid Cables: Bin Awadh Phone: 2742
JOHN ROBSON (SHIPLEY) LTD Alwataary General Trading & Agricultural Development Co. P.O. Box 61-2207 Maidan Altahrir Sanaa	
RUGGERINI MOTORI S.p.A. Abdullah H. Alsonidar & Sons P.O. Box 11 Sanaa	Telex: 226 Snidar

Yemen (Peoples Democratic Republic of)

CATERPILLAR OVERSEAS SA National Company for Home Trade P.O. Box 90 Crater Aden	Telex: ADN 211 Cables: Hometrade Phone: 51133-9, 51632-3
DORMAN DIESELS LIMITED Salem Saleh Bahaj P.O. Box 379 Aden	Cables: Bahaj Aden Phone: Crater 52533, 535909
L. GARDNER & SONS LIMITED The Arab General Trading Company P.O. Box 4195 Aden	Phone: 52479, 53450, 22976, 82266
LEYLAND INTERNATIONAL National Company for Foreign Trade Crater Aden	Telex: ADN 211 Cables: Hometrade Phone: 51134-8
LUCAS SERVICE OVERSEAS LIMITED National Company for Foreign Trade P.O. Box 90 Crater Aden	Telex: ADN 211 Cables: Foreigntrade Phone: 51133-9, 51632-3
PERKINS ENGINES LIMITED National Company for Home Trade Crater Aden	Telex: ADN 211 Phone: 51134

Yugoslavia

BRIGGS & STRATTON CORPORATION Elektrotehna Sektor Zastopstev Titova 51	Telex: 31-184 Phone: 061-311233/ 320241
CATERPILLAR OVERSEAS SA Omnikomerc Batajnicki put bb B.P. 637 Belgrade	Telex: 12223 Cables: Omnikomerc Phone: 608-322/3/4
DETROIT DIESEL ALLISON INTERNATIONAL – EUROPE c/o General Motors Suisse S.A. Salzhausstrasse 21 2501 Biele Bienne Switzerland	Telex: 34217 GMS CH Cables: GMS CH Bienne Phone: 032.26161
KLOCKNER-HUMBOLDT-DEUTZ AG TAM Tovarna Avtomobilov in Motorjev Maribor	Telex: 33111 Cables: Avtoma Maribor Phone: 32321
LEYLAND INTERNATIONAL Interpromet P.O. Box 328 11001 Belgrade	Telex: 11164 Phone: 435-132, 435-193 435-931, 435-834 435-895, 434-226 435-980
PERKINS ENGINES LIMITED Industrija Motora Rakovica Partijarha Dimitriha 7-13 Rakovica Belgrade	Telex: 11341 YU IMR Cables: Indmotor Beograd Phone: 562-043/322/992
PETTERS LIMITED Interkomerc Terezij 1100 Belgrade	Telex: 11 434 Uy Intercom Phone: 340-301
RUGGERINI MOTORI S.p.A. Agroservis Export-Import Amruseva 8 Zagreb	Telex: 21498 Agros
VOLVO PENTA Jugoslavenska Tankerska Plovidba Oour Jugotanker Borisa Kidrica 3 57000 Zadar	Telex: 27127 Phone: 22-377 24-255

Zaire

CATERPILLAR OVERSEAS SA Chanic Direction Générale B.P. 8512 Kinshasa	Telex: 300 Cables: Chamat Kinshasa Phone: 59815, 59819
KLOCKNER-HUMBOLDT-DEUTZ AG Magirus-Deutz-Zaire S.Z.A.R.L. 10me Rue B.P. 8616 Kinshasa – Limete	Telex: 23058 Phone: 77588
KOHLER INTERNATIONAL LIMITED Diomi Imports B.P. 2499 Kinshasa	

LEYLAND INTERNATIONAL
Industrie Nationale Zairoise Des Automobiles Leyland (Inzal)
B.P. 7121
Kinshasa 1
Telex: 382 Inzal Kin.
Cables: Inzal Kinshasa
Phone: 77748, 77768

LUCAS SERVICE OVERSEAS LIMITED
Industrie Nationale Zairoise des Automobiles Leyland (Inzal)
B.P. 7121
Kinshasa
Telex: 382 Inzal Kinshasa
Phone: 77748, 77768

NISSAN DIESEL MOTOR CO. LIMITED
A.C.A.
B.P. 1997
Kinshasa 1

PERKINS ENGINES LIMITED
Sedec Matforce
B.P. 13599
Kin. 1
Kinshasa
Telex: 348
Phone: 23972

PETTERS LIMITED
Affima
17 Avenue de l'Industrie
B.P. 2200
Kinshasa
Telex: Immarkin 974-38
Cables: Afrima Kinshasa
Phone: 22338-22194

WHITE ENGINES INCORPORATED
Africauto-Zaire
B.P. 2999
Kinshasa
Telex: 223
Phone: 22548-49

Zambia

CATERPILLAR OVERSEAS SA
Mazembe Tractor Co. Ltd.
Chibote House
ChaChaCha Road
P.O. Box 3450
Lusaka
Telex: ZA 4292 Lusaka
Cables: Assail Lusaka
Phone: 75168/75941

DETROIT DIESEL ALLISON
Velebilt Import-Export (Branch)
P.O. Box 870
Marsala Tolbuhina 79/1
YU 11000 Beograd
Yugoslavia
Telex: 11499
Phone: 444847, 443538

DORMAN DIESELS LIMITED
Northland Engineering Ltd.
P.O. Box 1640
Ndola
Telex: ZA 3331 A/B Ramine
Cables: Northland Ndola
Phone: 3712/3/4/5

L. GARDNER & SONS LIMITED
Samuel Osborn (Zambia) Ltd.
P.O. Box 1496
Kitwe
Phone: 2473

KIRLOSKAR OIL ENGINES LIMITED
Transocean Ltd.
P.O. Box 1778
Ndola
Telex: ZA 3368
Cable: Transocean
Phone: 4017/2422

KLOCKNER-HUMBOLDT-DEUTZ AG
Eimco (Interafrica) Ltd.
Arkwright Road
P.O. Box 885
Ndola
Telex: ZA 33621
Cables: Eimco
Phone: 4257/8/9

LEYLAND MOTORS (ZAMBIA) LIMITED
P.O. Box 1238
Lusaka
Cables: Leymotors Lusaka
Phone: 76082

LUCAS SERVICE OVERSEAS LIMITED
Stansfield Ratcliffe Ltd.
P.O. Box 780
Chachacha Road
Lusaka
Cables: Stancliffe
Phone: 72016-7

NISSAN DIESEL MOTORS LIMITED
National Transport Corporation (Zambia) Ltd.
P.O. Box 2607
Lusaka

PERKINS ENGINES LIMITED
Power Equipment Limited
P.O. Box 2699
Cairo Road (North End)
Lusaka
Telex: ZA 4174
Phone: 74951

PETTERS LIMITED
Power Equipment Ltd.
P.O. Box 2699
Cairo Road
Lusaka
Telex: ZA 4174
Phone: 74951/2/3/4

Zimbabwe-Rhodesia

CATERPILLAR OVERSEAS SA
Barlow's Tractor & Equipment Co. Ltd.
Cnr. Harrow Road & Martin Drive
Beverley East
Msasa
P.O. Box 1537
Telex: 3152
Phone: 67781

CATERPILLAR OVERSEAS S.A.
Barlow's Tractor & Equipment Co. Ltd.
5 Dunlop Road
Donnington
Bulawayo
Telex: 3152
Phone: 67781

12

Index of Manufacturers

This index refers to the main entries for manufacturers' equipment and excludes references in Section 10 to miscellaneous suppliers and in Section 11, the list of manufacturers' agents and distributors by country.